# Communications
# in Computer and Information Science     1949

## Rationale

The CCIS series is devoted to the publication of proceedings of computer science conferences. Its aim is to efficiently disseminate original research results in informatics in printed and electronic form. While the focus is on publication of peer-reviewed full papers presenting mature work, inclusion of reviewed short papers reporting on work in progress is welcome, too. Besides globally relevant meetings with internationally representative program committees guaranteeing a strict peer-reviewing and paper selection process, conferences run by societies or of high regional or national relevance are also considered for publication.

## Topics

The topical scope of CCIS spans the entire spectrum of informatics ranging from foundational topics in the theory of computing to information and communications science and technology and a broad variety of interdisciplinary application fields.

## Information for Volume Editors and Authors

Publication in CCIS is free of charge. No royalties are paid, however, we offer registered conference participants temporary free access to the online version of the conference proceedings on SpringerLink (http://link.springer.com) by means of an http referrer from the conference website and/or a number of complimentary printed copies, as specified in the official acceptance email of the event.

CCIS proceedings can be published in time for distribution at conferences or as postproceedings, and delivered in the form of printed books and/or electronically as USBs and/or e-content licenses for accessing proceedings at SpringerLink. Furthermore, CCIS proceedings are included in the CCIS electronic book series hosted in the SpringerLink digital library at http://link.springer.com/bookseries/7899. Conferences publishing in CCIS are allowed to use Online Conference Service (OCS) for managing the whole proceedings lifecycle (from submission and reviewing to preparing for publication) free of charge.

## Publication process

The language of publication is exclusively English. Authors publishing in CCIS have to sign the Springer CCIS copyright transfer form, however, they are free to use their material published in CCIS for substantially changed, more elaborate subsequent publications elsewhere. For the preparation of the camera-ready papers/files, authors have to strictly adhere to the Springer CCIS Authors' Instructions and are strongly encouraged to use the CCIS LaTeX style files or templates.

## Abstracting/Indexing

CCIS is abstracted/indexed in DBLP, Google Scholar, EI-Compendex, Mathematical Reviews, SCImago, Scopus. CCIS volumes are also submitted for the inclusion in ISI Proceedings.

## How to start

To start the evaluation of your proposal for inclusion in the CCIS series, please send an e-mail to ccis@springer.com.

Nguyen Thai-Nghe · Thanh-Nghi Do ·
Peter Haddawy
Editors

# Intelligent Systems and Data Science

First International Conference, ISDS 2023
Can Tho, Vietnam, November 11–12, 2023
Proceedings, Part I

Springer

*Editors*
Nguyen Thai-Nghe (iD)
Can Tho University
Can Tho, Vietnam

Thanh-Nghi Do (iD)
Can Tho University
Can Tho, Vietnam

Peter Haddawy (iD)
Mahidol University
Salaya, Thailand

ISSN 1865-0929                    ISSN 1865-0937 (electronic)
Communications in Computer and Information Science
ISBN 978-981-99-7648-5          ISBN 978-981-99-7649-2 (eBook)
https://doi.org/10.1007/978-981-99-7649-2

This Springer imprint is published by the registered company Springer Nature Singapore Pte Ltd.
The registered company address is: 152 Beach Road, #21-01/04 Gateway East, Singapore 189721, Singapore

Paper in this product is recyclable.

# Preface

This proceedings contains the papers from the first International Conference on Intelligent Systems and Data Science (ISDS 2023), held at Can Tho University, Vietnam from November 11–12, 2023. ISDS 2023 provided a dynamic forum in which researchers discussed problems, exchanged results, identified emerging issues, and established collaborations in related areas of Intelligent Systems and Data Science.

We received 123 submissions from 15 countries to the main conference and a special session. To handle the review process, we invited 100 expert reviewers. Each paper was reviewed by at least three reviewers. We followed the single-blind process in which the identities of the reviewers were not known to the authors. This year, we used the EasyChair conference management service to manage the submission and selection of papers. After a rigorous review process, followed by discussions among the program chairs, 35 papers were accepted as long papers and 13 as short papers, resulting in acceptance rates of 28.46% and 10.57% for long and short papers, respectively.

We are honored to have had keynote talks by Kazuhiro Ogata (Japan Advanced Institute of Science and Technology, Japan), Masayuki Fukuzawa (Kyoto Institute of Technology, Japan), Dewan Md. Farid (United International University, Bangladesh), Mizuhito Ogawa (JAIST, Japan), and Nguyen Le Minh (JAIST, Japan).

The conference program included four sessions: Applied Intelligent Systems and Data Science for Agriculture, Aquaculture, and Biomedicine; Big Data, IoT, and Cloud Computing; Deep Learning and Natural Language Processing; and Intelligent Systems.

We wish to thank the other members of the organizing committee, the reviewers, and the authors for the immense amount of hard work that has gone into making ISDS 2023 a success. The achievement of the conference was also contributed to by the kind devotion of many sponsors and volunteers.

We hope you enjoyed the conference!

Nguyen Thai-Nghe
Thanh-Nghi Do
Peter Haddawy
Tran Ngoc Hai
Nguyen Huu Hoa

# Organization

## Honorary Chairs

Ha Thanh Toan  Can Tho University, Vietnam
Nguyen Thanh Thuy  University of Engineering and Technology, VNU
  Hanoi, Vietnam

## General Chairs

Tran Ngoc Hai  Can Tho University, Vietnam
Nguyen Huu Hoa  Can Tho University, Vietnam

## Program Chairs

Nguyen Thai-Nghe  Can Tho University, Vietnam
Do Thanh Nghi  Can Tho University, Vietnam
Peter Haddawy  Mahidol University, Thailand

## Publication Chairs

Tran Thanh Dien  Can Tho University, Vietnam
Nguyen Minh Khiem  Can Tho University, Vietnam

## Technical Program Committee

Atsushi Nunome  Kyoto Institute of Technology, Japan
An Cong Tran  Can Tho University, Vietnam
Bay Vo  HCMC University of Technology, Vietnam
Bob Dao  Monash University, Australia
Binh Tran  La Trobe University, Australia
Bui Vo Quoc Bao  Can Tho University, Vietnam
Cu Vinh Loc  Can Tho University, Vietnam
Dewan Farid  United International University, Bangladesh

| | |
|---|---|
| Daniel Hagimont | Institut de Recherche en Informatique de Toulouse, France |
| Diep Anh Nguyet | University of Liege, Belgium |
| Duc-Luu Ngo | Bac Lieu University, Vietnam |
| Duong Van Hieu | Tien Giang University, Vietnam |
| Hiroki Nomiya | Kyoto Institute of Technology, Japan |
| Hien Nguyen | University of Information Technology - VNU HCMC, Vietnam |
| Huynh Quang Nghi | Can Tho University, Vietnam |
| Ivan Varzinczak | Université Paris 8, France |
| Lam Hoai Bao | Can Tho University, Vietnam |
| Le Hoang Son | Vietnam National University Hanoi, Vietnam |
| Luong Vinh Quoc Danh | Can Tho University, Vietnam |
| Maciej Huk | Wroclaw University of Science and Technology, Poland |
| Mai Xuan Trang | Phenikaa University, Vietnam |
| Masayuki Fukuzawa | Kyoto Institute of Technology, Japan |
| Nicolas Schwind | AIST - Artificial Intelligence Research Center, Japan |
| Nghia Duong-Trung | German Research Center for Artificial Intelligence (DFKI), Germany |
| Ngo Ba Hung | Can Tho University, Vietnam |
| Nguyen Chanh-Nghiem | Can Tho University, Vietnam |
| Nguyen Chi-Ngon | Can Tho University, Vietnam |
| Nguyen Dinh Thuan | University of Information Technology, VNU HCMC, Vietnam |
| Nguyen Huu Hoa | Can Tho University, Vietnam |
| Nguyen Minh Khiem | Can Tho University, Vietnam |
| Nguyen Minh Tien | Hung Yen University of Technology and Education, Vietnam |
| Nguyen Thai-Nghe | Can Tho University, Vietnam |
| Nguyen Thanh Thuy | University of Engineering and Technology, VNU Hanoi, Vietnam |
| Nguyen-Khang Pham | Can Tho University, Vietnam |
| Nhut-Khang Lam | Can Tho University, Vietnam |
| Peter Haddawy | Mahidol University, Thailand |
| Pham Truong Hong Ngan | Can Tho University, Vietnam |
| Phan Anh Cang | Vinh Long University of Technology Education, Vietnam |
| Phan Phuong Lan | Can Tho University, Vietnam |
| Quoc-Dinh Truong | Can Tho University, Vietnam |
| Salem Benferhat | CRIL, CNRS & Artois University, France |
| Si Choon Noh | Namseoul University, South Korea |

| | |
|---|---|
| Thai Tran | Lincoln University, New Zealand |
| Thanh-Hai Nguyen | Can Tho University, Vietnam |
| Thanh-Nghi Do | Can Tho University, Vietnam |
| Thanh-Nghi Doan | An Giang University, Vietnam |
| Thanh-Tho Quan | University of Technology - VNU HCMC, Vietnam |
| The-Phi Pham | Can Tho University, Vietnam |
| Thi-Lan Le | Hanoi University of Science and Technology, Vietnam |
| Thuong-Cang Phan | Can Tho University, Vietnam |
| Thai Minh Tuan | Can Tho University, Vietnam |
| Thi-Phuong Le | EBI School of Industrial Biology, France |
| Tomas Horvath | Eötvös Loránd University, Hungary |
| Tran Cao Son | New Mexico State University, USA |
| Tran Hoang Viet | Can Tho University, Vietnam |
| Tran Khai Thien | HCMC University of Foreign Languages and Information Technology, Vietnam |
| Tran Nguyen Minh Thu | Can Tho University, Vietnam |
| Trung-Hieu Huynh | Industrial University of Ho Chi Minh City, Vietnam |
| Truong Minh Thai | Can Tho University, Vietnam |
| Truong Xuan Viet | Can Tho University, Vietnam |
| Van-Hoa Nguyen | An Giang University, Vietnam |
| Vatcharaporn Esichaikul | Asian Institute of Technology, Thailand |
| Vinh Nguyen Nhi Gia | Can Tho University, Vietnam |
| Van-Sinh Nguyen | International University, VNU HCMC, Vietnam |
| Wu-Yuin Hwang | National Central University, Taiwan |
| Yuya Yokoyama | Kyoto Prefectural University, Japan |
| Yoshihiro Mori | Kyoto Institute of Technology, Japan |

## Finance and Secretary

| | |
|---|---|
| Phan Phuong Lan | Can Tho University, Vietnam |
| Lam Nhut Khang | Can Tho University, Vietnam |
| Dinh Lam Mai Chi | Can Tho University, Vietnam |

## Local Committee

| | |
|---|---|
| Le Nguyen Doan Khoi | Can Tho University, Vietnam |
| Le Van Lam | Can Tho University, Vietnam |

| | |
|---|---|
| Huynh Xuan Hiep | Can Tho University, Vietnam |
| Ngo Ba Hung | Can Tho University, Vietnam |
| Nguyen Nhi Gia Vinh | Can Tho University, Vietnam |
| Pham Nguyen Khang | Can Tho University, Vietnam |
| Pham The Phi | Can Tho University, Vietnam |
| Thuong-Cang Phan | Can Tho University, Vietnam |
| Tran Nguyen Minh Thu | Can Tho University, Vietnam |
| Truong Minh Thai | Can Tho University, Vietnam |
| Quoc-Dinh Truong | Can Tho University, Vietnam |

# Contents – Part I

# Contents – Part II

**Deep Learning and Natural Language Processing**

## Intelligent Systems

# Applied Intelligent Systems and Data Science for Agriculture, Aquaculture, and Biomedicine

# A Signal-Processing-Assisted Instrumentation System for Nondestructive Characterization of Semiconductor Wafers

Mio Maeda and Masayuki Fukuzawa[✉] [iD]

Graduate School of Science and Technology, Kyoto Institute of Technology, Matsugasaki, Sakyo-Ku, Kyoto 606-8585, Japan
fukuzawa@kit.ac.jp

**Abstract.** A signal-processing-assisted instrumentation system has been developed for nondestructive resistivity mapping in semiconductor wafers. With the recent performance improvement and market expansion on power electronic devices with gallium nitride (GaN) and silicon carbide (SiC), there is a strong demand for ultra-high-resolution resistivity mapping based on measurement informatics by combining narrow-pitch scanning and super resolution techniques, but it was still difficult to realize narrow-pitch scanning with high measurement stability. The developed system combined a special technique of signal processing with a conventional resistivity-measurement technique based on frequency response of metal-insulator-semiconductor-metal (MISM) structure consisting of two electrodes, air gap and semiconductor. This system enabled us to obtain the gap-dependent and resistivity-dependent components by straightforward peak estimation instead of parametric model fitting from the measured frequency response of the MISM structure to a square wave, which improves the reproducibility of gap control and stability of resistivity evaluation. The system performance was examined with commercial 6-inch $\phi$ GaAs wafers and it exhibited good reproducibility in gap control finer than the minimum translation step of 4 $\mu$m, and a small coefficient of variation of 1.7% for 100 repeated measurement of resistivity at the order of $10^8$ $\Omega$cm, which demonstrates a high applicability of the system in narrow-pitch scanning much finer than the probe diameter of 2 mm required for further study on ultra-high-resolution mapping with super resolution technique for GaN and SiC wafers.

**Keywords:** Signal Processing · Measurement Informatics · Frequency Response Analysis · Semiconductor Wafer · Resistivity · Narrow-Pitch Scanning

## 1 Introduction

Nondestructive characterization of semiconductor wafers had long been an essential research field in electronics, but recently it also becomes one of promising application fields of measurement informatics. Since the electronic and physical properties such as resistivity, carrier mobility, crystal defects and residual strains are indispensable for studying physical phenomena in semiconductors, there have been many attempts in academia for a long time to develop various characterization techniques [1, 2]. Many practical apparatuses have also been commercialized to inspect the commercial wafers such as silicon (Si) for large-scale integrated circuits (LSIs) and solar-cells, gallium arsenide (GaAs) and indium phosphide (InP) for opto-electronic devices because the nondestructive technique does not decrease the production yield. In recent years, there is a strong demand to improve the measurement performance and to cover novel materials such as gallium nitride (GaN) and silicon carbide (SiC) for power-electronic devices [3], but there are theoretical limits in electronic and physical measurements. One of the promising approaches to overcome these limitations is that based on measurement informatics. It is expected to drastically improve the performance or significantly expand the range of measurement condition by adopting computational techniques such as statistics, data mining, and machine learning techniques.

In the case of resistivity measurement, conventional techniques have already been established in semi-insulating type semiconductor wafers. Stibal et al. developed a nondestructive technique of resistivity measurement by forming a metal-insulator-semiconductor-metal (MISM) structure in a small part of the wafer, applying a unipolar square wave and analyzing its time-dependent charge response [4, 5]. Fukuzawa et al. proposed an alternative technique based on the frequency response of the MISM structure [6, 7]. It has some advantages such as low dependence on the measurement circuit and wide resistivity range, but still has some problems in stability due to the parametric model fitting. With the recent performance improvement and market expansion on GaN/SiC power devices, there is a strong demand on nondestructive resistivity mapping of commercial large-diameter wafers, but it is still difficult to achieve both high spatial-resolution and measurement stability.

This study aims to realize narrow-pitch scanning and stability improvement in the resistivity measurement based on the frequency response by developing a signal-processing-assisted instrumentation system. A new technique for analyzing the frequency response of MISM structure is proposed and applied to this system in order to improve the reproducibility of gap control and the stability of resistivity evaluation. This system is designed to cover novel GaN and SiC wafers, but it also provides narrow-pitch scanning data much finer than the probe diameter for further study on ultra-high-resolution mapping using a super resolution technique.

The remainder of this paper is as follows. In Sect. 2, we introduce the principle of resistivity evaluation with MISM structure and propose a technique to improve the estimation stability with signal processing. Section 3 describes the functional design and special implementation of the developed system in this study. In Sect. 4, the experimental results on system performance are provided and the system usefulness is discussed for further study on super resolution.

# 2   Signal-Processing-Assisted Resistivity-Evaluation Technique

## 2.1   MISM Structure and its Frequency Response

Figure 1 shows (a) probe geometry and (b) corresponding MISM structure with its equivalent circuit, adopted in this study. The probe consists of a back electrode connected to the waveform driver and an upper electrode with a charge amplifier mounted on a Z stage for approaching the semiconductor wafer and measuring the charge response of the MISM structure.

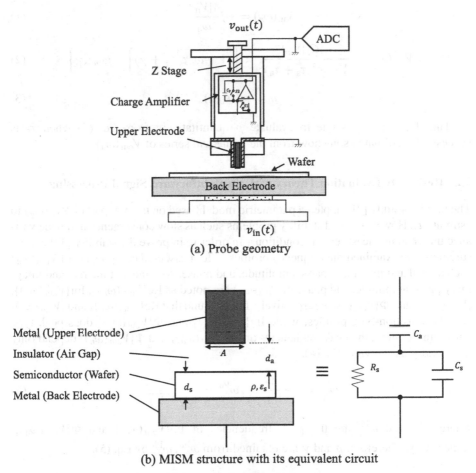

(a) Probe geometry

(b) MISM structure with its equivalent circuit

**Fig. 1.**  (a) Probe geometry and (b) corresponding MISM structure

An equivalent circuit of MISM structure consists of two capacitors $C_a$ and $C_s$, and a resistor $R_s$. They are formulated as follows: $C_a = \varepsilon_0 A/d_a$, $C_s = \varepsilon_0 \varepsilon_s A/d_s$ and $R_s = \rho d_s/A$, where $A$ is area of upper electrode, $d_a$ is a thickness of air gap between the upper electrode and the wafer surface, and $\rho$, $d_s$ and $\varepsilon_s$ are resistivity, thickness and relative permittivity of semiconductor wafer, respectively.

A bipolar square wave $v_{in}(t)$ with a period $T$ and magnitude $|V_{in}|$ is utilized in the resistivity measurement. It is applied to a MISM structure formed by the probe at a certain positon of the wafer. A charge response of the MISM structure is detected by the charge amplifier and acquired as an output voltage $v_{out}(t)$. Assume that the $n^{th}$ harmonics of complex spectrum for $v_{in}(t)$ and $v_{out}(t)$ are noted as $V_{in}(\omega_n)$ and $V_{out}(\omega_n)$, respectively. Some timing variables are also introduced as follows: $\tau_s = R_s C_s$, $\tau_a = R_s C_a$, $\gamma = \tau_a/\tau_s = d_s/(\varepsilon_s d_a)$. Their relationship is formulated as Eqs. (1) to (3), where $j = \sqrt{-1}$ and $n = 2m-1 (m = 1, 2, 3, \ldots)$ due to the symmetry of square wave.

$$V_{in}(\omega_n) = -j\frac{4|V_{in}|}{\omega_n} \tag{1}$$

$$V_{out}(\omega_n) = \frac{4C_a|V_{in}|}{1+\omega_n^2(\tau_s+\tau_a)^2} \times \left\{\left[1+\omega_n^2\tau_s^2(1+\gamma)\right] - j\omega_n\tau_s\gamma\right\} \tag{2}$$

$$\rho = \tau_s/(\varepsilon_0\varepsilon_s) \tag{3}$$

Therefore, it is possible to evaluate $\rho$ quantitatively from Eq. (3) when $\tau_s$ is successfully estimated somehow from the numerical series of $V_{out}(\omega_n)$.

## 2.2 Resistivity Evaluation Theory with Straightforward Signal Processing

The previous study [7] adopted a parametric model based on the real part of $V_{out}(\omega_n)$ to estimate $\tau_s$. However, it had stability problems such as slow convergence and wide variance under some measurement conditions. In order to improve the stability of the measurement, we examined the frequency profiles of four essential components of $V_{out}(\omega_n)$ such as real and imaginary parts, amplitude, and phase. Assume that the real and imaginary parts, amplitude, and phase of $V_{out}(\omega_n)$ are noted as $\text{Re}\{V_{out}(\omega_n)\}$, $\text{Im}\{V_{out}(\omega_n)\}$, $|V_{out}(\omega_n)|$, and $\Phi\{V_{out}(\omega_n)\}$, respectively. It was found that $\text{Re}\{V_{out}(\omega_n)\}$ and $|V_{out}(\omega_n)|$ reveals transition-type profiles, while $\text{Im}\{V_{out}(\omega_n)\}$ and $\Phi\{V_{out}(\omega_n)\}$ does peak-type ones. Furthermore, the peak frequencies of $\text{Im}\{V_{out}(\omega_n)\}$ and $\Phi\{V_{out}(\omega_n)\}$ depend only on $\tau_s$ and $\gamma$ as shown in Eq. (4),

$$\omega_{n_p}^{Im} = \frac{1}{\tau_s(1+\gamma)}, \quad \omega_{n_p}^{Ph} = \frac{1}{\tau_s\sqrt{1+\gamma}} \tag{4}$$

where $\omega_{n_p}^{Im}$ and $\omega_{n_p}^{Ph}$ are the peak frequencies of $\text{Im}\{V_{out}(\omega_n)\}$ and $\Phi\{V_{out}(\omega_n)\}$, respectively. Therefore, $\rho$ and $\gamma$ are obtained from $\omega_{n_p}^{Im}$, $\omega_{n_p}^{Ph}$ by Eq. (5).

$$\rho = \frac{\omega_{n_p}^{Im}}{\varepsilon_0\varepsilon_s\omega_{n_p}^{Ph\,2}}, \gamma = \left(\frac{\omega_{n_p}^{Ph}}{\omega_{n_p}^{Im}}\right)^2 - 1 \tag{5}$$

In this study, we proposed to estimate $\tau_s$ by implementing a non-parametric and straightforward algorithm of signal processing to convert $v_{out}(t)$ to $V_{out}(\omega_n)$ and to search or fit $\omega_{n_p}^{Im}$ and $\omega_{n_p}^{Im}$. This technique includes several advantages.

1. The stability problems of the previous technique will be avoidable. It is well known that the peak search or peak fitting is simple and robust compared with general curve fitting for transition-type function.
2. Since a special frequency range near $1/\tau_s$ is examined intensively in the peak search, unwanted noise effects in other frequency ranges will be suppressive.
3. The air gap $d_a$ can be retrospectively estimated from $\gamma$ after acquiring $v_{out}(t)$ because $\varepsilon_s$ and $d_s$ is regarded as constant at each measurement in the wafer. It is very helpful for keeping a constant $d_a$ throughout the measurement over the whole wafer.

The influence of $\gamma$ on resistivity evaluation was also considered. Equation (4) implies that the difference between $\omega_{n_p}^{Im}$ and $\omega_{n_p}^{Ph}$ will be decreased if $\gamma$ becomes much less than 1. Since the order of harmonics of $V_{out}(\omega_n)$ is finite in the resistivity evaluation, the error of peak order $n_p$ will be increased if the frequency difference between $\omega_{n_p}^{Im}$ and $\omega_{n_p}^{Ph}$ becomes small, which may result in poor estimation accuracy. Therefore, a large value of $\gamma$ is desirable to keep the accuracy in resistivity evaluation. On the other hand, to keep a large $\gamma$ requires a small air gap $d_a$, which increases the risk to hit the upper electrode to the wafer surface. In this study, we selected to control $d_a$ so as to keep $\gamma$ close to 1 by considering a trade-off between the risk of electrode hit and accuracy of resistivity evaluation.

## 3   System Design and Implementation

Figure 2 shows a functional block diagram of the system constructed in this study.

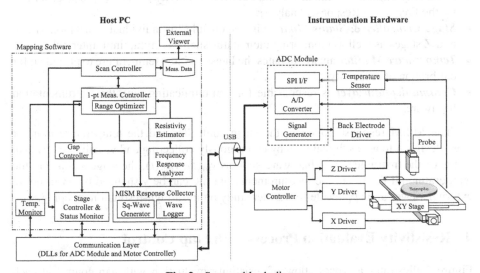

**Fig. 2.** System block diagram

The system consists of an instrumentation hardware and a host PC. The instrumentation hardware was constructed to handle translation of XY and Z stages, application of square wave to the back electrode, and capturing of the charge response waveform from the probe. The motor controller and the ADC module were connected to the host PC via USB and perform driving-pulse generation to the motor driver of each stage, square-wave generation, capturing of charge response signal and temperature measurement. An in-house back-electrode driver with a low output impedance was adopted to improve stability of resistivity evaluation. The diameter of the upper electrode was 2 mm and its structure with the charge amplifier was optimized to suppress the noises.

A mapping software in the host PC was developed to hierarchically handle everything from communication with the instrumentation hardware to overall control of scanning for obtaining the resistivity map. The essential functional blocks are as follows.

- **Scan Controller** is responsible for generation and execution of a scan procedure according to a configured measurement area, and recording the measurement result into a measurement data file.
- **1-pt Measurement Controller** is responsible for everything to be performed at each measurement such as invoking gap control, range optimization and resistivity evaluation.
- **Gap Controller** is responsible for bringing $d_a$ closer to a target gap by repeating probe translation and $d_a$ estimation. Details will be explained below.
- **MISM response collector** contains a square wave generator and a wave logger. It generates a square wave $v_{in}(t)$ to be applied to the back electrode, while it makes the ADC module to acquire $v_{out}(t)$ synchronized with $v_{in}(t)$.
- **Frequency Response Analyzer** generates $V_{out}(\omega_n)$ by fast Fourier transform (FFT) after filtering $v_{out}(t)$ as preprocessing.
- **Resistivity estimator** estimates $\tau_s$ and deduced resistivity $\rho$ from $V_{out}(\omega_n)$ generated by the Frequency Response Analyzer.
- **Stage Controller & Status Monitor** is responsible for individual translation of XY and Z stages as well as monitoring their status such as origin, limit, interlock, etc.
- **Temperature Monitor** always updates the latest ambient temperature to provide it for calibration of resistivity.
- **Communication layer** is responsible for communicating with the instrumentation hardware.

The MISM response collector, the stage controller and the temperature monitor were designed to act as digital twin by continuously collecting and visualizing physical status of the instrumentation hardware such as $v_{in}(t)$, $v_{out}(t)$, the stage status and the temperature all the time. The measurement data file can be visualized by an in-house external viewer and applicable for further study on ultra-high-resolution mapping.

## 4    Resistivity Evaluation Process with Gap Control

Figure 3 illustrates a process flow of resistivity evaluation with gap control at each measurement. In this flow, a gap control process precedes other things first, and then a series of resistivity measurement steps, including waveform acquisition, frequency analysis and resistivity evaluation, is repeated until a validity criterion is satisfied or the number of repetitions exceeds an upper limit.

### 4.1 Gap Control Process

The gap control process is to bring $d_a$ closer to a target gap by repeating probe translation and $d_a$ estimation. This system implemented a primitive version that raises the probe by a fixed amount $\Delta z_0$ after measuring each point and then gradually lowers it to the target gap before measuring the next point. Each control step is as follows.

1. $v_{in}(t)$ with a fixed period $T_0$ is applied to the back electrode and $v_{out}(t)$ is synchronously acquired.
2. The current gap is estimated from $v_{out}(t)$.
3. The process is finished if the current gap falls below a target gap.
4. The probe is lowered by an optimal $\Delta z$, then the process is repeated from the first step.

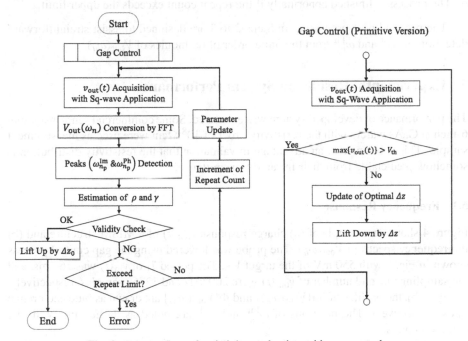

**Fig. 3.** Process flow of resistivity evaluation with gap control

As mentioned in the previous section, $\gamma$ is also available for estimating the current gap, but it is not appropriate before lowering the probe because a large gap leads to increase the error of $\gamma$. In this system, the amplitude of $v_{out}(t)$ was obtained from $\max\{v_{out}(t)\}$ and was used as an inverse measure of the current gap. Target gap was also configured as an inverse measure of $V_{th}$.

In order to reduce the number of repetition, $\Delta z$ was optimized to be proportional to the current gap. Since the amplitude of $v_{out}(t)$ is inversely proportional to the current gap, $\Delta z$ was determined to be inversely proportional to $\max\{v_{out}(t)\}$ and to be minimum $\Delta z_{min}$ near $V_{th}$. In this study, $\Delta z_{min}$ was configured as 4 μm.

### 4.2 Resistivity Evaluation

The resistivity evaluation process was designed to perform the following steps.

1. $v_{in}(t)$ with a tentative period $T_t$ is applied to the back electrode and $v_{out}(t)$ is synchronously acquired.
2. $V_{out}(\omega_n)$ is obtained by FFT after averaging of $v_{out}(t)$ and windowing as pre-processing.
3. Selective smoothing is applied to the odd harmonics of $V_{out}(\omega_n)$ as post-processing.
4. $\omega_{n_p}^{Im}$ and $\omega_{n_p}^{Ph}$ are detected and consequently $\rho$ and $\gamma$ are evaluated.
5. The validity of the current measurement is examined using $\gamma$ as a criterion. If it is valid, the process is finished normally after lifting up the probe by a fixed amount $\Delta z_0$ in preparation for the next measurement. Otherwise, re-measurement is repeated after updating some parameters such as $T_t$.
6. The process is finished abnormally if the repeat count exceeds the upper limit.

The pre- and post-processing in Steps 2 to 3 are designed to assist straightforward detection of $\omega_{n_p}^{Im}$ and $\omega_{n_p}^{Ph}$ from the finite order of harmonics of $V_{out}(\omega_n)$.

## 5 Experimental Results on System Performance

The performance of developed system was examined. Some commercial semi-insulating 6-inch $\phi$ GaAs wafers with the resistivity order of $10^8$ $\Omega$cm were used as measurement sample. This is because of an advantage in validation that the resistivity distribution is somehow predictable from their infrared transmittance.

### 5.1 Frequency Response

Figure 4 shows an example of (a) charge response $v_{out}(t)$ of a MISM structure and (b) its frequency spectrum $V_{out}(\omega_n)$. The probe was lowered using the gap control process shown in Fig. 3 with 550 mV of the target $V_{th}$. The period $T_t$ of $v_{in}(t)$ was 164 ms, and the sampling rate and number of $v_{out}(t)$ were 200 kHz and 32768 samples, respectively. In Fig. 4(b), the profiles of Im$\{V_{out}(\omega_n)\}$ and $\Phi\{V_{out}(\omega_n)\}$ are drawn as blue and orange lines, respectively. The positions of $\omega_{n_p}^{Im}$ and $\omega_{n_p}^{Ph}$ are noted as dashed line with the common colors.

$v_{out}(t)$ revealed a symmetric integral waveform known as charge response of MISM structure, and Im$\{V_{out}(\omega_n)\}$ and $\Phi\{V_{out}(\omega_n)\}$ clearly revealed peak-type profiles, which demonstrates the validity of the measurement and the signal processing. $\omega_{n_p}^{Im}$ and $\omega_{n_p}^{Ph}$ were 766 Hz and 1163 Hz, respectively and consequently $\tau_s$ and $\gamma$ were estimated as 88.2 $\mu$s and 1.3. $\rho$ was evaluated from $\tau_s$ as $7.9 \times 10^7$ $\Omega$cm and agreed with the value in the specification sheet of the wafer.

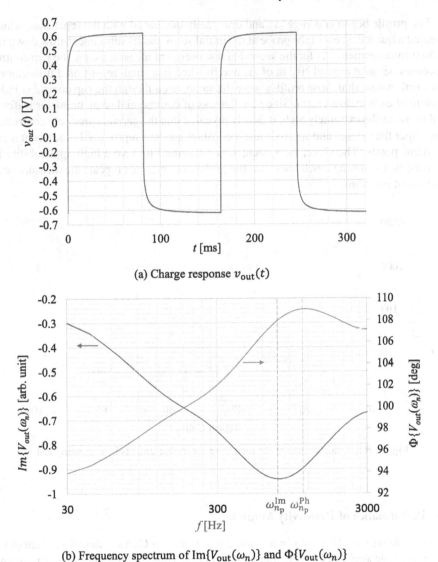

(a) Charge response $v_{out}(t)$

(b) Frequency spectrum of $Im\{V_{out}(\omega_n)\}$ and $\Phi\{V_{out}(\omega_n)\}$

**Fig. 4.** (a) Charge response and (b) its frequency spectrum of a MISM structure

## 5.2 Performance of Gap Control

Figure 5 shows a repetition profile of $\Delta z$ including six measurements at the beginning of a scanning test. It was performed under the same conditions as the experiment shown in Fig. 4. The initial position of the probe was 8 mm above the sample and $\Delta z_0$ was configured as 80 $\mu$m.

The profile began at a large $\Delta z$ and drastically decreased with the repetition, which revealed a quick descent of the probe at the initial steps and its subsequent slow down for the first measurement. As for the second and subsequent measurements, the gap control processes began at around 10 μm of $\Delta z$ and finished in a small repetition. Furthermore, it is worth noting that those profiles were the same, even though the gap control is independent at each measurement. Since the flatness of commercial semiconductor wafer is well known to be extremely high, it clearly revealed that the gap was precisely controlled much finer than $\Delta z_{min}$ and approximately constant gap was reproduced at adjacent measurement points. Therefore, the system was confirmed to have a high applicability in narrow-pitch scanning much finer than the probe diameter that repeats the measurement at adjacent positions.

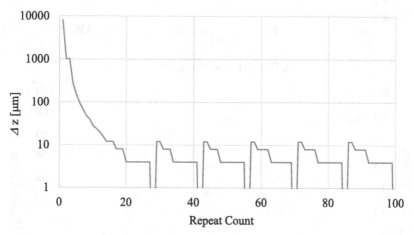

**Fig. 5.** Repetition profile of $\Delta z$ including six measurements in a scanning test

## 5.3 Performance of Resistivity Mapping

Figure 6 shows (a) 2D $\rho$ map in a commercial 6-inch $\phi$ GaAs wafer, (b) $\rho$ histogram over the whole wafer and (c) IR transmission map. The scanning pitch was 1 mm, and the scanning area was configured to a circle 10 mm inward from the wafer edge. The wafer outline and the area of IR map were indicated as black circle and blue rectangle, respectively in Fig. 6(a).

The $\rho$ map showed inhomogeneous distribution in the whole wafer and the distribution pattern exhibited a strong correlation with the contrast of IR transmission. This correlation was commonly found in commercial GaAs wafers, which supports the validity of this measurement. The averaged value of resistivity distribution was $3.6 \times 10^{8}$ $\Omega$cm and the coefficient of variation (CV) was 28%, which corresponds to the standard level of commercial wafer. On the other hand, the maximum CV was 1.7% for 100 repeated measurements at a center of this sample, which is sufficient in evaluating such spatial distribution of resistivity of the wafer.

Since the CV for repeated measurement in the conventional system developed in the previous study [7] was about 5%, it was also found that this system revealed high measurement stability about three times as high as that of the conventional one. Therefore, it can be concluded that the developed system provided not only sufficient stability in evaluating the spatial distribution of resistivity of a semiconductor wafer, but also high applicability in narrow-pitch scanning required for further study on ultra-high-resolution mapping with super resolution technique.

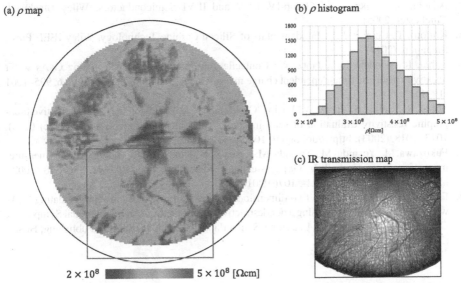

**Fig. 6.** (a) 2D $\rho$ map in a commercial 6-inch $\phi$ GaAs wafer, (b) $\rho$ histogram over the whole wafer and (c) IR transmission map

## 6 Conclusion

A signal-processing-assisted instrumentation system has been developed for nondestructive mapping of resistivity in semiconductor wafers. In order to realize narrow-pitch scanning and stability improvement, a special technique of signal processing was adopted to obtain the gap-dependent and resistivity-dependent components by straightforward peak estimation instead of parametric model fitting from the measured frequency response of the MISM structure to a square wave, which improves the reproducibility of gap control and stability of resistivity evaluation.

The system performance was examined with commercial 6-inch $\phi$ GaAs wafers and it exhibited good reproducibility in gap control finer than the $\Delta z_{min}$ of 4 μm, and a small CV of 1.7% for 100 repeated measurements of resistivity at the order of $10^8$ Ωcm, which demonstrates a high applicability of the system in narrow-pitch scanning much finer than the probe diameter required for further study on ultra-high-resolution mapping with super resolution technique.

**Acknowledgement.** This work was partly supported by JSPS Core-to-Core Program (grant number: JPJSCCB20230005).

# References

1. Schroder, D.K.: Semiconductor Material and Device Characterization, 3rd edn. John Wiley and Sons, Hoboken (2006)
2. Adachi, S.: Properties of Group-IV, III-V and II-VI Semiconductors, Wiley Blackwell, Chichester (2005)
3. Kimoto, T., Cooper, J.A.: Fundamentals of Silicon Carbide Technology, Wiley-IEEE Press, Singapore (2014)
4. Stibal, R., Windshelf, J., Jantz, W.: Contactless evaluation of semi-insulating GaAs wafer resistivity using the time-dependent charge measurement. Semiond. Sci. Technol. **6**, 995–1001 (1991)
5. Stibal, R., Müller, S., Jantz, W., Pozina, G., Magnusson, B., Ellison, A.: Nondestructive topographic resistivity evaluation of semi-insulating SiC substrates. Phys. Stat. Sol. (c) 0, (3), 1013–1018, (2003). https://doi.org/10.1002/pssc.200306237
6. Fukuzawa, M., Yoshida, M., Yamada, M., Hanaue, Y., Kinoshita, K.: Nondestructive measurement of resistivity in bulk $In_xGa_{1-x}As$ crystals. Mater. Sci. Eng. **B91–92**, 376–378 (2002). https://doi.org/10.1016/S0921-5107(01)01074-1
7. Fukuzawa, M., Yamada, M.: Two-dimensional mapping of resistivity in semi-insulating GaAs wafers with large diameter using a nondestructive technique. In: 29th International Symposium Compound Semiconductors, Lausanne, Switzerland 2002, pp. 85–88, IOP Publishing, Bristol (2003)

# SDCANet: Enhancing Symptoms-Driven Disease Prediction with CNN-Attention Networks

Thao Minh Nguyen Phan[1], Cong-Tinh Dao[1], Tai Tan Phan[2], and Hai Thanh Nguyen[2(✉)]

[1] National Yang Ming Chiao Tung University, Hsinchu City, Taiwan
pnmthaoct@gmail.com
[2] Can Tho University, Can Tho City, Vietnam
ntthai.cit@ctu.edu.vn

**Abstract.** Deep learning algorithms have revolutionized healthcare by improving patient outcomes, enhancing diagnostic accuracy, and advancing medical knowledge. In this paper, we propose an approach for symptom-based disease prediction based on understanding the intricate connections between symptoms and diseases by accurately representing symptom sets, considering the varying importance of individual symptoms. This framework enables precise and reliable disease prediction, transforming healthcare diagnosis and improving patient care. By incorporating advanced techniques such as a one-dimensional convolutional neural network (1DCNN) and attention mechanisms, our model captures the unique characteristics of each patient, facilitating personalized and accurate predictions. Our model outperforms baseline methods through comprehensive evaluation, demonstrating its effectiveness in disease prediction.

**Keywords:** disease prediction · healthcare · symptom

## 1 Introduction

Artificial Intelligence (AI)-based tools can support physicians in the diagnostic process, increasing the chances of timely healthcare care for a broader population [13]. AI provided solutions to key challenges, including reducing diagnosis time for critical conditions, improving access to comprehensive care, and reducing healthcare expenses. In addition, AI took advantage of Computer-Aided Diagnosis (CAD) systems, primarily to address healthcare crises such as the COVID-19 pandemic. The implementation of AI-driven solutions has witnessed extensive adoption across various fields.

However, several factors contributed to the limitations in implementing AI solutions in healthcare. One major limitation was the stringent compliance regulations imposed by the Health Insurance Portability and Accountability Act (HIPAA), which restricts the public availability of healthcare data, even for

N. Thai-Nghe et al. (Eds.): ISDS 2023, CCIS 1949, pp. 15–30, 2024.
https://doi.org/10.1007/978-981-99-7649-2_2

research purposes [16]. As a result, the development of new AI-driven health-care systems faced significant challenges. Moreover, the medical data tended to be imbalanced and skewed due to disparities in the number of participants between patient groups and healthy individuals in medical studies. As a result, the sample sizes for data collection were typically small, making rectifying the data imbalance issue complicated. Additionally, the need for more transparency in data collection posed a critical concern when developing AI-based health-care systems. In a published report by the World Health Organization (WHO) [8], with the lack of physicians for patients in many developing countries, pro-viding timely care and medical services to needy individuals has become an enormous challenge. The scarcity of healthcare professionals created a situation where it is nearly impossible for medical professionals to cater to the growing healthcare demands promptly. This disparity in the healthcare workforce posed significant barriers to accessing essential medical care. Despite these challenges, the benefits of an AI-led healthcare system outweighed the obstacles. Develop-ing a robust and accurate predictive system that provides initial diagnoses and fulfills the minimum healthcare support required by individuals becomes crucial. Deep learning (DL) techniques demonstrated immense promise and potential in medicine, particularly in accurate prognosis and diagnosis [16]. DL algorithms transformed healthcare using large volumes of medical data and uncovering hid-den patterns, leading to better patient outcomes, better diagnostic accuracy, and advancements in medical knowledge.

Symptoms were primary clinical phenotypes that are significant resources but often neglected. They played a pivotal role in clinical diagnosis and treatment, serving as essential indicators of an individual's health status. For instance, when considering a heart attack, key symptoms encompass diverse manifestations such as chest pain, discomfort radiating to the arms, shoulders, jaw, neck, or back, feelings of weakness, lightheadedness, fainting, and shortness of breath. This broad spectrum of symptoms vividly demonstrated the interconnectedness of homeostatic mechanisms, whose disruptions ultimately cause the emergence of a disease. Community health professionals and general practitioners' knowledge regarding symptoms of specific diseases was primarily derived from their obser-vations within hospital environments [5]. It should be noted that symptoms, the most directly observable disease characteristics, served as the foundational elements for classifying clinical diseases.

Given the limitations of doctors in avoiding diagnostic errors due to restricted expertise and potential human negligence, a symptom-based prognosis model held significant potential in providing valuable assistance [17]. This model could assist physicians by providing a systematic approach to the prognosis based on a given set of symptoms, offering convenience and accessibility in the diagnostic process. However, developing and implementing this novel approach presented several unique challenges that must be addressed to ensure its effectiveness. The first challenge revolved around modeling the complex relationships between symptoms and diseases. The intricate interplay between symptoms and their cor-responding diseases required a robust framework that accurately captured these relationships. This involved understanding the various factors that influence the manifestation of symptoms and the underlying mechanisms that link them to specific diseases. Developing a comprehensive model that effectively captures and represents these relationships could significantly improve prognostic accuracy, leading to better patient outcomes and healthcare decision-making. The second challenge focused on improving the accuracy of the representation of the symptom set. Accurately representing the collective information became crucial to a reliable prognosis when dealing with symptoms. It was essential to consider the varying importance and relevance of individual symptoms.

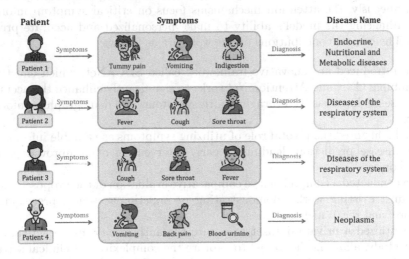

**Fig. 1.** An Illustrative Demonstration of Symptom-Based Disease Prediction

Addressing these challenges, DL methodologies, with their ability to uncover latent and hidden patterns within data, identified underlying connections between symptoms and diseases, thereby facilitating the development of an AI-based healthcare system [11]. The goal of this study was to predict diseases using their symptoms. The goal is to give doctors helpful resources to use while making diagnoses. We proposed a simplified approach that involves only presenting symptoms rather than relying on complex historical clinical records. Doctors may manually enter these symptoms, or computer programs may automatically extract them from medical records. The symptom-based disease prediction system ensures privacy and can be widely used to help doctors prescribe the proper medications using symptoms, which can provide information on a patient's physical condition while protecting their personal information. Figure 1 illustrates that patients 2 and 3 with identical symptoms, such as fever accompanied by cough and sore throat, are more likely to have the same disease diagnosed by medical professionals, specifically respiratory system diseases.

In this paper, we presented our proposed framework, known as **S**ymptoms-based **D**isease Prediction with **CNN**-**A**ttention **Net**works (SDCANet), with the primary goal of predicting diseases by leveraging both patient symptoms and demographics. Our framework addresses existing limitations in healthcare diagnosis, aiming to reduce diagnostic errors and enhance patient outcomes. Our model effectively analyzes and interprets the intricate relationships between symptoms and diseases by harnessing the power of 1D convolutional neural networks (1DCNN) and attention mechanisms. It overcomes the challenges posed by diverse symptom sets and considers the varying importance of individual symptoms, resulting in precise and reliable disease prediction. The incorporation of 1DCNN allows our model to capture sequential patterns and local dependencies within the symptom data, enabling a deeper understanding of disease dynamics. Simultaneously, the attention mechanisms focus on critical symptom information, enhancing the model's ability to make personalized and accurate predictions. The critical contributions of our work are as follows:

- We introduced an innovative framework called SDCANet, which effectively combines CNN and Attention Networks to significantly enhance the accuracy of disease diagnosis by leveraging both symptoms and demographics information.
- We highlighted the pivotal role of utilizing symptoms as valuable information for disease prediction, leading to early intervention and improved patient outcomes.
- We conducted a comprehensive comparative evaluation of the proposed model against existing works, demonstrating its superior accuracy, precision, F1 score, and recall metrics.
- We utilized a private dataset to ensure the authenticity and practicality of our study, allowing the model to capture the complexities of clinical settings and thereby increasing the relevance and applicability of the proposed disease prediction framework

The subsequent sections of this chapter are organized as follows. Section 2 provides a concise overview of current state-of-the-art approaches and the application of DL in disease prediction. Section 3 details the problem formulation relevant to this study. Section 4 presents an in-depth exploration of a DL-based approach for disease prediction. Subsequently, in Sect. 5, the proposed solution is evaluated and assessed. Finally, Sect. 6 concludes the chapter by summarizing the essential findings and implications of the research.

## 2   Related Work

### 2.1   Text Classification

There has been significant research on text classification and analysis methods, which are essential tasks in natural language processing (NLP). Several approaches have been proposed, including traditional machine learning (ML) methods such as support vector machines and decision trees and deep learning techniques such as CNNs and RNNs. In recent years, there has been increasing interest in using pre-trained language models, such as BERT and GPT-2, for text classification and analysis tasks. These models have achieved state-of-the-art performance in various NLP tasks, including sentiment analysis, named entity recognition, and question-answering. Significant progress has been made in developing text representation methods, including contextualized word embeddings. The text classification and analysis field rapidly evolved, with new methods and techniques being developed and tested regularly.

In medical diagnosis, a growing trend in recent years has been using NLP techniques to analyze text data. This approach has facilitated the development of novel methodologies to predict a diverse range of medical conditions based on information derived from textual data, including symptoms, demographics, and medical history. In particular, there were three broad categories of methods: rule-based, traditional ML-based, and DL-based. Rule-based approaches utilize expert knowledge to create rules to identify symptoms and infer diagnoses. Traditional ML-based approaches require labeled data sets to train statistical models, which are then used to predict diagnoses based on new text data. In contrast, DL-based approaches automatically use neural networks to learn relevant features from raw input data. Each method presented unique advantages and limitations, with the choice of which approach to use depending on the specific task requirements and available data. However, several challenges remained to be overcome, including dealing with imbalanced data sets and ensuring the reliability and interpretability of the results. Nonetheless, NLP techniques continued to promise to enhance medical diagnosis and patient outcomes, thereby reducing healthcare costs.

## 2.2 Symptoms-Based Disease Prediction Models

Many studies have proposed disease prediction models using patient-collected symptoms. These studies stressed the significance of utilizing the abundant symptom data to enhance diagnostic accuracy, enable early disease detection, and support proactive healthcare interventions. Researchers aimed to leverage advanced data analysis and machine learning to build robust models that could revolutionize disease prediction and enhance patient outcomes.

Kanchan et al. [2] introduced a comprehensive system for predicting diseases based on the symptoms exhibited by patients. To achieve accurate predictions, the researchers employed two machine learning algorithms, K Nearest Neighbor (KNN) and CNN. The CNN algorithm demonstrated an overall disease prediction accuracy of 84.5%, surpassing the performance of the KNN technique. However, it was observed that KNN required more time and memory resources than CNN. Besides, Keniya et al. [10] employed the KNN algorithm to predict diseases by assigning data points to the class containing most of the K closest data points. However, this method was susceptible to noise and missing data. Similarly, Taunk et al. [15] also utilized the KNN method and demonstrated its high Precision in various cases, including predicting diabetes and heart risks. However, there needed to be more data for disease classification. Moreover, Cao et al. [3] proposed a methodology that utilizes a Support Vector Machine (SVM) to classify diseases based on symptoms. The SVM model was effective in disease prediction but required more time for accurate predictions. The method had the potential for improved accuracy. However, it relied on classifying objects using a hyperplane, which was only partially effective. In the medical context, where symptoms correspond to multiple diseases, this binary classification approach was limited as it could only handle two classes, which needs to be improved for accurate diagnosis.

The work [14] proposed a method that utilized the Naïve Bayes algorithm to predict a limited number of diseases, including Diabetes, Malaria, Jaundice, Dengue, and Tuberculosis. However, their research did not involve working with an extensive data set to predict a broader range of diseases. Another approach [4] also utilized the Naïve Bayes classifier. However, their disease prediction model yielded poor accuracy, and they did not employ a standard dataset for training purposes. The work in [9] presented an approach to automate patient classification during hospital admission, focusing on symptoms extracted from text data. They leveraged the Bag of Words (BOW) model to generate word features from the textual information on diagnosing ten common diseases based on the set of symptoms, utilizing various machine learning algorithms such as Random Forest, SVM, Decision Tree, Multinomial Naive Bayes, Logistic Regression.

The approaches mentioned earlier have explored various machine-learning techniques for disease prediction. However, these existing works did not address factors such as efficiency, accuracy, the limited size of the data set used for model training, and the consideration of a restricted set of symptoms for disease diagnosis.

## 3   Problem Formulation

We aimed to tackle this challenge by developing a predictive model that effectively utilizes patients' symptoms and demographic information. All algorithms were presented using a single admission scenario to simplify the process. Our ultimate objective was to leverage this predictive model to provide a specific disease as a suitable treatment option based on a comprehensive analysis of the patient's symptoms and demographic factors. Hence, we strived to enhance the accuracy and efficiency of diagnosing patients and guided them toward the most appropriate course of treatment. The input and output were defined as follows:
**Input:** For each admission, the input data of our model consists of a patient demographics $p$ and set of symptoms $(s_1, s_2, ..., s_N)$

- The patient's demographic data, denoted as $p$, encompasses various attributes of the patient, such as age and sex, commonly documented in electronic health records (EHR). The demographic $p$ is a generated sentence; for instance, if the patient is a man and 24 years old, the demographic $p$ would be "He is 24 years old".
- A patient's set of symptoms $s$ contains multiple symptoms. We construct $s$ by discretizing each symptom.

**Output:** The predicted disease, denoted $\hat{y}$, represents the result of our model, which is determined based on the symptoms a patient exhibits and their demographic information.

## 4    Methodology

In this section, we present our proposed model **S**ymptoms-based **D**isease Prediction with **CNN**-**A**ttention **N**etworks (SDCANet) for predicting diseases by utilizing the symptoms and demographics of patients.

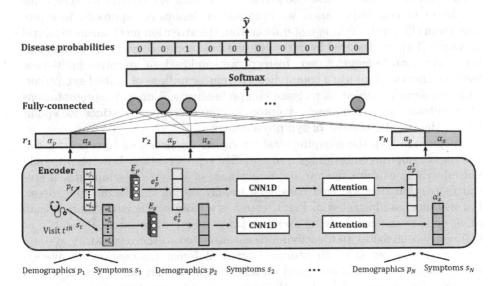

**Fig. 2.** The proposed model for symptom-based disease prediction

### 4.1    SDCANet Overview

Fig. 2 shows that the encoder block plays a vital role in our approach. SDCANet, specifically designed for predicting future diagnoses, incorporates two primary components: a patient representation encoder and a disease predictor. The patient representation encoder module teaches us intricate representations of symptoms and patient demographics. The accuracy and reliability of our predictions are enhanced by comprehensively understanding symptoms and demographic characteristics employing 1DCNN. Incorporating an attention mechanism further strengthens our ability to learn patient representations based on exhibited symptoms. This approach focuses on the relevant symptoms and their significance, capturing the unique characteristics of each patient. The attention-based approach ensures adaptability to individual patients, resulting in personalized and accurate disease predictions. Finally, disease probabilities are computed using the representations obtained from previous modules. Leveraging these learned representations and considering various factors based on attention scores, our model provides valuable insights into the likelihood of a patient having a specific disease.

## 4.2    Patient Representation Encoder

The objective of patient representation was to acquire a concise and informative vector that captures the patient's condition. During a clinical visit, doctors diagnose diseases by considering symptoms and demographic information. Our model also incorporated these two features in its analysis. Given that symptoms can be of varying importance, we proposed an innovative approach to represent them effectively. Taking inspiration from the attention mechanism in neural networks [1,6], we deviated from the conventional approach of merging all symptoms into a single lengthy text. Instead, we employed an attentive multi-view learning framework to learn comprehensive representations of unified symptoms. This framework enabled us to learn comprehensive and unified representations of symptoms by treating each symptom's information as a distinct viewpoint within the broader context of symptomatology.

Inspired in [17], the grouping strategy can be implemented to convert a collection of symptom embeddings into a unified representation. However, it could not effectively capture the varying importance of individual symptoms, a crucial factor that doctors must consider carefully during the diagnostic process. For instance, as illustrated in Fig. 1, cough is a more significant symptom than other symptoms (such as fever and sore throat) for Patient 2 and Patient 3, as it directly contributes to identifying diseases related to the respiratory system. Furthermore, given that symptoms occur at different frequencies, an averaging strategy alone would disregard unusual symptoms and fail to detect certain diseases. For example, Blood in the urine is a rare symptom associated with kidney stone disease, necessitating greater attention when representing a set of symptoms to accurately recommend diseases specific to the kidneys.

**Symptoms and Demographics Encoders.** The initial layer was word embedding, which transformed a symptom category from a sequence of words into a series of semantic vectors with lower dimensions. We denoted the word sequence of a category of demographics and symptoms as $[w_{p_1}^t, w_{p_2}^t, ..., w_{p_n}^t]$, $[w_{s_1}^t, w_{s_2}^t, ..., w_{s_n}^t]$, where $p_n$, $s_n$ represent the number of words of demographics and symptoms, respectively. Through the utilization of a word embedding look-up table $W_e \in \mathbb{R}^{V \times N}$, this sequence was converted into a sequence of word vectors $[e_1^t, e_2^t, ..., e_N^t]$ where $V$ represents the vocabulary size, while $N$ denoted the dimension of the word embeddings.

**Patient Representation.** The second layer consists of 1DCNN, which considers the importance of local word contexts within demographics for acquiring their representations. Similarly, in the case of symptom sets, specific symptoms held significant contextual information for determining the corresponding disease. For example, we considered the set of symptoms such as fever, cough, and sore throat. In this case, the local contexts of symptoms such as sore throat and cough were crucial for recognizing their relevance to respiratory-related conditions. Hence, we employed CNNs to effectively learn contextual representations of words by capturing their local contexts. We denoted the contextual representation of the $i$-th word as $c_i^t$, calculated as Eq. 1.

$$c_i^t = ReLU(F_t \times e_{(i-K):(i+K)}^t) + b_t \qquad (1)$$

where $e_{(i-K):(i+K)}^t$ is the concatenation of word embeddings from position $(i-K)$ to $(i + K)$. $F_t \in \mathbb{R}^{N_f \times (2K+1)D}$ and $b_t \in \mathbb{R}^{N_f}$ are the kernel and bias parameters of the CNN filters, $N_f$ is the number of CNN filters and $2K + 1$ is their window size. ReLU is the nonlinear activation function. The output of this layer is a sequence of contextual word representations, i.e., $[c_1^t, c_2^t, ..., c_N^t]$.

Then, we applied the third layer, which encompasses a word-level attention network [7], which addressed the fact that various words within the same symptom category tend to possess varying levels of informativeness when learning representations of symptoms. Certain symptoms held greater informativeness within a symptom category than others in representing the underlying disease. For example, in a set of symptoms such as fever, cough, and sore throat, we observed that the sore throat may carry more significance in indicating a specific disease than the Fever symptom. Acknowledging the significance of identifying crucial words within different symptom categories offered the potential to acquire more informative symptom representations. To accomplish this, we introduced a word-level attention network that selectively highlights important words within the context of each symptom category. The attention weight of the $i$-th word in a symptom category is denoted as $\alpha_i^t$, and formulated as Eqs. 2 and 3:

$$a_i^t = q_t^T tanh(V_t \times c_i^t + v_t) \qquad (2)$$

$$\alpha_i^t = \frac{exp(a_i^t)}{\Sigma_{j=1}^N exp(a_j^t)} \qquad (3)$$

where $V_t$ and $v_t$ are the projection parameters, $q_t$ denotes the attention query vector. The ultimate representation of a category of symptoms is determined by the summation of contextual representations of its words, each representation weighted by its corresponding attention weight. In other words, the final representation can be calculated as the sum of the contextual representations of the words multiplied by their respective attention weights. It is formulated as Eq. 4:

$$r_t = \Sigma_{j=1}^N \alpha_j^t c_j^t \qquad (4)$$

Within our SDCANet approach, the symptoms encoder module played a crucial role in acquiring representations for both historical symptoms reported by patients and the candidate symptoms that are to be recommended. This module was responsible for learning and capturing the essential features of these symptoms, enabling accurate representation and subsequent analysis.

## 4.3   Disease Predictor

In the disease prediction stage, attention scores played an important role in improving the accuracy and reliability of diagnoses. Attention scores are calculated by assigning importance scores to various demographic and symptom representation features. This allowed the model to focus on the most relevant information and capture its contributions to the disease prediction task. Different attention mechanisms, such as dot-product attention or self-attention, are employed depending on the model's architecture. By incorporating attention scores, the model gained the ability to extract meaningful patterns and relationships from input data effectively. The combination and integration of attention scores are achieved by concatenating or merging them to form a unified representation. These representations were then passed through a fully connected layer, where each neuron applies a weight to the corresponding attention score. This transformation and integration process enabled the model to incorporate the significance of different features and enhanced the overall understanding of the input data. The model became more adept at capturing the critical information necessary for accurate disease prediction by giving higher weights to essential features.

The softmax layer played a pivotal role in disease prediction by converting the transformed representation from the fully connected layer into probability distributions. Each disease was assigned a probability, indicating the likelihood of a patient having that specific disease based on their symptoms and demographic information. This enabled the model to quantify the confidence of its predictions and prioritize the most probable diseases. The model determined the predicted diagnosis by selecting the disease with the highest probability or applying a threshold. The utilization of attention scores and the softmax layer enhanced the accuracy of disease prediction and provided valuable insights for medical professionals in making informed decisions and improving patient outcomes.

## 5   Experiments

### 5.1   Dataset

In this investigation, an in-depth analysis was performed on the Patient Admission dataset [9], comprising 230,479 samples such as age, gender, and patient clinical symptoms. We used data from March 2016 to March 2021 from the Medical Center of My Tho City, Tien Giang in Vietnam, from the admissions and discharge office, outpatient department, accident, emergency department, and related reports. Patient information was collected through manual or semi-automatic retrieval, primarily utilizing the QRCode embedded in the patient's health insurance card. The dataset has fields including the patient's age (captured in the AGE field), gender (represented as 1 for male and 0 for female in the SEX field), an extensive compilation of clinical symptoms (stored within the CLINICAL SYMPTOMS field), and the ID DISEASES field showed the type of diagnosed disease by the patient's ICD10 disease code. The data set encompasses ten commonly encountered disease types in Vietnamese hospitals, as listed in Table 1. Documenting the clinical symptoms falls upon the medical staff stationed at the admission and discharge office. These healthcare professionals meticulously record the observed clinical symptoms upon the patient's declaration. The clinical symptom data is comprehensive, covering various aspects such as physical fitness, abnormal vital signs, and the manifestation of symptoms before and during the patient's arrival at the hospital.

**Table 1.** Statistical Analysis of a Patient Admission Dataset for Disease Prediction

| No. | Disease name | #samples |
|---|---|---|
| 1 | Neoplasms | 16271 |
| 2 | Endocrine, Nutritional and metabolic diseases | 38672 |
| 3 | Diseases of the eye and adnexa | 18443 |
| 4 | Diseases of the circulatory system | 37782 |
| 5 | Diseases of the respiratory system | 41888 |
| 6 | Diseases of the skin and subcutaneous tissue | 7044 |
| 7 | Diseases of the musculoskeletal system and connective tissue | 35427 |
| 8 | Diseases of the genitourinary system B212 | 17503 |
| 9 | Pregnancy, childbirth, and the puerperium | 3666 |
| 10 | Injury, poisoning and certain other consequences of external | 13783 |

## 5.2   Experimental Setup

The experimental results were obtained by testing an Ubuntu 18.04172 operating system server. The server had 20 different CPU configurations and boasted a substantial 64GB RAM capacity. The convolutional neural network in the experiments incorporated an attention mechanism constructed using the Keras library. The experiments are evaluated through 5-fold cross-validation.

## 5.3   Results

Table 2 demonstrates a comparison of different metrics, such as Accuracy (ACC), Precision, Recall, F1 score, and Area Under the Curve (AUC), for various optimizers, including Adam, SGD, and RMSProp, across different values of the MAX_SYMPTOM_LEN parameter (100, 150, and 200), which represents the maximum length of symptoms used. Besides these hyper-parameters, some others are a batch size of 64, a learning rate of 0.001, and the ReduceLROnPlateau scheduler, which are also utilized. Overall, the results show that the optimizers achieve comparable performance across most metrics and values of the number of symptoms. In terms of accuracy, all optimizers perform reasonably well, with

**Table 2.** Performance Evaluation (mean ± standard deviation) of SDCANet Model. The best results are marked in bold, and the second-highest results are underlined.

| Metrics | Optimizers | Maximum number of symptoms | | |
|---|---|---|---|---|
| | | 100 | 150 | 200 |
| ACC | Adam | 0.879 ± 0.0044 | 0.880 ± 0.0030 | 0.879 ± 0.0037 |
| | SGD | 0.879 ± 0.0052 | 0.879 ± 0.0048 | 0.878 ± 0.0045 |
| | RMSProp | 0.882 ± 0.0007 | 0.882 ± 0.0008 | **0.883 ± 0.0024** |
| Precision | Adam | 0.900 ± 0.0008 | 0.898 ± 0.0017 | 0.898 ± 0.0017 |
| | SGD | 0.899 ± 0.0020 | 0.899 ± 0.0018 | 0.899 ± 0.0027 |
| | RMSProp | 0.901 ± 0.0015 | 0.902 ± 0.0025 | **0.906 ± 0.0018** |
| Recall | Adam | 0.860 ± 0.0095 | 0.864 ± 0.0064 | 0.862 ± 0.0071 |
| | SGD | 0.860 ± 0.0094 | 0.860 ± 0.0086 | 0.859 ± 0.0082 |
| | RMSProp | **0.866 ± 0.0027** | 0.864 ± 0.0024 | 0.865 ± 0.0046 |
| F1 | Adam | 0.879 ± 0.0055 | 0.880 ± 0.0038 | 0.878 ± 0.0040 |
| | SGD | 0.878 ± 0.0061 | 0.878 ± 0.0055 | 0.878 ± 0.0057 |
| | RMSProp | **0.882 ± 0.0010** | 0.882 ± 0.0011 | 0.881 ± 0.0027 |
| AUC | Adam | 0.985 ± 0.0011 | 0.985 ± 0.0010 | 0.985 ± 0.0006 |
| | SGD | 0.986 ± 0.0009 | 0.986 ± 0.0007 | 0.986 ± 0.0010 |
| | RMSProp | 0.986 ± 0.0005 | 0.986 ± 0.0002 | **0.987 ± 0.0005** |

scores ranging from 0.879 to 0.883. RMSProp consistently achieves the highest accuracy across all values of the number of symptoms, with the highest score obtained for the number of symptoms equal to 200 being 0.883. Regarding the Precision metric, all optimizers achieve scores greater than 0.89, indicating a relatively high proportion of correctly predicted positive instances. Similar to accuracy, RMSProp tends to perform slightly better, with the highest Precision of 0.906 for the number of symptoms equal to 200.

Moreover, Recall values vary between 0.859 and 0.866, suggesting that the optimizers have slightly more difficulty identifying all positive instances correctly. However, the differences in the recall scores are relatively small, and RMSProp again tends to perform slightly better for the number of symptoms equal to 100. Besides, the F1-score value ranges from 0.878 to 0.882. Similarly to the previous metrics, RMSProp achieves the highest F1 scores, particularly for the number of symptoms equal to 100, obtaining a value of 0.882. The AUC measures the classifier's overall discriminative power. All optimizers achieve high AUC scores above 0.98, indicating strong performance. RMSProp consistently outperforms the other optimizers, with the highest AUC of 0.987 for the number of symptoms equal to 200. In summary, while all optimizers demonstrate competitive performance across most metrics, RMSProp performs slightly better, especially in ACC, Precision, F1, and AUC. RMSProp may be the preferred optimizer, particularly for the number of symptoms equal to 200.

**Table 3.** Comparison results of the proposed model SDCANet with other baselines in percent. The best results are marked in bold

| Method | ACC | Precision | Recall | F1 | AUC |
|---|---|---|---|---|---|
| Logistic Regression [9] | 0.791 | 0.799 | 0.791 | 0.795 | – |
| Deep Bidirectional LSTM with Tokenizer combined with Sequences [12] | 0.873 | 0.892 | 0.857 | 0.874 | 0.982 |
| **SDCANet (Ours)** | **0.883** | **0.906** | **0.865** | **0.881** | **0.987** |

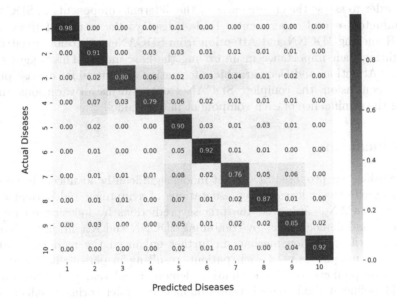

**Fig. 3.** Confusion matrix on Patient Admission Dataset predicted by SDCANet. Disease name numbers (from 1 to 10) are mapped in Table 1.

Table 3 compares the performance of the proposed model SDCANet with several baseline methods across different evaluation metrics, namely Accuracy (ACC), Precision, Recall, and F1 score. Regarding Accuracy, the proposed model outperforms all the baseline methods from previous studies, achieving an impressive accuracy of 88.3%. From the result reports as shown in Fig. 3, we calculate the confusion matrix, where each predicted outcome is derived from averaging the results across five different folds. Our observations show that the methods demonstrate strong performance in predicting conditions related to Neoplasms (No. 1), Endocrine, Nutritional, and metabolic diseases (No. 2), Diseases of the respiratory system (No. 5), Diseases of the skin and subcutaneous tissue (No. 6), as well as Injury, poisoning and certain other consequences of external (No. 10). All these mentioned diseases get performance over 90%.

**Table 4.** Ablation Study Results for SDCANet

| Method | ACC | Precision | Recall | F1 | AUC |
|---|---|---|---|---|---|
| SDCANet w/o Demographics | 0.873 | 0.899 | 0.849 | 0.872 | 0.982 |
| SDCANet w/o 1DCNN | 0.863 | 0.887 | 0.840 | 0.862 | 0.983 |
| SDCANet w/o Attention | 0.871 | 0.894 | 0.849 | 0.870 | 0.983 |
| **SDCANet** | **0.883** | **0.906** | **0.865** | **0.881** | **0.987** |

In order to assess the effectiveness of the different components in SDCANet, we conducted several experiments. Table 4 displays the results of these variations. Removing 1DCNN and Attention from SDCANet led to lower outcomes, highlighting their importance in improving the base model. This suggests that 1DCNN, Attention, and Demographic information are crucial for disease prediction. In conclusion, the complete SDCANet outperforms all variations, underscoring the significance of each component in our model.

## 6    Conclusion

In conclusion, our proposed SDCANet model significantly advanced disease prediction by effectively learning representations of symptoms and patient demographics. SDCANet improved the precise predictions by focusing on relevant symptoms and demographics by combining 1DCNN and Attention mechanism. Leveraging the attention mechanism further enhanced the model's ability to capture the unique traits of each patient, resulting in personalized and accurate disease predictions. Our evaluation demonstrated that SDCANet outperformed baseline methods across evaluation metrics, underscoring its effectiveness in disease prediction. Future research can explore integrating additional features like medications, procedures, and lab tests to enhance the model's accuracy. Additionally, incorporating longitudinal and genomics data can provide a better understanding of disease progression, enabling personalized treatment strategies. These advancements hold the potential to enhance healthcare decision-making and improve patient outcomes in the field of disease prediction.

## References

1. Vaswani, A., et al.: Attention is all you need. In: Proceedings of the 31st International Conference on Neural Information Processing Systems (NIPS 2017). Curran Associates Inc., Red Hook, NY, USA, pp. 6000–6010 (2017) .https://doi.org/10.5555/3295222.3295349
2. Kanchan, B.D., Kishor, M.M.: Study of machine learning algorithms for special disease prediction using principal of component analysis. In: IEEE International Conference on Global Trends in Signal Processing Information Computing and Communication (ICGTSPICC) (2016). https://doi.org/10.1109/ICGTSPICC.2016.7955260

3. Cao, J., Wang, M., Li, Y., Zhang, Q.: Improved support vector machine classification algorithm based on adaptive feature weight updating in the Hadoop cluster environment. PloS ONE **14**(4), e0215136 (2019). https://doi.org/10.1371/journal.pone.0215136
4. Chhogyal, K., Nayak, A.: An empirical study of a simple Naive Bayes classifier based on ranking functions. In: Kang, B.H., Bai, Q. (eds.) AI 2016. LNCS (LNAI), vol. 9992, pp. 324–331. Springer, Cham (2016). https://doi.org/10.1007/978-3-319-50127-7_27
5. Chen, J., Li, D., Chen, Q., Zhou, W., Liu, X.: Diaformer: automatic diagnosis via Symptoms Sequence Generation. In: AAAI Conference on Artificial Intelligence (2021). https://arxiv.org/abs/2112.10433
6. Wu, C., et al.: Neural news recommendation with attentive multi-view learning. In: Proceedings of the 28th International Joint Conference on Artificial Intelligence (IJCAI 2019), AAAI Press, pp. 3863–3869 (2019)
7. Wu, C., Wu, F., Liu, J., He, S., Huang, Y., Xie, X.: Neural demographic prediction using search query. In: Proceedings of the Twelfth ACM International Conference on Web Search and Data Mining (WSDM 2019). Association for Computing Machinery, New York, NY, USA, pp. 654–662 (2019). https://doi.org/10.1145/3289600.3291034
8. Guilbert, J.J.: The world health report 2006: working together for health. Educ. Health (Abingdon, England) **19**(3), 385–387 (2006). https://doi.org/10.1080/13576280600937911
9. Le, K.D.D., Luong, H.H., Nguyen, H.T.: Patient classification based on symptoms using machine learning algorithms supporting hospital admission. In: Cong Vinh, P., Huu Nhan, N. (eds.) ICTCC 2021. LNICST, vol. 408, pp. 40–50. Springer, Cham (2021). https://doi.org/10.1007/978-3-030-92942-8_4
10. Keniya, R., et al.: Disease prediction from various symptoms using machine learning. SSRN 3661426 (2020). https://doi.org/10.2139/ssrn.3661426
11. Kao, H.-C., Tang, K.-F., & Chang, E. (2018). Context-Aware Symptom Checking for Disease Diagnosis Using Hierarchical Reinforcement Learning. Proceedings of the AAAI Conference on Artificial Intelligence, 32(1). https://doi.org/10.1609/aaai.v32i1.11902
12. Nguyen, H.T., Dang Le, K.D., Pham, N.H., et al.: Deep bidirectional LSTM for disease classification supporting hospital admission based on pre-diagnosis: a case study in Vietnam. Int. J. Inf. Tecnol. **15**, 2677–2685 (2023). https://doi.org/10.1007/s41870-023-01283-x
13. Milella, F., Minelli, E.A., Strozzi, F., Croce, D.: Change and innovation in healthcare: findings from literature. ClinicoEconomics Outcomes Res. **2021**, 395–408 (2021). https://doi.org/10.2147/CEOR.S301169
14. Pingale, K., Surwase, S., Kulkarni, V., Sarage, S., Karve, A.: Disease prediction using machine learning. Int. Res. J. Eng. Technol. (IRJET) **6**(2019), 831–833 (2019). https://doi.org/10.1126/science.1065467
15. Taunk, K., De, S., Verma, S., wetapadma, A.: A brief review of the nearest neighbor algorithm for learning and classification. In: 2019 International Conference on Intelligent Computing and Control Systems (ICCS), pp. 1255–1260. IEEE (2019). https://doi.org/10.1109/ICCS45141.2019.9065747

16. Islam, S.R., Sinha, R., Maity, S.P., Ray, A.K.: Deep learning on symptoms in disease prediction. Mach. Learn. Healthcare Appl. (2021). https://doi.org/10.1002/9781119792611.ch5
17. Tan, Y., et al.: 4SDrug: symptom-based set-to-set small and safe drug recommendation. In: Proceedings of the 28th ACM SIGKDD Conference on Knowledge Discovery and Data Mining (KDD 2022), New York, NY, USA, pp. 3970–3980. Association for Computing Machinery (2022). https://doi.org/10.1145/3534678.3539089

# Development of a New Acoustic System for Nondestructive Internal Quality Assessment of Fruits

Nhut-Thanh Tran[1]([✉]), Cat-Tuong Nguyen[1], Huu-Phuoc Nguyen[1],
Gia-Thuan Truong[1], Chanh-Nghiem Nguyen[1], and Masayuki Fukuzawa[2]

[1] Faculty of Automation Engineering, Can Tho University, Can Tho 94000, Vietnam
nhutthanh@ctu.edu.vn
[2] Graduate School of Science and Technology, Kyoto Institute of Technology, Kyoto 606-8585,
Japan

**Abstract.** A new acoustic analysis system based on MEMS microphones has been
developed for assessing the internal quality of fruits. Although acoustic analysis
is one of the effective techniques for assessing the internal qualities such as firmness, maturity, hollowness and sweetness of fruits, it has not been widely applied
for fruits quality assessment in practice because the conventional apparatus was
complexed and bulky to suppress the effects of surrounding noise and to maintain a good frequency response of the system. This study aimed to achieve both
noise suppression and good frequency response with a simple system structure
using an ultra-low-cost MEMS microphone. By placing the MEMS microphone
inside a specially-designed fruit holder and optimizing its arrangement with a tapping module, the effect of ambient noise was significantly suppressed compared
with the MEMS microphone alone, realizing a drastic simplification of the system. The simple correlation analysis of watermelon sweetness and the received
acoustic frequency revealed a high correlation coefficient of –0.82. This result
not only demonstrates a high potential of the proposed system in non-destructive
internal quality assessment of fruits, and also strongly suggests the possibility of
further study on acoustic analysis based on machine-learning with large-scale data
collected effectively by using the developed system.

**Keywords:** Acoustic Analysis System · Fruit Quality Assessment · MEMS
Microphone · Watermelon Sweetness

## 1 Introduction

Watermelon is a popular edible fruit with high moisture content and vitamins. It is
grown in many countries in the world, with more than 1,000 varieties. The quality of
watermelon is assessed through several parameters. They are divided into two groups –
external qualities such as size, shape, and skin defects, and internal qualities such as
maturity, sweetness, nutrient, and hollowness [1]. The external quality parameters are
exactly assessed by physical measurements or visual inspections. Although internal

ones can be exactly evaluated by sampling and measuring destructively with specific devices, it is time-consuming and costly. Besides, an empirical method is also applied by thumbing or tapping on watermelons and listening to the reflected sound to predict the maturity or hollowness of watermelons. This empirical method is also applied to check the internal quality of other fruits such as durian, melon, coconut. However, this method is inaccurate and not suitable for applications of inline inspection.

Some non-destructive technologies for evaluating the internal quality of watermelons and other fruits have received considerable research attention in recent years, such as acoustic analysis, near-infrared (NIR) spectroscopy, machine vision, electronic nose, dielectric properties, nuclear magnetic resonance, and laser Doppler vibrometer [2, 3]. Among these technologies, acoustic analysis and NIR spectroscopy have received the most attention. Acoustic analysis is primarily used to evaluate some internal quality parameters of watermelon such as maturity, internal defects, firmness, while NIR spectroscopy is mainly used for assessing sweetness, maturity, and firmness [1]. Although NIR spectroscopy technology has been successfully applied to some thin-skin fruits, its application to watermelons and other thick-skin fruits is still limited due to light source penetration and overheating problems. Therefore, acoustic analysis technology is still more popular for evaluating the internal quality of thick-skin fruits because of cost-effectiveness and simple operation.

Various previous studies have successfully applied acoustic analysis technology for classifying the watermelon ripeness [4–7] and firmness [8]. This technology was also applied for evaluating the maturity or ripeness of durians [9, 10]. Khoshnam et al. implemented this acoustic technology to test the ripeness of the melon at different stages [11]. Although these studies have demonstrated the applicability of acoustic analysis in assessing the internal quality of watermelons and other fruits, the hardware design for the acoustic acquisition and insulator in these studies is not really suitable for practical applications. Therefore, it is necessary to renovate the hardware setup for practical application of inline inspection. In this study, a new acoustic analysis system based on a combination of an ultra-low-cost MEMS microphone and an optimal system arrangement is proposed for practical, high-reliability and cost-effective applications. Besides, the potential of the proposed system was demonstrated through the simple correlation analysis of watermelon sweetness and the received acoustic frequency.

## 2 Materials and Method

### 2.1 Proposed System

The proposed system was designed to include three main modules: a tapping module, a fruit holder with a MEMS microphone and acoustic insulators, and a sound recorder. The schematic diagram of the system and a picture of a prototype are shown in Fig. 1.

The tapping module aims to apply a constant impact force to a single point on the surface of a watermelon sample. It consisted of a wooden bar with a rubber ball (32 g) at one end, a motor with a disk cam on the shaft, and a support base. The constant impact force was produced by the free fall of the wooden bar lifted by the disk cam mechanism rotated by the motor.

The fruit holder is responsible both for holding the watermelon sample in the desired position and for preventing surrounding noises. It was made from a plastic bucket pasted with conventional plastic foam sheets as acoustic insulators on both sides of the bucket. A MEMS microphone was placed in an optimized position in the holder. One of the features of this system is to employ an ultra-low-cost MEMS microphone (Invensense INMP441) instead of expensive conventional microphones as in previous studies. Since this MEMS microphone integrated a signal conditioning circuit, a band-pass filter, and an analog-to-digital converter inside the package, it has clear advantages of giving a good performance of the analog system (i.e., 61 dB of SNR) and allowing the direct transfer of acoustic signal to the recorder as high-precision (24 bit) digital data. Some key features of this microphone are shown in Table 1.

The sound recorder is a device to record the acoustic signal. In this study, it was constructed using a single board computer (Raspberry Pi) because it equips a I2S digital interface which is required to connect to the MEMS sensor directly. A recording application was used to record the acoustic signal, while a software for acoustic analysis was originally developed to be performed on the Raspberry Pi.

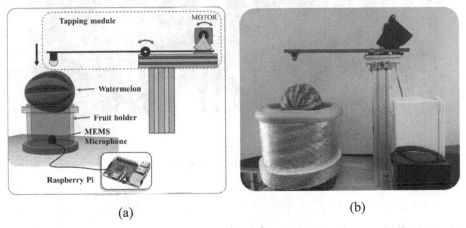

(a)                                  (b)

**Fig. 1.** (a) Schematic diagram of the proposed acoustic analysis system and (b) a system prototype

**Table 1.** Key features of INMP441 MEMS Microphone

| Feature | Description |
| --- | --- |
| Supply voltage | 1.6–3.3 V |
| Frequency response | 60 Hz to 15 kHz |
| Signal-to-noise ratio | 61 dB (A-weighted) |
| Data Precision | 24-bit |
| Current consumption | 1.4 mA |

## 2.2  Sample Preparation

A total of 18 watermelons (*Thanh Long F1 T522*) were harvested in Hau Giang province, Vietnam (Fig. 2). These fruits were 60 days after planting at harvest time. The average weight of these samples was around 3.0 kg. They were stored in a room with the temperature of 28 to 30 °C for 24 h before performing the experiments.

**Fig. 2.** Watermelon samples.

## 2.3  Acoustic Signal Acquisition and Analysis

In the experiment of acoustic signal acquisition, a watermelon was placed on the fruit holder with the horizontal stem–bloom orientation. Each watermelon was tapped six times at three positions around the middle area to get the acoustic signals of six watermelon samples. The recorded signals of the watermelon samples were converted to the frequency domain by the fast Fourier transformation algorithm. From the frequency-domain signal, a frequency with the largest amplitude of each watermelon sample, called $f_{max}$, was extracted.

## 2.4  Watermelon Sweetness Measurement

Figure 3 presents a procedure for measuring the sweetness of watermelon samples. After the acoustic signal acquisition process, the watermelons were cut into a slice with a thickness of 4 cm at the tapping region (at the middle of the watermelons). The watermelon samples were then peeled (about 1 cm) and squeezed to get the sample juice. Finally, the sweetness value of each sample was obtained by calculating the average of three measurement values of Brix with a pocket Brix-Acidity meter (model PAL-BX|ACID15 Master Kit, Atago Co., Ltd., Tokyo, Japan). This pocket meter has a sweetness measurement range from 0 to 90 °Bx with the resolution and accuracy of 0.1 and ± 0.2 °Bx respectively.

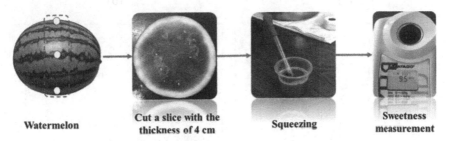

| Watermelon | Cut a slice with the thickness of 4 cm | Squeezing | Sweetness measurement |

**Fig. 3.**  The procedure for watermelon sweetness measurement

# 3  Results and Discussion

## 3.1  Functional Performance Testing

### Performance of Acoustic Insulator

The performance of the acoustic insulator in the developed system was evaluated through two experiments (without and with insulator) under the common condition of surrounding noises. They include the wind noises of a ceiling fan and the sound of music being played at a high volume in a closed room. In the experiment without the insulator, the watermelon was placed on a table and next to the microphone, while the watermelon was placed on the designed fruit holder with the microphone inside in the experiment with the insulator. Figure 4 shows the recorded watermelon signals of the experiments without and with the insulator.

The result of the experiment without the insulator showed that the recorded signal exhibited the peaks at frequencies around 190, 245, 363, and 379 Hz, as shown in Fig. 4b. By using the designed fruit holder, most of the high-frequency peaks in the previous experiment were significantly suppressed. The recorded signal had only one dominant signal at a frequency of around 190 Hz (Fig. 4d). This frequency can be interpreted as the watermelon frequency when tapping and others can be done as noises. The experimental results of these two experiments demonstrated a good performance of acoustic insulation in the system using the designed fruit holder

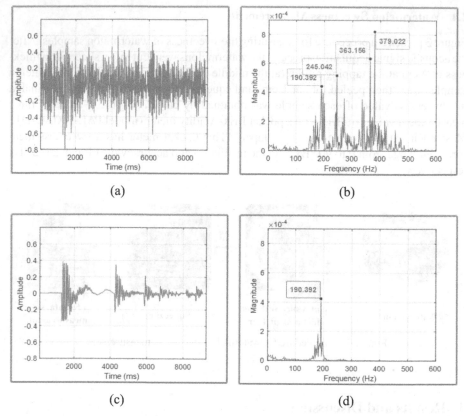

Fig. 4. Recorded watermelon signals in time and frequency domains (a) without and (b) with acoustic insulator.

## Performance of INMP441 MEMS Microphone

To check the performance of the MEMS microphone, a conventional microphone (*Rode SmartLav Plus*) used in previous studies was used to compare the recording results. Figure 5 presents the acoustic signals of a watermelon in time and frequency domains, recorded with the two microphones. Although the recorded signal amplitude of the MEMS microphone was reduced by half compared to that of the conventional one (Figs. 5a and 5c), the $f_{max}$ values were almost the same (Figs. 5b and 5d).

This result demonstrated that the MEMS microphone can completely be used to replace conventional microphones in assessing the internal quality of fruits based on acoustic analysis. The amplitude difference between the two microphones could be due to the position of the microphone installation and the tapping.

For evaluating the stability of the system, the acoustic signals of a watermelon sample were collected simultaneously by the MEMS and conventional microphones while this sample was tapped in 6 times at one position. The $f_{max}$ values of these signals are shown in Table 2.

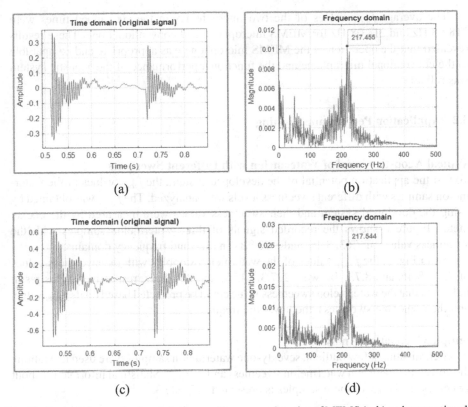

**Fig. 5.** Recorded acoustic signals in time and frequency domains of MEMS (a, b) and conventional (c, d) microphones.

**Table 2.** The $f_{max}$ values of the sample tapped in six times at a one position

| No | MEMS microphone (Hz) | Conventional microphone (Hz) | Difference (%) |
|---|---|---|---|
| 1 | 220.7 | 217.2 | 1.59 |
| 2 | 216.8 | 217.9 | 0.51 |
| 3 | 220.2 | 217.9 | 1.04 |
| 4 | 217.4 | 217.5 | 0.05 |
| 5 | 217.3 | 217.9 | 0.28 |
| 6 | 219.7 | 220.1 | 0.18 |
| **Average** | **218.68** | **218.08** | **0.61** |

Table 2 shows that the recorded frequencies of the MEMS and conventional microphones in each tapping were almost the same because their differences were 1.59, 0.05, and 0.61% for the largest, smallest, and average ones, respectively.

The average $f_{max}$ values of the two microphones in the six tapped times were 218.68 Hz and 218.08 Hz for MEMS microphone and conventional one. These results revealed that the operation of the MEMS microphone was appropriate and comparable to the conventional microphone and the functional performance of the acoustic system was reliable.

## 3.2 Application Performance Testing

**Typical Acoustic Signal of Watermelon with Different Sweetness**
To test the application potential of the developed system, the $f_{max}$ values of the watermelon samples with different sweetness levels were analyzed. The $f_{max}$ was obtained by using the developed system and their sweetness values were measured by the pocket meter. Figure 6 shows the recorded signals of three watermelon samples with the sweetness values of 7.63, 8.46, and 8.73 °Bx in time and frequency domains.

From Fig. 6, the $f_{max}$ values of the watermelon samples with the sweetness values of 7.63, 8.46, and 8.73 °Bx were 227.3, 215.7, and 204.1 Hz, respectively. This result indicated that the watermelon sweetness value could be predicted based on its $f_{max}$ value, and the higher the sweetness, the lower the frequency.

**Simple Correlation Analysis**
After eliminating a few outliers, seventy-one watermelon samples were used to examine the correlation between watermelon sweetness and its $f_{max}$. Statistical information about sweetness and $f_{max}$ of these samples is presented in Table 3.

**Table 3.** Statistical information of 71 watermelon samples

| Measure | Sweetness (°Bx) | $f_{max}$ (Hz) |
|---|---|---|
| Min | 7.73 | 206.05 |
| Max | 9.13 | 261.72 |
| Average | 8.47 | 226.41 |
| Standard deviation | 0.36 | 13.45 |

The $f_{max}$ values of the samples were from 206.05 to 226.41 Hz. The corresponding sweetness values of these samples ranged from 7.73 to 9.13 °Bx. There was an inverse correlation between the watermelon sweetness and its $f_{max}$ value with a high correlation coefficient of −0.82, as shown in Fig. 7. This result was also consistent with the results of previous studies on predicting watermelon sweetness by $f_{max}$ [2]. These results suggested that the proposed system had a great potential for predicting or classifying watermelon sweetness.

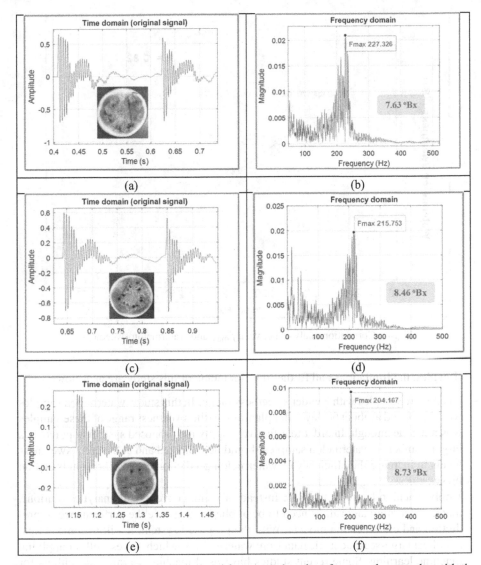

**Fig. 6.** Typical acoustic signals in time and frequency domains of watermelon samples with the sweetness values of 7.63 °Bx (a), 8.46 °Bx (b), and 8.73 °Bx (c).

### 3.3 Discussion

A cost-effective and high-performance acoustic analysis system was successfully developed. By placing the ultra-low-cost MEMS microphone inside the fruit holder made of common materials, the recorded acoustic signal was isolated from the noise of the environment. Moreover, with the optimized arrangement of the tapping module and the fruit holder, this developed system can be easily integrated into the conveyor system for automatic and inline assessment of fruit quality.

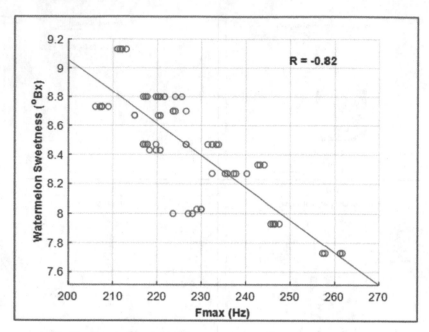

**Fig. 7.** Correlation analysis between $f_{max}$ and watermelon sweetness

Besides, the limitations and further developments of this study are as follows:

− Need more samples with a wider sweetness range: In this study, watermelon samples were harvested at about 60 days after planting, so the sweetness range of these samples was not wide enough. In order to effectively apply the proposed system in practice, a larger number of watermelon samples should be collected and at different sweetness levels to increase the efficiency of watermelon prediction or classification based on its sweetness.
− Apply machine learning analysis: Instead of using conventional analytic solutions, machine learning algorithms need to be applied to improve the performance of prediction and classification models. With the proposed system, it will become simple to collect large-scale acoustic data from watermelons, which serves well for applying machine learning. Some recent studies have used machine learning algorithms for quality assessment of watermelons [6, 12, 13] and coconuts [14, 15].

## 4  Conclusion

In this study, a simple and low-cost acoustic analysis system based on an ultra-low-cost MEMS microphone and the special fruit holder for internal quality assessment of fruits was proposed and validated. By optimizing the arrangement of the system components, the received acoustic signal was not affected by the surrounding noise. The performance of the MEMS microphone was also comparable to that of the conventional microphone used in previous studies. A strong correlation result of watermelon sweetness and the received acoustic frequency not only demonstrated the feasibility of the proposed system for assessing the internal quality of fruits in practice, but also strongly suggested the possibility of further study on acoustic analysis based on machine-learning with large-scale data collected effectively by using the developed system.

**Acknowledgments.** This work was supported by JSPS Core-to-Core Program (grant number: JPJSCCB20230005).

# References

1. Mohd Ali, M., Hashim, N., Bejo, S.K., Shamsudin, R.: Rapid and nondestructive techniques for internal and external quality evaluation of watermelons: a review. Sci. Hortic. (Amsterdam) **225**, 689–699 (2017). https://doi.org/10.1016/j.scienta.2017.08.012
2. Jie, D., Wei, X.: Review on the recent progress of non-destructive detection technology for internal quality of watermelon. Comput. Electron. Agric. **151**, 156–164 (2018). https://doi.org/10.1016/j.compag.2018.05.031
3. Abbaszadeh, R., Rajabipour, A., Delshad, M., Mahjub, M., Ahmadi, H., Laguë, C.: Application of vibration response for the nondestructive ripeness evaluation of watermelons. Aust. J. Crop. Sci. **5**, 920–925 (2011)
4. Chen, X., Yuan, P., Deng, X.: Watermelon ripeness detection by wavelet multiresolution decomposition of acoustic impulse response signals. Postharvest Biol. Technol. **142**, 135–141 (2018). https://doi.org/10.1016/j.postharvbio.2017.08.018
5. Baki, S.R., Yassin, I.M., Hasliza, A.H., Zabidi, A.: Non-destructive classification of watermelon ripeness using mel-frequency cepstrum coefficients and multilayer perceptrons. In: The 2010 International Joint Conference on Neural Networks (IJCNN), pp. 1–6. IEEE (2010)
6. Chawgien, K., Kiattisin, S.: Machine learning techniques for classifying the sweetness of watermelon using acoustic signal and image processing. Comput. Electron. Agric. **181**, 105938 (2021). https://doi.org/10.1016/j.compag.2020.105938
7. Zeng, W., Huang, X., Müller Arisona, S., McLoughlin, I.V.: Classifying watermelon ripeness by analysing acoustic signals using mobile devices. Pers. Ubiquitous Comput. **18**, 1753–1762 (2014). https://doi.org/10.1007/s00779-013-0706-7
8. Liza Pintor, A.C., Anthony Magpantay, M.A., Santiago, M.R.: Development of an android-based maturity detector mobile application for watermelons [Citrullus Lanatus (Thunb.) Matsum. & Nakai] using acoustic impulse response. Philipp. e-Journal Appl. Res. Dev. **6**, 44–56 (2016)
9. Phoophuangpairoj, R.: Durian ripeness striking sound recognition using N-gram models with N-best lists and majority voting. In: Boonkrong, S., Unger, H., Meesad, P. (eds.) Recent Advances in Information and Communication Technology. AISC, vol. 265, pp. 167–176. Springer, Cham (2014). https://doi.org/10.1007/978-3-319-06538-0_17
10. Terdwongworakul, A., Neamsorn, N.: Non-destructive maturity measurement of 'Montong' durian using stem strength and resonant frequency. Eng. Appl. Sci. Res. **33**(5), 555–563 (2013). https://ph01.tci-thaijo.org/index.php/easr/article/view/6017
11. Khoshnam, F., Namjoo, M., Golbakhshi, H.: Acoustic testing for melon fruit ripeness evaluation during different stages of ripening. Agric. Conspec. Sci. **80**(4), 197–204 (2015).https://hrcak.srce.hr/en/clanak/231272
12. Rajan, R., Reshma, R.S.: Non-destructive classification of watermelon ripeness using acoustic cues. In: Emerging Trends in Engineering, Science and Technology for Society, Energy and Environment, pp. 739–744. CRC Press (2018)

13. Choe, U., Kang, H., Ham, J., Ri, K., Choe, U.: Maturity assessment of watermelon by acoustic method. Sci. Hortic. (Amsterdam). **293**, 110735 (2022). https://doi.org/10.1016/j.scienta.2021.110735
14. Fadchar, N.A., Dela Cruz, J.C.: A non-destructive approach of young coconut maturity detection using acoustic vibration and neural network. In: 2020 16th IEEE International Colloquium on Signal Processing & Its Applications (CSPA), pp. 136–137. IEEE (2020). https://doi.org/10.1109/CSPA48992.2020.9068723
15. Caladcad, J.A., et al.: Determining Philippine coconut maturity level using machine learning algorithms based on acoustic signal. Comput. Electron. Agric. **172**, 105327 (2020). https://doi.org/10.1016/j.compag.2020.105327

# MR-Unet: Modified Recurrent Unet for Medical Image Segmentation

Song-Toan Tran[1]([✉]) [iD], Ching-Hwa Cheng[2], Don-Gey Liu[2]([✉]), Phuong-Thao Cao[1], and Tan-Hung Pham[1]

[1] Tra Vinh University, Tra Vinh 87000, Vietnam
{tstoan1512,cpthao,pthung}@tvu.edu.vn
[2] Feng Chia University, Taichung 40724, Taiwan, R.O.C.
{chengch,dgliu}@fcu.edu.tw

**Abstract.** In recent years, there has been significant interest in medical image segmentation. Traditional methods are being surpassed by deep learning, which has demonstrated their superiority. However, one drawback of using standard convolutional neural networks (CNNs) in deep learning is their large model size, which leads to excessive memory consumption and longer computation time. To address this issue, a novel model called MR-Unet was introduced in this study. MR-Unet combines the advantages of Recurrent Convolutional Neural Networks (RCNN) with the multiple layers Unet ($U^n$-Net) architecture. By incorporating RCNN and leveraging all output feature maps in the convolution units of the network nodes, the overall network size was reduced. To assess the effectiveness of our proposed model, we conducted experiments on liver segmentation using the LiTS 2017 dataset, spleen segmentation using the Medical Segmentation Decathlon Challenge 2018 datasets, and skin lesion segmentation on dataset is supplied by the ISIC-2018 Challenge. The experimental results clearly demonstrate that our model not only achieves smaller size but also improves performance compared to existing models.

**Keywords:** Convolution neural network · Multiple layers Unet · Liver segmentation · Spleen segmentation · Skin lesion segmentation

## 1 Introduction

In recent years, the field of medical image segmentation has faced numerous challenges. Various methods have been proposed to address these challenges, including traditional image processing, basic machine learning, and deep learning [1]. Among these methods, deep learning has emerged as the most widely used approach, with the number of published studies increasing significantly from 2014 to 2018 [2]. Convolutional neural networks (CNNs) have played a crucial role in medical image segmentation applications. They have been employed in the segmentation of various anatomical structures, such as the liver and liver tumors [3–5], the brain and brain tumors [6–8], the lungs and lung nodules [9, 10], and many others.

Unet has emerged as the predominant deep learning architecture [11]. However, both Unet itself [12] and architectures based on Unet tend to be large and computationally intensive. Modifying the number of filters or nodes typically leads to an increase in model size. Some approaches have sought to address this limitation by employing dense architectures [13, 14], altering the convolution function structure [15], or utilizing recurrent convolutional neural networks (RCNN) [16, 17]. Multi-layer Unet architecture ($U^n$-Net) [18], combined with dense skip connection (TMD-Unet) [19], and modified RCNN [20] have been studied and applied to the medical image segmentation problem. Dense architectures have successfully reduced the number of parameters, but at the cost of increased network complexity and computation time. While modifying the convolution structure has been proposed to enhance model performance, it does not effectively reduce the network size. RCNN, on the other hand, tackles the issue by reusing the weights of convolutions, resulting in a reduction in network size.

In our research, we deviate from the conventional CNN structure and employ a recurrent convolutional neural network (RCNN). The utilization of RCNN offers a key advantage in reducing the number of parameters while maintaining the effectiveness of the convolutional function. Consequently, this leads to a reduction in the overall network size. Additionally, we introduce a novel approach by combining the extraction of all feature maps from the convolutional functions within the nodes of the Unet architecture.

To summarize our research contributions, we have addressed the following technical challenges:

1. We have introduced a novel network model called MR-Unet, which incorporates the RCNN structure while effectively utilizing feature maps in the network nodes.
2. We have conducted evaluations of the proposed model in the context of liver segmentation, spleen segmentation and skin lesion segmentation.
3. Our experimental findings demonstrate notable improvements in the model. Specifically, we have achieved a significant reduction in model size, and the segmentation performance has surpassed that of existing models.

## 2 Modified Recurrent Unet (MR-Unet)

Figure 1 illustrates the distinction between the traditional CNN and the RCNN proposed in our study. In the proposed network, the output feature maps play a crucial role in the subsequent layers and the decoder side of the Unet structure. For the proposed model, MR-Unet, the convolution units from the second onwards will be reused. This is the strength of the Recurrent structure. In the proposed model, we used all the outputs of the convolutional units, this is the advantage of the $U^n$-Net model. Therefore, the number of parameters of the model will not change as the convolution unit depth is increased.

The overview of the MR- Unet models is presented in Fig. 2. The skip connection path, max-pooling, and up-convolution path encompass multiple paths. This is due to the utilization of all output feature maps from the encoder nodes. Both the encoder nodes and decoder nodes are comprised of RCL blocks. The output states of the RCL block depend on the number of times the convolutional unit is reused.

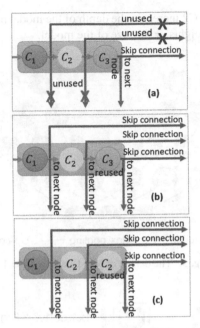

**Fig. 1.** The node structure of (a) traditional Unet (b) $U^n$-Net, and (c) the proposed model

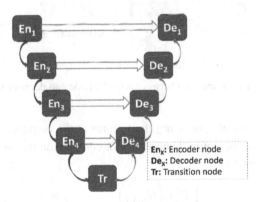

**Fig. 2.** The MR-Unet architecture

In this study, we explore two cases: $t = 1$ and $t = 2$, indicating the number of times the convolution unit is reused in the Recurrent Convolutional Layer (RCL). When $t = 1$, MR-Unet consists of 3 sub-Unet models, as illustrated in Fig. 3. Each convolution unit within the RCL block consists of a $3 \times 3$ kernel size convolutional function followed by a ReLU activation function. The number of filters in the RCL blocks is as follows: 16; 32; 64; 128; 256. Additional details regarding the connection between the encoder node and the decoder node can be found in Fig. 4. When $t = 2$, it means that the convolution block is reused twice. The convolution unit C2 block appears 3 times in the RCL block. Therefore, for the MR2-Unet model, the number of sub-Unets will be 4. The third appearance of block C2 increases the depth of the model but does not significantly increase the number of weights or the size of the model.

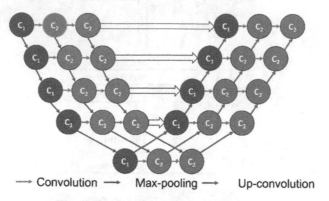

**Fig. 3.** The details of MR-Unet when t = 1

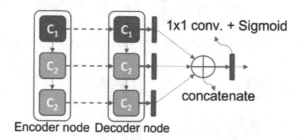

**Fig. 4.** The connection between encoder node and decoder node.

To clarify the details of connecting the features at the outputs at the model nodes. We have defined $U_i^j$ as the output feature map of convolution unit in encoder node. $U_i^j$ will be given by the formula:

$$U_i^j = \begin{cases} \mathbb{C}\left(P\left(U_{i-1}^j\right)\right) & j = 1 \\ \mathbb{C}\left(\left[P\left(U_{i-1}^j\right), U_i^{j-1}\right]\right) & j > 1 \end{cases} \tag{1}$$

where $\mathbb{C}(*)$, $[*]$, $P(*)$ denote the convolution function, the concatenate function, and the max-pooling function, respectively. Letter '$i$' indicates the node number and the '$j$' represents the convolution number.

On the decoder side, the inputs to the convolution unit are composed of three features: the feature obtained after applying up-convolution from the previous node, the feature from the encoder node at the same level, and the feature from the previous convolution unit within the same node. This relationship can be described by the following calculation equation:

$$D_i^j = \begin{cases} \mathbb{C}\left(\left[\mathbb{Q}\left(D_{i+1}^j\right), U_i^j\right]\right) & j = 1 \\ \mathbb{C}\left(\left[\mathbb{Q}\left(D_{i+1}^j\right), U_i^j, D_i^{j-1}\right]\right) & j > 1 \end{cases} \quad (2)$$

where $D_i^j$ is the output feature of the decoder node, $\mathbb{Q}(*)$ represents the transposed convolution.

The final output of the MR-Unet model is derived by concatenating all the output features from the first level decoder node. Subsequently, a $1 \times 1$ convolution and sigmoid function are applied to obtain the final output. Let's denote the final output as $O$, which can be expressed as follows:

$$O = S\left(\Delta\left(\left[S\left(\Delta\left(D_i^k\right)\right)\right]_{k=1}^j\right)\right) \quad (3)$$

where $S(*)$ and $\Delta(*)$ depict the sigmoid function and the $1 \times 1$ convolution.

## 3 Experiments and Results

### 3.1 Datasets and Preprocessing

To assess the performance of the model, we conducted evaluations on three different applications: liver segmentation, spleen segmentation, and skin lesion segmentation. This section provides comprehensive details regarding the datasets used and the preprocessing steps employed for these tasks. Figure 5 presents some image examples from the dataset used in this study.

For liver segmentation application, we utilized the dataset provided by the 2017 LiTS (Liver Tumor Segmentation) challenge for training purposes [23]. The dataset consists of 131 CT volumes, which were divided into training/validation (131 volumes) and testing (70 volumes). However, it's important to note that the annotations were available only for a subset of 131 volumes. Therefore, in our experiments, we exclusively used the volumes that included ground truth annotations. The primary objective of the LiTS challenge was to perform both liver segmentation and liver tumor segmentation. However, in this study, we focused solely on implementing the liver segmentation task, where the tumor labels were considered as part of the liver labels. The total number of slices across the 131 volumes amounted to 58,638, with each slice sized at $512 \times 512$. To prepare the dataset for training and testing, we divided it into three subsets: 90 volumes for training, 11 volumes for validation, and 30 volumes for testing. Preprocessing steps were applied as follows: Firstly, each slice was cropped to a size of $448 \times 448$ and then scaled down to $224 \times 224$. This resizing process was employed to reduce computation time. Secondly, a Hounsfield unit threshold ranging from $-200$ to 250 was applied to remove redundant areas around the liver and reduce noise. Finally, three adjacent slices were combined to form an RGB image. Additionally, slices without liver presence were excluded from the dataset. In summary, a total of 22,109 images were used in our study, with 4,494 images designated for validation and 7,059 images reserved for testing and validation purposes.

The spleen segmentation application employed a dataset obtained from the Medical Segmentation Decathlon Challenge 2018 [24]. The dataset comprised 63 CT volumes, out of which only 41 volumes included ground truth annotations. Across these 41 volumes, there were a total of 3,650 slices, each with a size of $512 \times 512$. To prepare the dataset for training, we randomly divided it into three subsets: the training set, which encompassed 2,920 slices, the validation set, consisting of 584 slices, and the testing set, comprising 730 slices. Similar to the liver segmentation application, we applied the Hounsfield Unit threshold in the range of $[-200, 250]$ as part of the preprocessing steps. Additionally, the images were cropped to a size of $224 \times 224$ before being used for training. In summary, the spleen segmentation dataset consisted of 41 volumes with 3,650 slices. These slices were randomly divided into a training set (2,920 slices), a validation set (584 slices), and a testing set (730 slices). Preprocessing involved the application of a Hounsfield Unit threshold and image cropping to a size of $224 \times 224$.

The skin lesion dataset is supplied by the ISIC-2018 Challenge [25] and consists of 2594 high-resolution dermoscopy images. There is substantial variance in the dimensions of the images in the dataset. To standardize them, we resize all images to a $224 \times 224$ resolution, then randomly divide the dataset into three subsets: a training set containing 1660 images, a validation set of 415 images, and a testing set with 519 images.

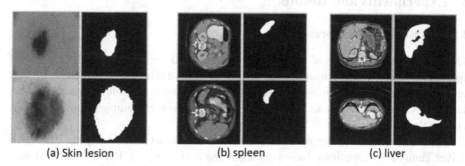

(a) Skin lesion          (b) spleen          (c) liver

**Fig. 5.** The images used in this study. The original images are in the first column and the ground truth are presented in the second one.

## 3.2   Experimental Setting

In this study, we utilized a hybrid loss function due to the presence of imbalanced data. The hybrid loss function combines weighted cross-entropy and dice loss. The expressions for the loss function are as follows:

$$\mathcal{L}_\tau = \mathcal{L}_\alpha + \mathcal{L}_\beta \tag{4}$$

$$\mathcal{L}_\alpha = -\frac{1}{N} \sum_1^N \left( (1-\alpha) y_i \log \widehat{y_i} + \alpha (1 - y_i) \log \left(1 - \widehat{y_i}\right) \right) \tag{5}$$

$$\mathcal{L}_\beta = 1 - \frac{2\sum_{i=1}^N (y_i \widehat{y_i}) + \delta}{\sum_{i=1}^N (y_i + \widehat{y_i}) + \delta} \tag{6}$$

where $\mathcal{L}_\tau$, $\mathcal{L}_\alpha$, and $\mathcal{L}_\beta$ represent the total loss, weighted cross-entropy loss, and the dice loss, respectively. $y_i$ and $\widehat{y_i}$ are the background pixel and the foreground pixel recognition probability. $N$ is the total pixels number, the $\alpha$ is the foreground weight. The $\delta$ is used to avoid the infinitive problem.

Our model was implemented using the Keras package. For the liver segmentation task, we trained the model for 100 epochs, while for spleen segmentation, we trained it for 200 epochs. The batch size was set to 8. To optimize the learning process, we employed a learning rate scheduler with an *initial* value of $3 \times 10^{-4}$. The learning rate was updated during training using an $initial*(0.9^{(epochs/10)})$ formula. To improve efficiency, we utilized the early-stopping mechanism, which helps save training time by stopping the training process if there is no further improvement in performance. The model was trained on a personal computer with NVIDIA GeForce RTX 3070 Graphics Processing Unit 6 GB of memory, Intel Core i7 11370H CPU, and 24 GB of RAM. For the spleen and skin lesion dataset, we applied augmentation techniques to enhance the training process and increase the model's ability to handle variations in the images.

### 3.3 The Metrics

The metrics used to evaluate the performance of our models include the Dice coefficient (*Dice*), F1-score (*F1Score*), and mean Intersection over Union (*mIoU*). These metrics are defined as follows:

$$Dice\left(Y, \widehat{Y}\right) = \frac{2 \times \left|Y \cap \widehat{Y}\right|}{|Y| + \left|\widehat{Y}\right|} \tag{7}$$

$$mIoU\left(Y, \widehat{Y}\right) = \frac{\left|Y \cap \widehat{Y}\right|}{\left|Y \cup \widehat{Y}\right|} \tag{8}$$

$$F1Score = \frac{TP}{TP + \frac{1}{2}(FP + FN)} \tag{9}$$

where TP represents the number of true positive pixels (pixels correctly classified as the positive class), FP represents the number of false positive pixels (pixels incorrectly classified as the positive class), FN represents the number of false negative pixels (pixels incorrectly classified as the negative class), and TN represents the number of true negative pixels (pixels correctly classified as the negative class). The terms and denote the total number of positive class pixels in the ground truth and the total number of pixels predicted as the positive class, respectively.

### 3.4   Results and Discussion

This study involved training a total of four network models: Unet, Unet++ [14], MR1-Unet (t = 1), and MR2-Unet (t = 2). Table 1 presents a comparison of the liver segmentation results and parameter numbers between the proposed model and Unet, Unet++, TMD-Unet, and $U^n$-Net (with n = 3). All models used for comparison in this study were trained and tested on the same setting and hardware.

**Table 1.** The liver segmentation results. The best one is in **bold**, the second is in *italic*.

| Models | Parameters (million) | Weight file size (MB) | Metrics (%) | | |
|---|---|---|---|---|---|
| | | | Dice | F1Score | mIoU |
| Unet | 7.8 | 93.4 | 91.26 | 91.31 | 84.07 |
| Unet++ | 9.5 | 115 | 94.75 | 95.19 | 91.08 |
| TMD-Unet | 9.1 | 110 | 95.48 | 95.71 | 91.40 |
| $U^3$-Net | 21.7 | 261 | **96.38** | – | – |
| MR1-Unet | **3.5** | **42.3** | 95.86 | *96.04* | *92.06* |
| MR2-Unet | *4.6* | *55.6* | *96.25* | **96.43** | **92.78** |

Table 1 demonstrates that our proposed models outperformed Unet and Unet++ in terms of results. For MR-Unet (t = 1), the metrics show improvements of 4.60%, 4.73%, and 7.99% in DC, F1, and IoU, respectively, compared to Unet. Similarly, for MR2-Unet, the improvements are 4.99%, 5.12%, and 8.71%, respectively. Additionally, our proposed models consume less memory than Unet and Unet++ (3.48 million and 4.64 million compared to 7.77 million and 9.51 million). The $U^3$-Net achieved the best results; however, the size of the models is more than five times bigger.

**Table 2.** The spleen segmentation results. The best one is in **bold**, the second is in *italic*.

| Models | Parameters (million) | Weight file size (MB) | Metrics (%) | | |
|---|---|---|---|---|---|
| | | | Dice | F1Score | mIoU |
| Unet | 7.8 | 93.4 | 90.28 | 90.30 | 82.93 |
| Unet++ | 9.5 | 115 | 95.00 | 95.42 | 90.49 |
| TMD-Unet | 9.1 | 110 | 95.41 | 95.59 | 91.32 |
| MR1-Unet | **3.5** | **42.3** | *96.30* | *96.56* | *92.88* |
| MR2-Unet | *4.6* | *55.6* | **96.44** | **96.62** | **93.15** |

In Table 2, the segmentation results on the spleen dataset are presented. It is evident that MR2-Unet achieved the best results across all metrics. Our proposed models exhibit significant improvements compared to the Unet and Unet++ model. Compared with

TMD-Unet, the proposed model not only has a smaller size (approximately 2 times) but also achieves better segmentation efficiency (1.03%).

The results of skin lesion segmentation are presented in Table 3. The proposed model achieves better performance than other models. Although the superiority is not significant compared to TMD-Unet (0.08%), it proves the optimality of the proposed model in terms of size (4.6 milion vs 9.1 million).

**Table 3.** The skin lesion segmentation results. The best one is in **bold**, the second is in *italic*.

| Models | Parameters (million) | Weight file size (MB) | Metrics (%) | | |
|---|---|---|---|---|---|
| | | | Dice | F1Score | mIoU |
| Unet | 7.8 | 93.4 | 83.15 | 84.28 | 71.22 |
| Unet++ | 9.5 | 115 | 82.84 | 85.29 | 71.16 |
| TMD-Unet | 9.1 | 110 | *87.27* | *88.42* | *77.66* |
| MR1-Unet | **3.5** | **42.3** | 86.64 | 88.16 | 76.67 |
| MR2-Unet | *4.6* | *55.6* | **87.35** | **88.89** | **77.79** |

In Fig. 6, examples of segmentation results for liver and spleen segmentations are represented. It is evident that our proposed models exhibit superior segmentation results for both liver and spleen compared to Unet and Unet++. This serves as evidence of the effectiveness of our proposed network models.

The Unet architecture has achieved remarkable success in medical image segmentation challenges. However, a major drawback of network models is their large size, which leads to high memory consumption and longer training times. This increase in network size is often attributed to changes in convolution architecture, improved skip connections, or the use of dense structures. To address this issue, the effectiveness of using Recurrent Neural Networks (RNNs) to reduce network size has been demonstrated. By replacing Unet's network nodes with Recurrent Convolutional Layers (RCLs), the network size has been significantly reduced. Additionally, exploiting all the output features of the convolutional units has enhanced model performance.

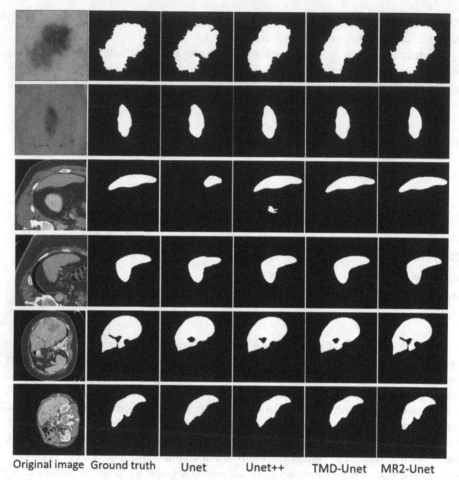

Original image  Ground truth      Unet          Unet++       TMD-Unet    MR2-Unet

**Fig. 6.** Examples of segmentation result for skin lesion (first and second rows), spleen slices (third and forth rows) and liver slices (fifth and sixth rows).

Table 4 presents a comparison of the results between the proposed model and popular models. Overall, the proposed model outperforms the existing models, except for Unet 3+ in liver segmentation. However, it's important to note that Unet 3+ has a significantly larger network size compared to the proposed models (26.97 million compared to 3.48 million and 4.64 million). Despite this difference in network size, the proposed models still demonstrate competitive performance in terms of segmentation results.

**Table 4.** Comparing with popular network models

| Models | Dice (%) | | |
|---|---|---|---|
| | Liver | Spleen | Skin |
| PSPNet [21] | 92.42 | 92.40 | – |
| Deeplab V3+ [22] | 91.86 | 92.90 | **87.70** |
| TMD-Unet [19] | 95.48 | 95.41 | 87.27 |
| $U^3$-Net [18] | 96.38 | – | – |
| Unet 3+ [13] | **96.75** | 96.20 | – |
| MR1-Unet | 95.86 | *96.30* | 86.64 |
| MR2-Unet | 96.25 | **96.44** | *87.35* |

## 4  Conclusion

In this study, a novel network model called MR-Unet was proposed for medical image segmentation. MR-Unet combines RCNN with the exploitation of output feature maps within the network nodes. Experimental results have demonstrated the effectiveness of the proposed network. MR-Unet not only reduces the network size but also surpasses popular networks in terms of segmentation results. Our study has made valuable contributions by introducing a new network model, evaluating it in specific segmentation tasks, and showcasing its enhanced performance in terms of reduced size and improved segmentation accuracy compared to current models. This innovative approach holds promise for future advancements in medical image segmentation.

## References

1. Wang, R., Lei, T., Cui, R., Zhang, B., Meng, H., Nandi, A.K.: Medical image segmentation using deep learning: a survey. arXiv200913120 Cs Eess (2020)
2. Zhou, T., Ruan, S., Canu, S.: A review: deep learning for medical image segmentation using multi-modality fusion. Array (2019)
3. Li, X., Chen, H., Qi, X., Dou, Q., Fu, C.-W., Heng, P.A.: H-DenseUNet: hybrid densely connected UNet for liver and tumor segmentation from CT volumes. IEEE Trans. Med. Imaging, 2663–2674 (2018)
4. Seo, H., Huang, C., Bassenne, M., Xiao, R., Xing, L.: Modified U-Net (mU-Net) with incorporation of object-dependent high level features for improved liver and liver-tumor segmentation in CT images. IEEE Trans. Med. Imaging (2020)
5. Xi, X.-F., Wang, L., Sheng, V.S., Cui, Z., Fu, B., Hu, F.: Cascade U-ResNets for simultaneous liver and lesion segmentation. IEEE Access, 68944–68952 (2020)
6. Menze, B.H., et al.: The multimodal brain tumor image segmentation benchmark (BRATS). IEEE Trans. Med. Imaging (2015)
7. Wang, L., et al.: Nested dilation networks for brain tumor segmentation based on magnetic resonance imaging. Front. Neurosci. (2019)
8. Zhang, J., Lv, X., Zhang, H., Liu, B.: AResU-Net: attention residual U-Net for brain tumor segmentation. Symmetry (2020)

9. Aresta, G., et al.: iW-Net: an automatic and minimalistic interactive lung nodule segmentation deep network. Sci. Rep. (2019)
10. Keetha, N.V., Parisapogu, S.A.B., Annavarapu, C.S.R.: U-Det: a modified U-Net architecture with bidirectional feature network for lung nodule segmentation. arXiv200309293 Cs Eess Stat, March 2020
11. Panayides, A.S., et al.: AI in medical imaging informatics: current challenges and future directions. IEEE J. Biomed. Health Inform., 1837–1857 (2020)
12. Ronneberger, O., Fischer, P., Brox, T.: U-Net: convolutional networks for biomedical image segmentation. In: Navab, N., Hornegger, J., Wells, W., Frangi, A. (eds.) MICCAI 2015. LNCS, vol. 9351, pp. 234–241. Springer, Cham (2015). https://doi.org/10.1007/978-3-319-24574-4_28
13. Huang, H., et al.: UNet 3+: a full-scale connected UNet for medical image segmentation. In: ICASSP 2020 - 2020 IEEE International Conference on Acoustics, Speech and Signal Processing (ICASSP), pp. 1055–1059, May 2020
14. Zhou, Z., Siddiquee, M.M.R., Tajbakhsh, N., Liang, J.: UNet++: redesigning skip connections to exploit multiscale features in image segmentation. IEEE Trans. Med. Imaging, 1856–1867 (2020)
15. Chen, Y., et al.: Channel-UNet: a spatial channel-wise convolutional neural network for liver and tumors segmentation. Front. Genet. (2019)
16. Liang, M., Hu, X.: Recurrent convolutional neural network for object recognition. In: 2015 IEEE Conference on Computer Vision and Pattern Recognition (CVPR), pp. 3367–3375 (2015)
17. Alom, M.Z., Hasan, M., Yakopcic, C., Taha, T.M., Asari, V.K.: Recurrent residual convolutional neural network based on U-Net (R2U-Net) for medical image segmentation. arXiv180206955 Cs, May 2018
18. Tran, S.-T., Cheng, C.-H., Liu, D.-G.: A multiple layer U-Net, $U^n$-Net, for liver and liver tumor segmentation in CT. IEEE Access, 3752–3764 (2021)
19. Tran, S.-T., Cheng, C.-H., Nguyen, T.-T., Le, M.-H., Liu, D.-G.: TMDUNet: triple-UNet with multi-scale input features and dense skip connection for medical image segmentation. Healthcare (2021)
20. Tran, S.-T., Nguyen, M.-H., Dang, H.-P., Nguyen, T.-T.: Automatic polyp segmentation using modified recurrent residual UNet network. IEEE Access, 65951–65961 (2022)
21. Zhao, H.S., Shi, J.P., Qi, X.J. , Wang, X.G., Jia, J.Y.: Pyramid scene parsing network. In: The IEEE Conference on Computer Vision and Pattern Recognition, pp. 2881–2890 (2017)
22. Chen, L.C., Zhu, Y., Papandreou, G., Schroff, F., Adam, H.: Encoder-decoder with atrous separable convolution for semantic image segmentation. In: Ferrari, V., Hebert, M., Sminchisescu, C., Weiss, Y. (eds.) ECCV 2018. LNCS, vol. 11211, pp. 833–851. Springer, Cham (2018). https://doi.org/10.1007/978-3-030-01234-2_49
23. Bilic, P., et al.: The liver tumor segmentation benchmark (LiTS). Med. Image Anal. (2023)
24. Simpson, A.L., et al.: A large annotated medical image dataset for the development and evaluation of segmentation algorithms. arXiv:1902.09063 (2019)
25. Codella, N.C.F., et al.: Skin lesion analysis toward melanoma detection: a challenge. In: 2018 IEEE 15th International Symposium on Biomedical Imaging (ISBI 2018), Washington, DC, April 2018

# Retrospective Analysis of a Large-Scale Archive of Ultrasonic Movies for Ischemic Diseases of Neonatal Brain

My N. Nguyen[1,2]([envelope]) [ORCID], Ryosuke Kawamidori[1], Yoshiki Kitsunezuka[1], and Masayuki Fukuzawa[1]([envelope]) [ORCID]

[1] Graduate School of Science and Technology, Kyoto Institute of Technology, Matsugasaki, Sakyo-ku, Kyoto 606-8585, Japan
nnmy@cit.ctu.edu.vn, fukuzawa@kit.ac.jp
[2] College of Information and Communication Technology, Can Tho University, 3-2 Street, Ninh Kieu District, Can Tho, Vietnam

**Abstract.** Retrospective analysis of neonatal cranial ultrasonic movies has been effectively realized by developing a novel movie analysis system. Despite the recognition of the diagnostic potential of neonatal cranial ultrasonic movies, several issues still persist in handling a large-scale archive of approximately 1000 cases. These issues include extensive, redundant computations and inconsistent data. The proposed system was designed to address these issues through a systematic organization of processes. It includes a primary layer for fully-automated preprocessing such as frequency spectrum analysis and several consequent layers for semi-automated retrospective analysis such as detection of probe-stabilized scene and assessment of pulsatile tissues. Four test cases were examined by the conventional and novel systems to evaluate the performance. The results demonstrated that the novel system performed essential tasks three to five times faster, while also providing new capabilities that the conventional system lacks. Overall, the proposed system confirmed a significant improvement of efficiency in retrospective analysis compared with the conventional process flow and the retrospective analysis has been realized to the entire archive for the first time, revealing its clinical applicability for further analysis of ischemic diseases.

**Keywords:** Large-Scale Archive · Ultrasonic Movie · Neonatal Brain · Ischemic Diseases · Time-Frequency Analysis

## 1 Introduction

Ultrasonic (US) movie is widely used by pediatricians at neonatal intensive care unit (NICU) to observe the brain tissues through the anterior fontanelle of neonates with potential risks of ischemic and related diseases such as Hypoxic-Ischemic Encephalopathy (HIE) [1] and Intraventricular Hemorrhage (IVH) [2, 3]. Since early detection of these diseases is essential for timely intervention, pediatricians have diagnosed such abnormalities from the shape and motion of brain tissues in brightness mode images and

© The Author(s), under exclusive license to Springer Nature Singapore Pte Ltd. 2024
N. Thai-Nghe et al. (Eds.): ISDS 2023, CCIS 1949, pp. 55–64, 2024.
https://doi.org/10.1007/978-981-99-7649-2_5

movies while taking care of some factors such as choice of suitable equipment, optimal settings, appropriate scanning protocols, and scanning experience to ensure its image quality [4].

One of cardinal viewpoints in the diagnosis of ischemic diseases is to observe a pulsatile motion of brain tissues due to blood flow of neighboring arteries. In order to assist such pediatric diagnosis, we have conducted several studies in detecting and visualizing the pulsatile tissues from US movies with time-frequency analysis [5–8]. It was based on a heartbeat-frequency component $I(f_{HB})$ in the frequency spectrum obtained from a time-variant echo intensity of each pixel over 64 frames in a movie fragment extracted from a scene without any motion of the US probe. $I(f_{HB})$ was provided to assist pediatric diagnosis by visualizing as a superimposed image while overlaying a color gradation proportional to $I(f_{HB})$ onto the original US image [8]. It was also given by statistically analyzing as a pulsatile tissue measures. These procedures are shown in Fig. 1. An extensive archive of neonatal cranial US movies was also prepared for a variety of studies, and its retrospective analysis was examined by using a conventional analysis system. However, it was problematic to handle the entire archive due to several issues in the conventional analysis system such as extensive, redundant computations and inconsistent data.

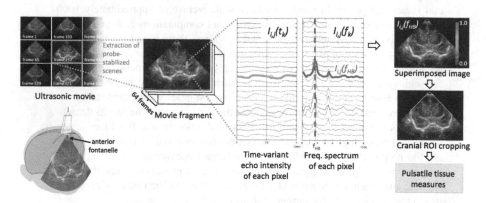

**Fig. 1.** Procedures for detection and visualization of pulsatile tissues from a movie fragment in a scene without probe-motion

In this study, we introduce an approach for handling the entire archive by developing a novel analysis system dedicated to neonatal cranial US movies of ischemic diseases while mitigating potential challenges associated with our conventional system. Additionally, several test cases are formulated to compare the performance of two systems.

The reminder of this paper is as follows. In Sect. 2, the archive associated with the conventional image analysis system and its issues are presented. Sect. 3 proposes a novel image analysis system to alleviate such issues. Next, Sect. 4 illustrates some experiments to evaluate the novel system performance. Finally, achievements of this paper are summarized in Sect. 5.

## 2  Movie Archive and Issues in Its Retrospective Analysis Using the Conventional System

The movie archive utilized in this study consists of an immense volume of cranial US movies with approximately 1000 cases recorded at NICUs in the previous affiliated institutions (Himeji Red Cross Hospital and Saiseikai Hyogo-ken Hospital) of the author (Kitsunezuka, Y.) and provided anonymously under the approval of the affiliated department, or the ethics committee after its establishment. It consists of around 213 h of movie and encompassing around 1000 cases. Table 1. shows the statistics of the examined archive. Since the US apparatus was different between the groups of different recording periods, the resolution and the fan-shaped field of view (FOV) should be considered for each group.

**Table 1.** Statistics of the US movie archive

| Group | A | B | C |
|---|---|---|---|
| Original Recording Media | Hi-8 | DV, HDD | HDD |
| Recording Period | 1994–2001 | 2005–2011 | 2011–2012 |
| Number of Cases | 693 | 205 | 108 |
| Capture Duration [hour] | 106 | 79 | 28 |
| Frame Resolution [pixels] | 640 × 480 | 720 × 480 | 640 × 480 |

The process flow in the previous studies using the conventional analysis system is illustrated in Fig. 2. In this figure, rounded and angular rectangles indicate the processes and their resultant data, respectively. The gray background indicates a process that requires manual intervention for each case or movie fragment, while the white one means what can be performed for multiple cases or fragments automatically at once.

In that conventional process flow, four types of manual process were required for each scene prior to the time-frequency analysis in order to select appropriate scenes for detecting pulsatile tissues. They are: 1) the anatomical identification of tomographic section and diagnosis annotation, 2) probe-stabilized scene detection to eliminate the visual incoherence due to the sway of probe, 3) $f_{HB}$ estimation according to the heartbeat frequency for each neonate, and 4) cranial ROI detection to focus on the area of brain tissues. All of them are performed by visual observation of the movie.

The case selection process was repeatedly invoked each time a certain purpose of study was considered for corresponding analysis and visualization. Therefore, it was difficult to investigate the relationship between pulsatile tissues and pathologic conditions. The problem of this process flow revolved around the uncertainty of the success or failure of all the manual processes until the visualization of the time-frequency analysis were observed. For example, the probe-stabilized scene was not valid until the $I(f_{HB})$ superimposed image was observed to confirm that it did not contain motion artifacts, as shown in Fig. 3. If it contained motion artifacts, we had to re-extract the scene and perform the time-frequency analysis again. In most cases, that process required some iterations with the feedback from the results of the time-frequency analysis. Since each of the other manual processes also required its own verification, a complete manual process included complex iterations. Therefore, the analysis of the entire archive was challenging because as more scenes were processed, various problems with iterative processes arose such as extensive, redundant computations and inconsistent data.

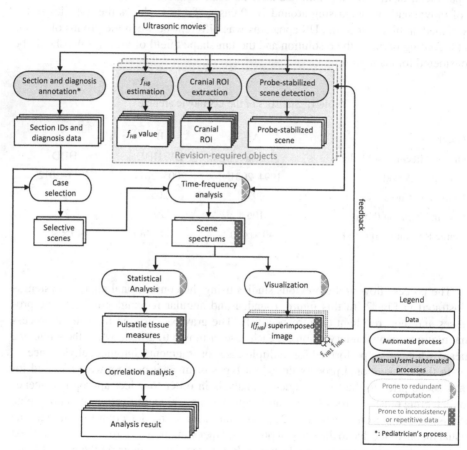

**Fig. 2.** Process flow in the previous studies using the conventional image analysis system

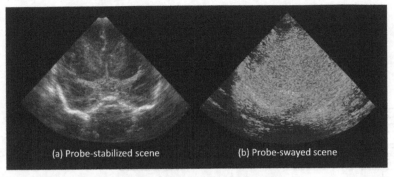

**Fig. 3.** Example of $I(f_{HB})$ superimposed images obtained from (a) probe-stabilized and (b) probe-swayed scenes

## 3   A Proposal of the Novel Image Analysis System

To tackle the issues in the conventional process flow, a novel image analysis system has been proposed. Figure 4 shows the systematic process flow enabled by that system. The shape and the background color of the rectangle in Fig. 4 indicate the same process type as those shown in Fig. 2. The novel system adopts a multilayered architecture to arrange processes into distinct layers including: 1) primary layer, which prepares extensive amount of intermediate data, 2) refinement layer, which filters or transforms intermediate data into more lightweight and concentrated data, and 3) exploratory layer, which discovers data patterns that are associated with the pathologic conditions.

One of the important design features of this system is that all the processes are performed on each movie fragment rather than on each movie scene. In addition, prior to any manual process, time-frequency analysis and its visualization are performed on the entire archive, regardless of the appropriateness for detecting pulsatile tissues, and the results are recorded as an intermediate data. An element of the intermediate data consists of a fragment spectrum that includes the whole spectrum $I(f)$ corresponding to a movie fragment, and an $I(f_{HB})$ superimposed image made from the initial frame of the fragment. Example of spectrum $I(f)$ and corresponding $I(f_{HB})$ superimposed image was reflected in Fig. 1. Since a huge number of intermediate data was generated to cover the entire archive, it can be directly used for subsequent processes without regeneration, which is a clear advantage over the conventional system. Therefore, this design feature allowed all the processes of intermediate data generation to be positioned as the primary, feedback-free layer in the process flow.

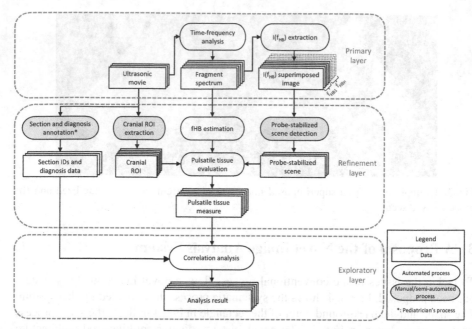

**Fig. 4.** Systematic process flow enabled by a novel image analysis system

Another important feature of this system is to efficiently perform the manual processes by using both intermediate data and the US movies. Among the manual processes, the section identification and ROI detection should be performed by visual observation of US movies since it is based on anatomical knowledge. However, the probe-stabilized scene can be effectively detected, by visual observation of a series of $I(f_{HB})$ superimposed images to ensure the absence of motion artifacts [9]. Furthermore, $f_{HB}$ can be effectively estimated from the peak frequency of the fragment spectrum. Since iteration was no longer required for all the manual processes, they were positioned as a refinement layer in the process flow to generate pulsatile tissue measures from the intermediate data. In other words, this layer refines the large amount of data from the primary layer into more focused data for further analysis.

The final output of this flow is the result of correlation analysis between the diagnosis data and the pulsatile tissue measures to assist pediatric diagnosis. Since it is expected to apply several pediatric diagnoses such as prognosis of HIE and risk assessment of IVH, this process was positioned as the exploratory layer in the process flow and designed to be performed for a certain subset of the archive.

In our system, a fragment was constructed from 64 frames (2 s), and Blackman was selected as the windowing function, by considering the appropriateness to handle the actual $f_{HB}$ of neonates (2 to 3 Hz) with the standard video rate (29.97 fps).

# 4  Evaluation of System Performance

In order to compare the performance between the conventional and the proposed systems, four test cases were designed and conducted as follows.

1. Time to extract the valid probe-stabilized scenes: This test aims to evaluate the time required to extract and validate all the probe-stabilized scenes from a 20-min US movie of a certain neonate.
2. Time to calculate the pulsatile tissue measures: This test aims to evaluate the time required to calculate the pulsatile tissue measures from the 1200 probe-stabilized scenes for 100 neonates with a common fan-shape ROI.
3. Applicability to the prospective studies in NICU: This test aims to examine the system applicability for a kind of prospective study in the NICU to compare an ongoing disease with the previous cases. We gave a new US movie of an ongoing case and specified 3 previous cases in the archive, and tested the ability to generate and visualize the pulsatile tissue images of them within 1 h, assuming an acceptable latency in NICU. The pulsatile tissue image means a special $I(f_{HB})$ superimposed image generated from the validated probe-stabilized scenes.
4. Applicability to the retrospective (case-control) studies: This test aims to examine the system applicability for retrospective studies to compare two case groups in the archive. We specified two case groups (HIE positive and negative) including around 10 patients for each in the archive and tested the ability to extract a certain pulsatile tissue measure and to visualize its sectional or time dependence.

Table 2. shows the results of four test cases. The extraction of valid probe-stabilized scenes was five times faster in the novel system. The primary reason for this result is the ability of the pre-generated $I(f_{HB})$ superimposed images to eliminate the iteration of manual extraction of probe-stabilized scenes. Likewise, the calculation of pulsatile tissue measures was three times faster in the novel system because it utilized the pre-generated fragment spectrum and the refinement data.

The novel system succeeded in instantly generating the pulsatile-tissue images for both ongoing and multiple previous cases, while the conventional system failed to do so due to time-consuming iterative processes. From this result, it was found that the applicability to prospective studies in the NICU was realized exclusively in the novel system.

**Table 2.** Performance evaluation contrasting conventional and novel archive systems

| # | Case Description | Target Subset | System Performance | |
|---|---|---|---|---|
| | | | Conventional | Novel |
| 1 | Time to extract the valid probe-stabilized scenes | A 20-min. US movie of certain neonate | 150 min | 30 min |
| 2 | Time to calculate the pulsatile tissue measures | 1200 valid probe-stabilized scenes with 100 neonates | 60 min | 20 min |
| 3 | Applicability to the prospective studies in NICU | 1 ongoing and 3 previous cases | No | Yes |
| 4 | Applicability to the retrospective studies | Two (normal and abnormal) case groups (N≈ 10 for each) | No | Yes |

The novel system also succeeded in extracting the pulsatile tissue measures and visualizing their sectional or time variation in specified two case groups, while it was impossible for the conventional system because the valid probe-stabilized scene was unknown. Figure 5 shows examples of pulsatile tissue images and corresponding measures obtained in HIE negative and positive case groups retrospectively. A normalized pulsatile area in a fan-shape ROI [10] was used as the pulsatile tissue measure in this experiment and noted at the upper right of each image. The pulsatile tissue image clearly revealed different distribution of pulsatile tissues slice by slice as well as case by case. The pulsatile tissue measure revealed a significant difference between HIE positive and negative groups. It should be emphasized that the symptom of HIE was diagnosed a few weeks after birth, while the US movie was taken just after birth. Therefore, this result strongly suggests the applicability of pulsatile tissue measures to predict the prognosis of HIE in the early postnatal stage.

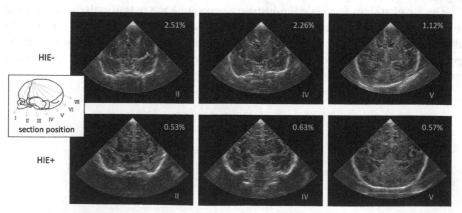

**Fig. 5.** Pulsatile tissue images and corresponding measures (normalized pulsatile area) obtained in HIE negative and positive case groups retrospectively

# 5 Conclusion

This research presented a layer-based analysis system to deal with a large-scale archive of neonatal cranial US movies while successfully eliminating the intensive and redundant computations and inconsistent data. The system was designed to include a primary layer for fully-automated preprocessing such as frequency spectrum analysis and several consequent layers for semi-automated retrospective analysis such as detection of probe-stabilized scene and assessment of pulsatile tissues. Four test cases were examined by the conventional and novel systems to evaluate the performance. The results demonstrated that the novel system performed essential tasks three to five times faster, while also providing new capabilities that the conventional system lacks. Besides, the solution also showcases the potential applications for analysis of periodical phenomena in other mass movie repositories than our archive. In summary, the proposed system successfully handled the large-scale archive of US movies, accelerating its retrospective analysis and revealing the clinical applicability of the system.

**Acknowledgement.** The authors would like to thank Mr. Daiki Terai for his significant contribution to the initial stages of this study.

This work was supported by JSPS Core-to-Core Program (grant number: JPJSCCB20230005).

# References

1. Kurinczuk, J.J., White-Koning, M., Badawi, N.: Epidemiology of neonatal encephalopathy and hypoxic-ischaemic encephalopathy. Early Hum. Dev. **86**(6), 329–338 (2010). https://doi.org/10.1016/j.earlhumdev.2010.05.010
2. Christian, E.A., et al.: Trends in hospitalization of preterm infants with intraventricular hemorrhage and hydrocephalus in the United States, 2000–2010. J. Neurosurg. Pediatr. **17**(3), 260–269 (2016). https://doi.org/10.3171/2015.7.PEDS15140
3. de Figueiredo Vinagre, L.E., et al.: Temporal trends in intraventricular hemorrhage in preterm infants: a Brazilian multicenter cohort. Eur. J. Paediatr. Neurol. **39**, 65–73 (2022). https://doi.org/10.1016/j.ejpn.2022.05.003
4. Leijser, L.M., Vries, L.S., Cowan, F.M.: Using cerebral ultrasound effectively in the newborn infant. Early Hum. Dev. **82**(12), 827–835 (2006). https://doi.org/10.1016/j.earlhumdev.2006.09.018
5. Yamada, M., et al.: Pulsation detection from noisy ultrasound-echo moving images of newborn baby head using Fourier transform. Jpn. J. Appl. Phys. **34**, 2854–2856 (1995). https://doi.org/10.1143/jjap.34.2854
6. Fukuzawa, M., Kitsunezuka, Y., Yamada, M.: A real-time processing system for pulsation detection in neonatal cranial ultrasonogram. Jpn. J. Appl. Phys. **37**, 3106–3109 (1998). https://doi.org/10.1143/jjap.37.3106
7. Fukuzawa, M., Kubo, H., Kitsunezuka, Y., Yamada, M.: Motion analysis of artery pulsation in neonatal cranial ultrasonogram. In: Proceedings of the SPIE 3661, Medical Imaging 1999: Image Processing, California, United States (1999). https://doi.org/10.1117/12.348519
8. Fukuzawa, M., Yamada, M., Nakamori N., Kitsunezuka, Y.: A new imaging technique on strength and phase of pulsatile tissue-motion in brightness-mode ultrasonogram. In: Proceedings of the SPIE 6513, Medical Imaging 2007: Ultrasonic Imaging and Signal Processing, 65130B, California, United States (2007). https://doi.org/10.1117/12.709354

9. Tabata, Y., Fukuzawa, M., Izuwaki, Y., Nakamori, N., Kitsunezuka, Y.: Time-frequency analysis of neonatal cranial ultrasonic movies for selective detection of pulsatile tissues by avoiding probe-motion artifact. In: Proceedings of the SPIE 9419, Medical Imaging 2015: Ultrasonic Imaging and Tomography, 941913, Florida, United States (2015). https://doi.org/10.1117/12.2081383
10. Fukuzawa, M., Takahashi, K., Tabata, Y., Kitsunezuka, Y.: Effect of echo artifacts on characterization of pulsatile tissues in neonatal cranial ultrasonic movies. In: Proceedings of the SPIE 9790, Medical Imaging 2016: Ultrasonic Imaging and Tomography, 979014, California, United States (2016). https://doi.org/10.1117/12.2216751

# Exam Cheating Detection Based on Action Recognition Using Vision Transformer

Thuong-Cang Phan[1(✉)] ⓘ, Anh-Cang Phan[2] ⓘ, and Ho-Dat Tran[2]

[1] Can Tho University, Can Tho, Vietnam
ptcang@cit.ctu.edu.vn
[2] Vinh Long University Of Technology Education, Vinh Long, Vietnam
cangpa@vlute.edu.vn, datth@vlute.edu.vn

**Abstract.** Cheating is the use of prohibited actions to illegally gain in the process of taking tests and exams. These actions cause negative consequences, making learners less capable of learning and creating, leading to unqualified individuals in the social workforce. In this study, we propose an action recognition method based on LSTM, VGG-19, Faster R-CNN, Mask R-CNN, and Vision Transformer architectures to classify fraudulent actions such as copying or pulling up to copy other students' works, communicating, or transferring test material. Experimental results show that Vision Transformer identifies fraudulent actions with an accuracy of above 98%. This contributes to supporting teachers in managing candidates in exams, creating fairness and transparency in education.

**Keywords:** Vision Transformer · Cheating in exams · Action recognition

## 1 Introduction

Training quality is always a top concern in every country in the world. In order to assess learners' ability, tests and exams are usually conducted in the middle and at the end of every semester. Learners are required to rely on the knowledge they have learned to answer the requirements posed in the test without any help from anyone else or materials that are not allowed to be brought into the exam room. Exam fraud is the act of bringing unauthorized documents into the exam room, communicating with other candidates, copying other candidates' work, etc. [1,2]. In Vietnam, Circular No. 15/2020/TT-BGDDT[1] issued on May 26, 2020, defines exam cheating as an act contrary to the regulations of the examination board such as copying and bring unauthorized materials into the exam room. Cheating

---

[1] https://thuvienphapluat.vn/van-ban/EN/Giao-duc/Circular-15-2020-TT-BGDDT-promulgation-of-Regulation-on-high-school-graduation-exam/445907/tieng-anh.aspxaccessedon22June2023.

© The Author(s), under exclusive license to Springer Nature Singapore Pte Ltd. 2024
N. Thai-Nghe et al. (Eds.): ISDS 2023, CCIS 1949, pp. 65–77, 2024.
https://doi.org/10.1007/978-981-99-7649-2_6

in exams causes bad effects on students, making them in a state of dependence and a lack of willingness to strive for good academic results.

The application of advanced technology to detect cheating in exams has been included in research. Rehab and Ali [14] proposed to identify cheating behaviors in online exams by three different classification algorithms: Support Vector Machine (SVM), Random Forest (RF) and K-nearest neighbor (KNN). Students' behaviors were classified as cheating or not with an accuracy of up to 87%. Yulita et al. [20] detected cheating in online exams using deep learning techniques with the MobileNetV2 architecture. Student's activities in web camera in an online exam were monitored and cheating actions were detected with a F1-score of 84.52%. Tiong and Lee [17] proposed a fraud detection model in online exams using deep learning techniques. Students' behaviors were monitored to detect and prevent learners' cheating. The system achieved an accuracy of 68% for deep neural networks (DNNs), 92% for Long Short Term Memory (LSTM), 95% for DenseLSTM, and 86% for recurrent neural networks (RNNs). Waleed Alsabhan [3] employed SVM, LSTM, and RNN classifiers to identify whether a student is cheating or not in higher education. The system achieved an accuracy of 90%. Dilini et al. [4] explored the use of eye tracking to detect cheating in exams. This study analyzed eye movement patterns and staring behaviors to identify suspicious activities, such as viewing unauthorized material. This approach achieved a positive results with an accuracy of 92.04%. Li Zhizhuang et al. [12] proposed a method to detect cheating in multiple-choice tests based on LSTM and RAE algorithm (combining of linear regression and expectation-maximization algorithm). This was to determine each student's mastery of knowledge points based on the exam problem solving. The system achieved an average accuracy of 81%. Ozdamli et al. [13] used computer vision algorithms and deep learning algorithms to detect emotions and feelings of cheating students in distance learning. The system achieved an accuracy of 87.5% in real-time student tracking head, face, and expressions of fear emotion during exams. Kamalov et al. [11] proposed an approach to detect potential cheating cases in final exams using machine learning techniques. This model applied recurrent neural networks along with anomaly detection algorithms and achieved an average true positive rate of 95%. Hussein et al. [10] proposed a method to automatically detect cheating by classifying video sequences. This helped to detect cheating behavior of students in paper-based exams. The authors achieved an average accuracy of 91%.

## 2   Background

### 2.1   Exam Fraud

Cheating in exams is an action contrary to the regulations of the examination board. Several types of cheating in exams are copying/pulling up to copy other students' works, viewing cheat sheets, and communicating/exchanging test materials with others. These actions negatively affect the fairness of the assessment process and lead to dishonest test results.

## 2.2   Network Models

**2.2.1   LSTM** Long Short Term Memory (LSTM) [7] is a type of recurrent neural network (RNN) architecture. LSTM was introduced by Hochreiter and Schmidhuber in 1997. It has been successfully applied to various sequential data tasks, such as natural language processing, speech recognition, machine translation, and video analysis. It is outstanding at capturing long-term dependencies in sequence, which is important for tasks involving context or temporal relationships. LSTM has played an important role in advancing the field of deep learning for sequential data analysis. The ability to model and capture complex temporal dependencies has resulted in improved performance in various domains, making them an influential and widely adopted architecture in the field of study.

**2.2.2   VGG-19** VGG-19 [16] is a convolutional neural network (CNN) architecture introduced by K. Simonyan and A. Zisserman. The VGG-19 model is a variant of the VGG network, which has 19 layers including 16 convolutional layers and 3 fully connected layers. It gained popularity for its simplicity and high performance in large-scale image recognition tasks. It was designed with the goal of exploring the effect of network depth on performance in image classification tasks. VGG family, including VGG19, has achieved excellent performance on a variety of benchmark datasets. Researchers often use VGG19 pre-trained weights, which are trained on large-scale image datasets, as a starting point for computer vision tasks.

**2.2.3   Faster R-CNN** Faster R-CNN [15] is an efficient and widely applied object detection model in computer vision. It was introduced by Shaoqing Ren et al. in 2017. Faster R-CNN is flexible and can be easily adapted to different object detection tasks. By modifying the architecture and training data, it can be used for different types of objects and scales, making it suitable for a wide range of applications, which in turn can handle detection tasks in real time or near real time. This helps to make it suitable for applications that require fast and efficient processing. It can process images in a batch-wise manner, allowing for scalability and efficient inference on large datasets.

**2.2.4   Mask R-CNN** Mask R-CNN [9] was introduced by Kaiming He et al. in 2017. Mask R-CNN achieves good performance on several benchmark datasets, consistently outperforming previous methods in terms of accuracy and durability. It can be applied to many computer vision tasks beyond segmentation, including object detection, object tracking, and semantic segmentation. Its flexibility and precision make it a suitable choice for various applications such as autonomous driving, robotics, and medical imaging.

**2.2.5   Vision Transformer** Vision Transformer (ViT) [16] is a neural network architecture introduced by A. Dosovitskiy et al. (2020). It is a successful extension of Transformers, originally designed for natural language processing, into the field of computer vision. Vision Transformer splits the input image into smaller patches and processes them as a token string. It applies the standard Transformer architecture, which includes self-attention mechanisms and feed-forward neural networks, to capture the local and global relationships between these image arrays.

ViT has demonstrated outstanding performance on various computer vision standards, competing with or even surpassing convolutional neural networks in certain settings [8]. Unlike traditional convolutional neural networks, ViT can process images of arbitrary size. This scalability makes it suitable for tasks that require different input resolutions without modifying the training architecture [5]. ViT's architecture allows it to be applied to various computer vision tasks without significant change. By fine-tuning the pre-trained ViT on tasks, it can adapt to image classification, object detection, and other related tasks that other neural networks force the user to reprocess. The accuracy of the Vision Transformer model compared to other models may vary depending on specific datasets, tasks, and test setup. According to Li Yuan et al. [19], ViT demonstrated competitive accuracy on the ImageNet dataset compared to traditional CNN models, demonstrating ViT's potential to achieve high accuracy without relying on transfer learning. Hugo Touvron et al. [18] indicate that the ViT model can achieve the same accuracy as a CNN while requiring fewer labeled training samples. This highlights ViT's potential for effective learning and generalization.

### 2.3 Evaluation Metrics

The simplest and most commonly used metric in evaluating network models is accuracy [6]. This evaluation simply calculates the ratio between the number of correctly predicted samples and the total number of samples in the dataset.

$$accuracy = \frac{TP + TN}{TP + TN + FP + FN} \tag{1}$$

where:

- TP: True positive
- TN: True negative

- FP: False positive
- FN: False negative

Mean average precision (mAP) is a commonly used method in multi-class classification problems. The mAP measure is calculated using Eq. 2 after obtaining the AP (average precision) where $N$ is the number of classes. The AP measure is calculated by Eq. 3, with $\rho_{interp}(r)$ performing 11-point interpolation to summarize the shape of the Precision x Recall curve by averaging the accuracy at a set of 11 equally spaced points [0, 0.1, 0.2,..., 1] and $\rho(\tilde{r})$ is the precision measured on $\tilde{r}$.

$$mAP = \frac{1}{N} \sum_{i=1}^{N} AP_i \tag{2}$$

$$AP = \frac{1}{11} \sum_{r \in 0,0.1,\ldots,1} \rho_{interp}(r) \quad \text{with} \quad \rho_{interp}(r) = \max_{\tilde{r}:\tilde{r} \geq r} \rho(\tilde{r}) \qquad (3)$$

## 3 Proposed Method

In this study, we use deep learning techniques and Vision Transformer to detect sequences of cheating actions in exams. The proposed approach consists of two phases: training and testing. The details of the phases are shown in Fig. 1.

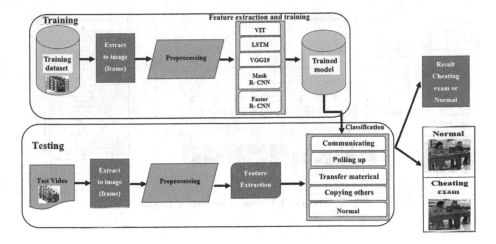

**Fig. 1.** The proposed approach in action identification and classification.

### 3.1 Training Phase

**3.1.1 Pre-processing** The dataset consists of videos recorded from the students' exams or tests. Each video is 5 to 10 s long at 30 frame per second (FPS). We normalize the input images to $224 \times 224$ pixels to maintain the ability compatible and consistent across different models. To enhance the dataset, image flipping and rotating were used to obtain more training observations. Deep learning models require large amounts of training data to perform well. When there is a lack of data, models can suffer from overfitting, where they perform exceptionally well on the training data but fail to generalize to new, unseen data. Image rotating and flipping are basic data augmentation techniques that helps mitigate this issue by generating new, diverse samples, effectively simulating a larger dataset.

**3.1.2    Feature Extraction and Training** With the advantages of deep learning networks presented earlier, we conduct feature extraction of action types based on extracted images with five models including LSTM, VGG-19, Faster R-CNN, Mask R-CNN, and Vision Transformer. These network models have a classification layer that helps in classifying fraudulent actions. Especially with the Vision Transformer model, we cut the input images into patches and passed them through the layers to conduct encryption, feature extraction, and classification (Fig. 2). The result of this process will be the classified and partitioned image.

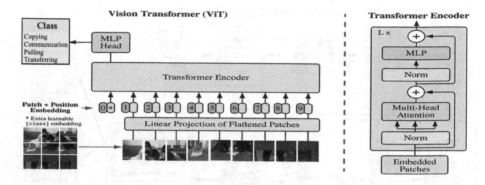

**Fig. 2.** Proposed Vision transformer model.

## 3.2    Testing Phase

To evaluate the performance of network models, we test the trained models using the testing dataset. This includes videos that we have collected combined with the segmentation of people on the YOLO algorithm. The test results will identify the actions in the video as normal actions or fraudulent actions.

## 4    Experiments

### 4.1    Installation Environment and Dataset Description

The system is installed in Python language and runs on the Colab environment with Windows 10 and the configuration of 12 GB RAM and Nvidia Geforce GPU. The libraries to support training network models are Tensorflow and Keras.

The dataset used in this study is video clips taken from students' exams or tests. These videos are divided into fraudulent actions such as copying other student works (1,480 images), exchanging test materials (1,850 images), pulling up to copy other students (1,560 photos), communicating with others (2,078 images), and acting normally (1,680 images). This dataset is divided into a training dataset and a validation dataset with a ratio of 80% and 20%, respectively.

For testing, we build a hypothetical scenario where the students were doing the exam, then recorded a video, and tested it with a number of 15 videos on a cheating action to check the accuracy of the network models.

## 4.2   Scenarios

To conduct the experiment, we perform five scenarios with training parameters as shown in Table 1.

**Table 1.** Proposed scenarios and training parameters

| Scenarios | Models | Epochs | Learning_rate | Batch_size | Image_size |
|---|---|---|---|---|---|
| 1 | Vision transformer | 200 | 0.001 | 9 | $224 \times 224$ |
| 2 | LSTM | 400 | 0.001 | 4 | $64 \times 64$ |
| 3 | VGG-19 | 400 | 0.001 | 4 | $224 \times 224$ |
| 4 | Mask R-CNN | 400 | 0.001 | 4 | $224 \times 224$ |
| 5 | Faster R-CNN | 400 | 0.001 | 4 | $224 \times 224$ |

## 4.3   Training Results

Figure 3 shows the loss values of five scenarios during the training phase. The validation loss values of scenarios 1 to 5 are 0.0048, 0.1872, 0.1426, 0.1989, and 0.1643, respectively. In scenarios 2 and 4, the validation loss is high and higher than the training loss, which shows that the models are not optimal and can lead to errors in the prediction process. Scenarios 3 and 5 show the loss values better than scenarios 2 and 4, but the validation loss values are still higher than the training loss. In scenario 1, the training loss and validation loss values are more stable and approaching 0 with 200 epochs, which is less by half than other models. With this result, scenario 1 has a very low loss value, less than 5% compared to other models after going through 200 training steps. This means that the error rate when predicting scenario 1 is the lowest compared to the remaining models. In the remaining scenarios, the validation loss value is still high, thus the models will have a higher error rate when making predictions.

Figure 4 shows the accuracy of five scenarios during the training phase. Scenarios 1 to 5 achieved an accuracy of 98.2%, 89.6%, 93.8%, 91.3%, and 92.7%, respectively. In scenarios 2 and 4, the validation accuracy is low and lower than the training accuracy, which can lead to incorrect identification of actions. Scenarios 1, 3, and 5 give more optimal results with train accuracy relatively close to validation accuracy. However, in scenarios 3 and 5, the validation accuracy value is still low. In scenario 1, the validation accuracy and train accuracy values gradually move towards 1. Thus, this model is better at prediction than the other models.

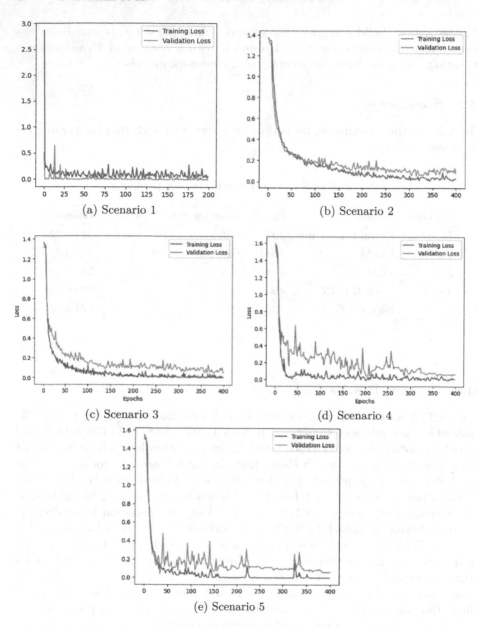

(a) Scenario 1    (b) Scenario 2

(c) Scenario 3    (d) Scenario 4

(e) Scenario 5

**Fig. 3.** Loss of five scenarios.

Figure 5 shows the training time of the proposed scenarios. The training time of scenario 1 is 117.6 min, scenario 2 is 93.6 min, scenario 3 is 125.6 min, scenario 4 is 114.3 min, and scenario 5 is 98.4 min. Through the above results, scenarios 2 and 5 give faster training time than the other scenarios. Scenario model 1 gives

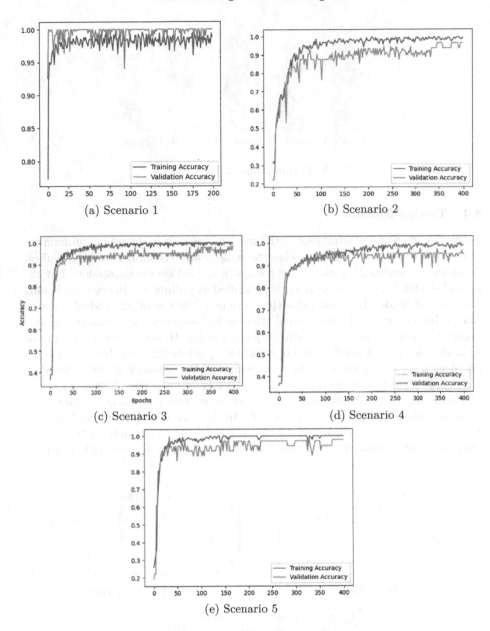

**Fig. 4.** Accuracy of five scenarios.

a close training time compared to other scenarios with less than half number of training steps.

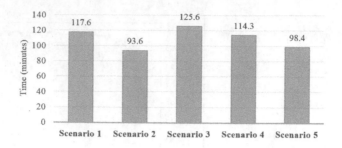

**Fig. 5.** Training time of five scenarios.

## 4.4   Testing Results

Table 3 shows some illustrations of the five scenarios. In the case of communication, all five scenarios can recognize the action, and scenario 1 has the highest accuracy. Scenarios 2, 3, and 4 can recognize the action of one student but the second student's action is incorrectly identified as pulling up. In scenario 5, only one student is identified with cheating actions while the other student's action cannot be recognized. In the case of transferring material, all scenarios can recognize the action and give relatively good results. However, compared to the scenarios, scenario 1 predicts the best outcome, nearly 99%. In the case of copying others, scenario 1 gives the correct result with an accuracy of 100%. Scenario 2 does not recognize fraudulent actions, possibly because the identification and classification activities give a low accuracy rate. The remaining scenarios can recognize and give relatively good results. In the case of pulling up and normal, scenario 2 cannot recognize the actions, possibly due to the actions being relatively unclear. Scenarios 1, 3, 4, and 5 all recognize the actions with relatively

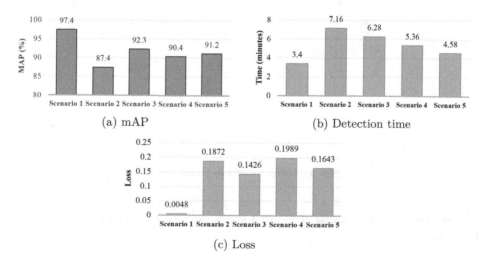

(a) mAP

(b) Detection time

(c) Loss

**Fig. 6.** mAP, Loss, and detection time on testing dataset.

**Table 2.** Average accuracy of the scenarios on fraudulent actions

| Scenarios | Communicating | Pulling up | Copying others | Transfer material |
|-----------|---------------|------------|----------------|-------------------|
| 1 | 98.2% | 97.1% | 96.5% | 97.8% |
| 2 | 87.1% | 87.8% | 86.2% | 88.5% |
| 3 | 92.7% | 91.5% | 90.8% | 94.2% |
| 4 | 90.5% | 89.6% | 91.8% | 89.7% |
| 5 | 91.6% | 92.1% | 90.8% | 90.3% |

**Table 3.** Illustration of classification results from the experimental dataset.

high accuracy. Scenario 1 gives the highest accuracy among the scenarios. The above results show that scenario 1 gives better performance than the remaining scenarios in identifying and classifying fraudulent actions.

The comparison results are summarized in Fig. 6. The mAP of the five scenarios is shown in Fig. 6a. Scenario 1 shows the highest mAP of 97.4% compared to scenario 2 (87.4%), scenario 3 (92.3%), scenario 4 (90.4%), and scenario 5 (91.2%). On the testing dataset, the shortest detection time is of scenario 1 with 3.4 min (Fig. 6b), which is almost two times faster than scenarios 2, 3, and 4 and almost 1.5 times faster than scenario 5. The loss value of scenario 1 is 0.0048 (Fig. 6b), which is the lowest among the five scenarios. From the above results, scenario 1 using Vision Transformer can identify and classify fraudulent actions more effectively than the remaining models.

Table 2 shows the average accuracy of the fraudulent actions in five scenarios. Scenario 1 provides better accuracy in four classes compared to the other four scenarios. Scenario 2 has the lowest accuracy in the four fraudulent classes.

## 5   Conclusion

The consequences of cheating in exams for students are enormous. It creates bad habits and bad qualities for students, affecting the process of being human. In this work, we bring our contributions to building a training dataset of action recognition with the use of advanced network architectures such as LSTM, VGG-19, Mask R-cNN, Faster R-CNN, and Vision Transformer. Fraudulent behaviors considered in this study are copying others, pulling up, communicating, and exchanging test materials. The detection of fraudulent actions all gives extremely positive results with an accuracy of above 85%, in which Vision Transformer has the best accuracy of above 98%. This helps to detect fraudulent action sequences in a timely and effective manner. Although bringing high results in the identification process, this work only stops at identifying common fraudulent behaviors, we will continue to recognize various cheating actions and more sophisticated cheating techniques. Future development could focus on addressing these limitations by incorporating additional features, exploring new architectures, and combining data from multiple sources, such as audio or oral gestures to identify cheating.

## References

1. https://www.adelaide.edu.au/student/academic-skills/cheating-in-exams. Accessed 22 June 2023
2. https://www.niu.edu/academic-integrity/faculty/types/index.shtml. Accessed 22 June 2023
3. Alsabhan, W.: Student cheating detection in higher education by implementing machine learning and LSTM techniques. Sensors 23(8), 4149 (2023)
4. Dilini, N., Senaratne, A., Yasarathna, T., Warnajith, N., Seneviratne, L.: Cheating detection in browser-based online exams through eye gaze tracking. In: 2021 6th International Conference on Information Technology Research (ICITR), pp. 1–8. IEEE (2021)

5. Dosovitskiy, A., et al.: An image is worth 16x16 words: transformers for image recognition at scale. arXiv preprint: arXiv:2010.11929 (2020)
6. Goutte, C., Gaussier, E.: A probabilistic interpretation of precision, recall and $F$-score, with implication for evaluation. In: Losada, D.E., Fernández-Luna, J.M. (eds.) ECIR 2005. LNCS, vol. 3408, pp. 345–359. Springer, Heidelberg (2005). https://doi.org/10.1007/978-3-540-31865-1_25
7. Graves, A.: Long short-term memory. In: Graves, A. (ed.) Supervised Sequence Labelling with Recurrent Neural Networks. Studies in Computational Intelligence, vol. 385, pp. 37–45. Springer, Berlin (2012). https://doi.org/10.1007/978-3-642-24797-2_4
8. Han, K., et al.: A survey on vision transformer. IEEE Trans. Pattern Anal. Mach. Intell. **45**(1), 87–110 (2022)
9. He, K., Gkioxari, G., Dollár, P., Girshick, R.: Mask R-CNN. In: Proceedings of the IEEE International Conference on Computer Vision, pp. 2961–2969 (2017)
10. Hussein, F., Al-Ahmad, A., El-Salhi, S., Alshdaifat, E., Al-Hami, M.: Advances in contextual action recognition: automatic cheating detection using machine learning techniques. Data **7**(9), 122 (2022)
11. Kamalov, F., Sulieman, H., Santandreu Calonge, D.: Machine learning based approach to exam cheating detection. PLoS ONE **16**(8), e0254340 (2021)
12. Li, Z., Zhu, Z., Yang, T.: A multi-index examination cheating detection method based on neural network. In: 2019 IEEE 31st International Conference on Tools with Artificial Intelligence (ICTAI), pp. 575–581. IEEE (2019)
13. Ozdamli, F., Aljarrah, A., Karagozlu, D., Ababneh, M.: Facial recognition system to detect student emotions and cheating in distance learning. Sustainability **14**(20), 13230 (2022)
14. Rehab, K.k., Ali, Z.H.: Cheating detection in online exams using machine learning. J. AL-Turath Univ. Coll. **2**(35) (2023)
15. Ren, S., He, K., Girshick, R., Sun, J.: Faster R-CNN: towards real-time object detection with region proposal networks. In: Advances in Neural Information Processing Systems, vol. 28 (2015)
16. Simonyan, K., Zisserman, A.: Very deep convolutional networks for large-scale image recognition. arXiv preprint: arXiv:1409.1556 (2014)
17. Tiong, L.C.O., Lee, H.J.: E-cheating prevention measures: detection of cheating at online examinations using deep learning approach-a case study. arXiv preprint: arXiv:2101.09841 (2021)
18. Touvron, H., Cord, M., Douze, M., Massa, F., Sablayrolles, A., Jégou, H.: Training data-efficient image transformers & distillation through attention. In: International Conference on Machine Learning, pp. 10347–10357. PMLR (2021)
19. Yuan, L., et al.: Tokens-to-token ViT: training vision transformers from scratch on ImageNet. In: Proceedings of the IEEE/CVF International Conference on Computer Vision, pp. 558–567 (2021)
20. Yulita, I.N., Hariz, F.A., Suryana, I., Prabuwono, A.S.: Educational innovation faced with COVID-19: deep learning for online exam cheating detection. Educ. Sci. **13**(2), 194 (2023)

# Big Data, IoT, and Cloud Computing

# Building a Health Monitoring System

Tri-Thuc Vo[1]([✉]) and Thanh-Nghi Do[1,2]

[1] College of Information Technology, Can Tho University, 92000 Cantho, Vietnam
{vtthuc,dtnghi}@cit.ctu.edu.vn
[2] UMI UMMISCO 209 (IRD/UPMC), Sorbonne University, Pierre and Marie Curie
University, Paris 6, France

**Abstract.** In this paper, we propose a health monitoring system, which leverages the Internet of Things (IoT) in conjunction with time-series deep learning techniques. The main purpose is to build the system for connecting to the MiBand smart watch device as well as to provide timely notifications about health abnormal signs of the wearer to their relatives, doctors to assist in timely handling. Cardiovascular disease is one of the dangerous diseases, causing patients to have a very high risk of death and rapid death. Our proposed system can monitor, predict and detect abnormal heart rhythms, so it plays an important role for early detection of cardiac dysfunction, timely treatment, and reducing the risk of death. To solve this problem, we have developed a system to collect data from the MiBand device, build a model for predicting heart rate and an abnormal alert system to relatives and doctors. The heart rate prediction module is trained on six popular deep learning models. The experimental results show that all models have the mean absolute error (MAE) in the range of 3.4 to 4.2. The results of this study can be the basis for further studies to develop other health monitoring initiatives with comparable objectives.

**Keywords:** Health monitoring · MiBand4 · Heart rate · Deep learning

## 1 Introduction

The conventional approach involved patients and doctors primarily interacting through in-person visits or communication via phone. However, this method had its limitations, making it inconvenient for doctors and healthcare providers to continuously monitor patients' health. Advancements in technology, particularly with the emergence of the Internet of Things (IoT), have revolutionized the healthcare industry [3,10,17,18]. These developments have made remote monitoring in healthcare a reality, empowering patients with greater control over their lives and treatment. These devices provide valuable insights into symptoms and trends, leading to reduced healthcare costs and improved treatment outcomes.

The advancements in smartwatches with health monitoring capabilities have significantly impacted the medical industry in recent years. These devices allow continuous real-time monitoring including heart rate, steps, runs, stairs climbed,

N. Thai-Nghe et al. (Eds.): ISDS 2023, CCIS 1949, pp. 81–94, 2024.
https://doi.org/10.1007/978-981-99-7649-2_7

calories burned through sensors. Wearable technology plays an important role for early detection of health abnormalities, providing health care benefits as well as limiting the risk of complicated disease progression. Early diagnosis enables more effective treatment during the initial stages of the disease. Numerous studies have focused on building health monitoring systems by leveraging data from smartwatch devices and developing corresponding applications on smartphones [9,11,23].

Heart disease is one of the leading causes of death in the world because the heart cannot perform its main function of pumping blood to other organs. Medically, heart rate is defined as the number of heartbeats per minute, and it varies across different age groups, with the average adult heart rate ranging from 60 to 100 beats per minute. Heart rate can be affected by many factors such as physical activity, mental state, gender, age, etc. The study and monitoring of heart rate variations hold significant importance in patient health monitoring, providing valuable insights into potential health issues and enabling timely interventions as needed.

This research proposed a health monitoring system that integrates the Miband 4 smartwatch with deep learning techniques. The system was designed to include MiBand device management, data collection and support to detect and predict health abnormalities, with a particular focus on heart rate. Moreover, our innovative approach also supports sending notifications to patients, their relatives and doctors about health indicators that exceed the threshold. The study holds significant importance for the prevention and management of heart-related diseases. The research involved collecting heart rate data using the MiBand 4 smartwatch. Subsequently, we train six deep learning models, including LSTM, GRU, Bi-LSTM, Bi-GRU, CNN-LSTM, and Conv-LSTM, for heart rate prediction. The experimental results reveal that the Bi-GRU model exhibits the most promising performance when compared to the other five methods.

The subsequent content of this article is structured as follows. Section 2 provides an overview of relevant studies. The architecture of the health monitoring system is presented in Sect. 3. Section 4 elaborates on the experimental results. Finally, in Sect. 5, we conclude the article and discuss future research directions.

## 2   Related Work

Research in [7] indicated that wearable IoT technologies, when integrated with remote medical infrastructure, have the potential to enhance the efficiency of treating chronic diseases like diabetes, obesity, and cardiovascular disorders. Through continuous tracking and detection of symptom patterns using sensor technology, these devices can record valuable data directly on the patient. Moreover, the integration of sensors, mobile phones, and medical infrastructure fosters seamless communication between physicians and patients. In [16], modern wearable devices, such as the Fitbit[1] and Pebble smartwatch[2] have sparked innovative

---

[1] https://www.fitbit.com/.Retrieved10July2023.
[2] https://www.pebblecart.com/.Retrieved10July2023.

approaches to harnessing the potential of IoT for the human body and beyond. These devices empower individuals with self-monitoring capabilities and offer insights into health and fitness beyond smartphones. They are valuable tools for proactive health condition prevention, including hypertension and stress monitoring.

In [13], authors provide a comprehensive synthesis of various studies investigating the application of smartwatch devices in healthcare. The findings reveal that six studies (25%) centered on smartwatch usage among the elderly, focusing on health monitoring and smart home applications. Additionally, five studies (21%) were dedicated to exploring the potential benefits of smartwatches for patients with Parkinson's disease. The third and fourth most prominent categories of studies addressed food and diet monitoring (4.17%) and medication adherence monitoring in patients with chronic diseases (3.13%). In [12], this research focuses on developing a health monitoring system that harnesses the capabilities of smartwatches and mobile phones, enabling remote monitoring of elderly individuals or patients. The system functions by collecting sensor data from the smartwatch, which is then transmitted to the mobile phone for processing and onward transmission to a server application. Caregivers can access the data at any time, providing timely notifications and valuable insights to improve the quality of life for the elderly, particularly those with age-related diseases or chronic conditions.

Many studies explored heart rate prediction utilizing machine learning and deep learning methodologies. The authors in [24] suggested a heart rate prediction approach based on heart rate data and physical activities. They employed the Feedforward Neural Network method, which yielded the evaluation of model with value Mean Absolute Error within 5. In [15], the authors proposed heart rate prediction from acceleration values using wearable devices for patients with heart valve intervention. The obtained experimental results show that the MAE value is 2.89. The research team [21] proposed a deep learning architecture to predict heart rate using PPG signals with input data be 8-second. The proposed model was evaluated on IEEE SPC 2015 dataset with MAE value of $3.46 \pm 4.1$ beats per minute. The research by Dwaipayan Biswas et al. [4] focused on a deep learning approach comprising 2 CNN layers, 2 LSTM layers, and 1 Densenet layer to predict heart rate from PPG signals in ambulance vehicles. The network architecture of this study was evaluated on the TROIKA dataset, which included 22 PPG samples, achieving a Mean Absolute Error (MAE) value of $1.47 \pm 3.37$ heartbeats per minute. Abdullah Alharbi et al. in [2] focused on realtime heart rate prediction. The training data was sourced from the Information Mart for Intensive Care (MIMIC-II) dataset and processed using various deep learning networks, including RNN, LSTM, GRU, and BI-LSTM. The experimental results revealed that GRU achieved the best performance based on the Root Mean Square Error (RMSE). In their work [14], the researchers conducted heart rate predictions using a dataset that encompassed five attributes: heart rate signal, gender, age, physical activity, and mental state. The outcomes demonstrated that Adam-LSTM is an effective method for predicting heart rate and capturing heart rate variability trends in daily life. The study conducted by the author in [22] compares three models, namely the Autoregressive Model, LSTM, and Convolutional LSTM, for heart rate prediction. The data was collected from 12 individuals using the Fitbit Versa device, which included heart rate information, sleep time, and physical activity data. The experimental results indicated that the Autoregressive Model outperformed the other two models, achieving the best performance with value MAE of 2.069.

# 3   A Health Monitoring System

## 3.1   System Architecture

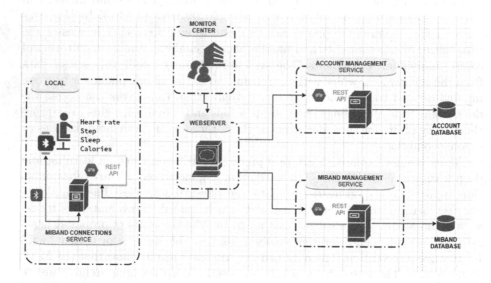

**Fig. 1.** The architecture of health monitoring system.

In our health monitoring system, the data collection was facilitated through the utilization of the MiBand smartwatch device. The system's architecture was designed by integrating various components, each playing a crucial role in gathering and processing the health-related information and sending timely notifications to patient's relatives and physicians. Through this seamless combination of components, the system effectively collected and managed data, ensuring the availability of valuable insights to the users. By utilizing the MiBand smartwatch device as a reliable data source, the health monitoring system presented comprehensive and user-friendly displays, enabling individuals to stay informed about their health status with ease and convenience. The core structure of our system encompasses five primary components: the local MiBand connection management service, webserver, account management service, MiBand management service, and monitor center. These interconnected components work in synergy to ensure the seamless functioning of the health monitoring system.

**The Local MiBand Connection Management Service.** Serves as an intermediate server, responsible for establishing connections and disconnections with the devices. It efficiently reads data from the MiBand devices and provides the data values to other services upon receiving REST API requests. This service acts as a bridge, facilitating smooth communication between the MiBand devices and the rest of the system components.

**Webserver.** Serves as a crucial cornerstone in our health monitoring system, playing a pivotal role in delivering an intuitive and user-friendly interface. Its primary responsibility is to display a visually appealing UI that enables users to access and interpret their health-related data effortlessly. Beyond the presentation layer, the webserver acts as a key communicator, facilitating interactions between users and the other integral services of the system. It achieves this by effectively sending API calls to the respective services in response to users' specific requests. This communication mechanism ensures that users can seamlessly access their personalized health information and receive real-time updates based on their individual needs. The webserver acts as a central hub for information dissemination, allowing users to interact with their data in a meaningful and insightful manner. The webserver component empowers users to actively engage with the health monitoring system, gaining valuable insights into their health.

**The Account Management Service.** Responsible for a critical pillar in the foundation of our system, providing robust functionality to handle and process a wide array of requests concerning user accounts. Upon receiving requests from users, the account management service performs essential operations such as user authentication, data storage, and access control. This service facilitates personalized access control, allowing users to customize their preferences, privacy settings, and data sharing permissions. The account management service serves as a trusted guardian of user data, maintaining data integrity, confidentiality, and compliance with privacy regulations. The account management service plays a fundamental role in streamlining user interactions and ensuring the seamless functioning of our health monitoring system.

**The MiBand Management Service.** Serves as a pivotal component in our health monitoring system, taking on the responsibility of handling and executing requests concerning the MiBand database. This service acts as a central hub for managing all interactions related to MiBand devices, ensuring their seamless integration within the system. Upon receiving requests, the MiBand management service efficiently processes and manages various operations related to MiBand devices. It facilitates device registration, configuration, and tracking, enabling smooth and reliable data collection from the connected MiBand devices. By effectively managing the MiBand database, this service ensures that data collection is streamlined and optimized for real-time health monitoring. The MiBand management service ensures that MiBand devices are synchronized with other system components. This seamless integration optimizes the user experience, allowing for comprehensive health data collection and analysis.

**The Monitor Center.** Assumes a central role in system, actively overseeing and evaluating users' health conditions. Its primary responsibility is to continuously monitor various health parameters and detect any abnormal situations that may arise. Particularly noteworthy are indicators that surpass safety thresholds, such as heart rates exceeding the typical range of 60 to 100 beats per minute for adults with normal heart rates. In the event of detecting abnormal health indicators or critical situations, the monitor center promptly initiates contact with the user's designated family members or emergency contacts by messages. This proactive approach ensures that timely support and assistance can be provided, enhancing the safety and well-being of the user. With its sophisticated monitoring capabilities, the monitor center acts as a vigilant guardian, safeguarding users' health and promoting early intervention in case of any potential health risks. By being equipped to monitor and respond to critical situations, the monitor center plays a vital role in enhancing user safety and fostering a sense of security and trust in the health monitoring system.

## 3.2    Xiaomi Mi Smart Band 4

Xiaomi Mi Smart Band 4[3] represents the generation of Xiaomi's smart bracelets, equipped with functions of tracking health and fitness. This impressive wearable is capable of monitoring a wide range of physical activities, including walking, running, swimming, and cycling. With its built-in heart rate sensor, the MiBand 4 effectively tracks heart rate, offering valuable insights into users' cardiovascular health. The device records various metrics such as speed, steps taken, and calories burned, enabling users to stay informed about their fitness progress and activity levels. To extract and analyze the data collected by the MiBand 4, a Python program has been developed, facilitating seamless Bluetooth connectivity with the wearable device. This program efficiently extracts essential data, including heart rate, steps, distance walked, and calories burned.

## 3.3    Heart Rate Prediction System

To evaluate the health monitoring system, we developed a module capable of predicting heart rate and detecting and alerting users about any abnormalities in heart rate. The heart rate prediction system using deep learning involves five key stages: data collection, data preprocessing, division of the dataset for training and testing, model training with six deep learning methods, and model evaluation. The data collection process is achieved through a Bluetooth-connected program with a MiBand 4 device, capturing four attributes: heart rate, number

---

[3] Xiaomi United Kindom. Retrieved 10 July 2023, from https://www.mi.com/uk/mi-smart-band-4/.

of steps, distance, and calories consumed. After collecting data from four individuals, it undergoes preprocessing, which includes applying a sliding window of 16 s. Following preprocessing, the dataset is divided into two subsets: a training set and a test set, serving to train and evaluate the deep learning model for heart rate prediction. Subsequently, 67% of the dataset is utilized to train the model using six deep learning techniques. Lastly, the performance of the deep learning model is assessed using the mean absolute error (MAE) on the remaining 33% of the test dataset.

## 3.4   Deep Learning Algorithms for Heart Rate Prediction

**Long Short-Term Memory (LSTM).** The concept of LSTM networks was introduced by Hochreiter and Schmidhuber in 1997 [8]. LSTM represents a specialized type of Recurrent Neural Network (RNN) with the unique ability to capture distant dependencies in sequences. It has proven to be highly effective for time series sequence prediction tasks. The LSTM architecture consists of a series of interconnected cells, each equipped with three essential gates: the forget gate, responsible for determining which information to discard; the input gate, responsible for updating the memory state based on relevant input information; and the output gate, which regulates the output value. The one-cell architecture of LSTM is illustrated in Fig. 2. LSTM networks have demonstrated remarkable efficacy in handling sequential data, making them a preferred choice for various predictive modeling tasks involving time series data. At each time step, the equations governing the LSTM cell's behavior are as follows:

$$
\begin{aligned}
\text{Input Gate } (i_t): \quad & i_t = \sigma(W_i \cdot x_t + U_i \cdot h_{t-1} + b_i) \\
\text{Forget Gate } (f_t): \quad & f_t = \sigma(W_f \cdot x_t + U_f \cdot h_{t-1} + b_f) \\
\text{Output Gate } (o_t): \quad & o_t = \sigma(W_o \cdot x_t + U_o \cdot h_{t-1} + b_o) \\
\text{Candidate Cell State } (\widetilde{C}_t): \quad & \widetilde{C}_t = \tanh(W_c \cdot x_t + U_c \cdot h_{t-1} + b_c) \\
\text{Current Cell State } (C_t): \quad & C_t = f_t \odot C_{t-1} + i_t \odot \widetilde{C}_t \\
\text{Current Hidden State } (h_t): \quad & h_t = o_t \odot \tanh(C_t)
\end{aligned}
\tag{1}
$$

**Gated Recurrent Units (GRU).** The GRU was introduced by Kyunghyun Cho et al. in 2014 [5]. Similar to LSTM, the GRU represents a specialized type of RNN architecture. In the GRU network (see Fig. 3), interconnected cells form an array, with each cell containing two essential gates: the update gate and the reset gate. The update gate determines the extent to which previous state information is incorporated into the new state, while the reset gate governs whether the prior hidden state is ignored or not. Notably, compared to LSTM, the GRU boasts a streamlined design with fewer gates, rendering it a more compact and parameter-efficient architecture. The GRU's simplified structure results in a reduced number of parameters, making it computationally more

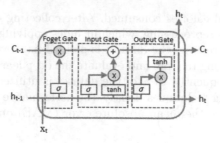

**Fig. 2.** Structure of the LSTM cell.

efficient than LSTM. Despite its reduced complexity, the GRU has proven to be highly effective in capturing temporal dependencies in sequential data. At each time step, the main equations governing the GRU's behavior are as follows:

$$
\begin{aligned}
\text{Reset Gate } (r_t): \quad & r_t = \sigma(W_r \cdot x_t + U_r \cdot h_{t-1} + b_r) \\
\text{Update Gate } (z_t): \quad & z_t = \sigma(W_z \cdot x_t + U_z \cdot h_{t-1} + b_z) \\
\text{Candidate Activation } (\tilde{h}_t): \quad & \tilde{h}_t = \tanh(W_h \cdot x_t + r_t \odot (U_h \cdot h_{t-1} + b_h)) \\
\text{Current Hidden State } (h_t): \quad & h_t = (1 - z_t) \odot h_{t-1} + z_t \odot \tilde{h}_t
\end{aligned}
\tag{2}
$$

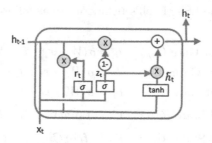

**Fig. 3.** Structure of the GRU cell.

**Bidirectional Long Short-Term Memory (Bi-LSTM).** The Bi-LSTM architecture was inspired by the concept of a bidirectional regression network (Bidirectional RNN) [19] that processes sequence data in both forward and backward directions. In this approach, prediction information at each time step is gathered from both the forward and backward directions, enhancing the model's ability to capture contextual dependencies in sequential data. LSTM, being a specialized type of RNN, is well-suited for integration into the Bi-RNN model. Each unit of the RNN is now replaced with an LSTM unit, comprising the essential components of a forget gate, an input gate, and an output gate. These components enable LSTM to efficiently capture and process temporal dependencies in both directions, enhancing the model's capacity for sequence prediction tasks. The Bi-LSTM architecture consists of two LSTMs in Fig. 4: one for processing the input data in the forward direction and another for processing it in the reverse direction. This bidirectional design allows the model to leverage information from both past and future time steps, significantly improving its ability to model and understand the sequential patterns present in the data.

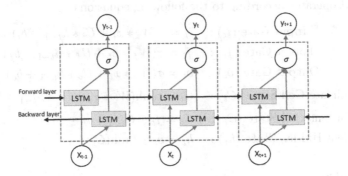

**Fig. 4.** Architecture of the Bi-LSTM.

**Bidirectional Gated Recurrent Units(Bi-GRU).** The Bi-GRU architecture draws inspiration from the idea of a Bidirectional Regression Network (Bidirectional RNN) [19]. Like Bi-LSTM, the Bi-GRU also incorporates corresponding GRU components for both forward and reverse directions. Each of these components includes two crucial gates: the update gate and the reset gate. The Bi-GRU architecture showcases the data flow in both forward and reverse directions, providing a comprehensive view of the input sequence.

**Convolutional Neural Network Long Short-Term Memory (CNN-LSTM).** The CNN-LSTM network architecture is a hybrid model that combines the strengths of both convolutional neural networks (CNN) for effective feature extraction and long short-term memory (LSTM) for sequence prediction. This powerful combination enables the CNN-LSTM to handle various tasks, including visual time series prediction. For instance, in the context of heart rate prediction, the CNN-LSTM is well-suited to tackle this form of the problem. It leverages the CNN's ability to extract meaningful visual features from the time series data and then utilizes the LSTM's memory and sequential modeling capabilities to make accurate predictions.

**Convolutional Long Short-Term Memory (ConvLSTM).** In 2015, Xingjian Shi et al. introduced the ConvLSTM [20], an innovative architecture that builds upon the LSTM framework. The ConvLSTM incorporates a convolution operator in the state-to-state and input-to-state transitions, making it a powerful variant of the traditional LSTM. The ConvLSTM architecture illustrates the integration of convolutions within the LSTM cells, which allows it to efficiently process spatial and temporal information simultaneously. This combination of LSTM and convolution operators empowers the ConvLSTM to excel in tasks involving both spatial and temporal dependencies, making it a compelling choice for applications that processing sequential data. At each time step t, the ConvLSTM operates according to the following equations:

$$
\begin{aligned}
\text{Input Gate } (i_t): \quad & i_t = \sigma(W_i * x_t + U_i * h_{t-1} + b_i) \\
\text{Forget Gate } (f_t): \quad & f_t = \sigma(W_f * x_t + U_f * h_{t-1} + b_f) \\
\text{Output Gate } (o_t): \quad & o_t = \sigma(W_o * x_t + U_o * h_{t-1} + b_o) \\
\text{Candidate Cell State } (\widetilde{C}_t): \quad & \widetilde{C}_t = \tanh(W_c * x_t + U_c * h_{t-1} + b_c) \\
\text{Current Cell State } (C_t): \quad & C_t = f_t \odot C_{t-1} + i_t \odot \widetilde{C}_t \\
\text{Current Hidden State } (h_t): \quad & h_t = o_t \odot \tanh(C_t)
\end{aligned}
\tag{3}
$$

## 4   Results

### 4.1   Dataset

Data was collected from four individuals wearing MiBand 4 devices during their daily activities. The participants consisted of two males (aged 48 and 13) and two females (aged 47 and 15). The data collection program established a Bluetooth connection with the MiBand 4 devices, capturing data every second while users engaged in various activities, including daytime tasks, studying, working, and sleeping. The data collection period coincided with the 2021 social distancing period, prompted by the Covid-19 outbreak. In total, 359.610 s of data were obtained through the MiBand 4 devices. To prepare the data for model training, we applied a sliding window approach, segmenting the data into 16-second intervals. This data preprocessing yielded 359.594 samples of 16 s each. Subsequently, the dataset was split into two subsets: a training dataset comprising 67% (240.928 samples) and a test dataset consisting of 33% (118.666 samples).

### 4.2   Experimental Results for Heart Rate Prediction

We implemented the data collection program for MiBand 4 and the heart rate prediction program using the Python programming language. For the heart rate prediction, we employed LSTM, GRU, Bi-LSTM, Bi-GRU, CNN-LSTM, and ConvLSTM algorithms, which were implemented in Python using the Keras library [6] and Tensorflow [1]. Mean absolute error (MAE) was used to evaluate the performance of six deep learning models as follows:

$$
\text{MAE} = \frac{1}{n} \sum_{i=1}^{n} |y_i - \hat{y}_i| \quad * \quad y_i\text{:true value, } \hat{y}_i\text{:the predicted value} \tag{4}
$$

**Table 1.** Experimental results on heart rate prediction

| Algorithm | Mean absolute error (MAE) |
|-----------|---------------------------|
| GRU | 3.53 |
| LSTM | 3.58 |
| Bi-GRU | 3.45 |
| Bi-LSTM | 3.48 |
| CNN-LSTM | 4.14 |
| Conv-LSTM | 3.60 |

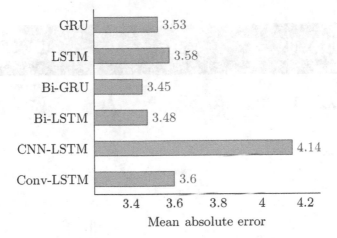

**Fig. 5.** Experimental results on heart rate prediction (MAE)

The experimental results of heart rate prediction obtained from six deep learning models (GRU, LSTM, Bi-GRU, Bi-LSTM, CNN-LSTM, and Conv-LSTM) are presented in Table 1 and depicted in Fig. 5. The graph illustrates that all models exhibit MAE values ranging from 3.4 to 4.2. Notably, five models (GRU, LSTM, Bi-GRU, Bi-LSTM, and Conv-LSTM) show no significant difference in their MAE values, which fall within the range of 3.4 to 3.6. Among the six models, the Bi-GRU model demonstrated the lowest MAE value of 3.45, signifying its superior accuracy in heart rate prediction. Conversely, the CNN-LSTM model recorded the highest MAE value of 4.14, indicating slightly less accuracy compared to the other models. These findings provide valuable insights into the performance of different deep learning models in predicting heart rates, with the Bi-GRU model emerging as the most effective among the tested approaches. The user interface of the health monitoring system is shown in Fig. 6.

Heart rate prediction outcomes are instrumental in bolstering human health monitoring efforts. They prove especially valuable when they anticipate heart rates that are poised to surpass established norms. For instance, when the predictive model signals the likelihood of heart rates exceeding the standard thresholds for various age groups, including preschoolers (3–5 years) with a typical range of 80–120 beats per minute, school-age children (6–12 years) falling within the range of 70–100 beats per minute, adolescents (13–18 years), adults (18+ years) and the elderly, particularly those aged 65 and above, whose typical range also spans from 60–100 beats per minute. Its pivotal role extends beyond mere observation. It promptly triggers notifications, reaching out to both concerned family members and healthcare professionals when the predicted heart rate exceeds the threshold. This proactive response ensures that timely measures can be taken to address potential health concerns, ultimately providing peace of mind and a vigilant approach to safeguarding well-being.

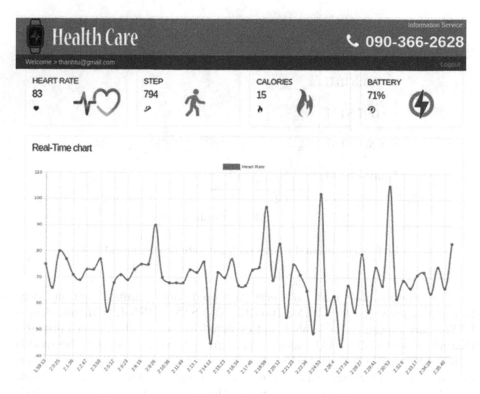

**Fig. 6.** The health monitoring system.

## 5   Conclusion and Future Works

We presented a health monitoring system by integrating the MiBand4 smartwatch with deep learning algorithms. Our proposed system is designed to receive data from the smartwatch device, analyze data, predict and detect any abnormal signs of the wearer's health when it exceeds the threshold of a healthy person.

Additionally, the system is equipped to send health status messages to patients, their relatives, and medical professionals. This integration aims to bridge the gap between patients and the medical team, facilitating easier disease management and providing timely interventions. To evaluate the system's performance, we collected physical activity and heart rate data from the wearers through MiBand4. We then trained six deep learning models to predict heart rates. Our primary contribution lies in proposing a health monitoring model and assembling a dataset comprising 359.594 samples from the MiBand4 device. Furthermore, we have successfully implemented a system capable of predicting and warning against abnormal heart rhythms when it goes beyond the boundary of a person in good health. Through experimentation and evaluation using MAE, we observed that the Bi-GRU model outperformed the others with a superior MAE value of 3.45. Our proposed system has promising applications in similar diseases with IoT devices, smartwatches, and enhancing collaboration between patients, their relatives and healthcare providers for better outcomes.

**Acknowledgements.** This work has been funded by the STARWARS project (STormwAteR and WastewAteR networkS heterogeneous data AI-driven management). The authors would like to particularly thank for the STARWARS' support.

# References

1. Abadi, M., et al.: TensorFlow: large-scale machine learning on heterogeneous distributed systems. arXiv preprint arXiv:1603.04467 (2016)
2. Alharbi, A., Alosaimi, W., Sahal, R., Saleh, H.: Real-time system prediction for heart rate using deep learning and stream processing platforms. Complexity **2021**, 1–9 (2021)
3. Alshamrani, M.: IoT and artificial intelligence implementations for remote healthcare monitoring systems: a survey. J. King Saud Univ. Comput. Inf. Sci. **34**(8), 4687–4701 (2022)
4. Biswas, D., et al.: CorNET: deep learning framework for PPG-based heart rate estimation and biometric identification in ambulant environment. IEEE Trans. Biomed. Circuits Syst. **13**(2), 282–291 (2019)
5. Cho, K., et al.: Learning phrase representations using RNN encoder-decoder for statistical machine translation. arXiv preprint arXiv:1406.1078 (2014)
6. Chollet, F., et al.: Keras. https://keras.io (2015)
7. Hiremath, S., Yang, G., Mankodiya, K.: Wearable internet of things: concept, architectural components and promises for person-centered healthcare. In: 2014 4th International Conference on Wireless Mobile Communication and Healthcare-Transforming Healthcare Through Innovations in Mobile and Wireless Technologies (MOBIHEALTH), pp. 304–307. IEEE (2014)
8. Hochreiter, S., Schmidhuber, J.: Long short-term memory. Neural Comput. **9**(8), 1735–1780 (1997)
9. Jat, A.S., Grønli, T.M.: Smart watch for smart health monitoring: a literature review. In: International Work-Conference on Bioinformatics and Biomedical Engineering. pp. 256–268. Springer (2022). https://doi.org/10.1007/978-3-031-07704-3_21

10. Khan, M.M., Alanazi, T.M., Albraikan, A.A., Almalki, F.A., et al.: IoT-based health monitoring system development and analysis. Secur. Commun. Netw. **2022** (2022)
11. King, C.E., Sarrafzadeh, M.: A survey of smartwatches in remote health monitoring. J. Healthc. Inf. Res. **2**, 1–24 (2018)
12. Kosanovic, M., Stosovic, S., Stojanovic, D.: Smartwatch-based wellbeing monitoring system for the elderly (2018)
13. Lu, T.C., Fu, C.M., Ma, M.H.M., Fang, C.C., Turner, A.M.: Healthcare applications of smart watches. Appl. Clin. Inform. **7**(03), 850–869 (2016)
14. Luo, M., Wu, K.: Heart rate prediction model based on neural network. In: IOP Conference Series: Materials Science and Engineering, vol. 715, p. 012060. IOP Publishing (2020)
15. McConville, R., et al.: Online heart rate prediction using acceleration from a wrist worn wearable. arXiv preprint arXiv:1807.04667 (2018)
16. Metcalf, D., Milliard, S.T., Gomez, M., Schwartz, M.: Wearables and the internet of things for health: wearable, interconnected devices promise more efficient and comprehensive health care. IEEE Pulse **7**(5), 35–39 (2016)
17. Nancy, A.A., Ravindran, D., Raj Vincent, P.D., Srinivasan, K., Gutierrez Reina, D.: IoT-cloud-based smart healthcare monitoring system for heart disease prediction via deep learning. Electronics **11**(15), 2292 (2022)
18. Sangeethalakshmi, K., Preethi, U., Pavithra, S., et al.: Patient health monitoring system using IoT. Mater. Today Proc. **80**, 2228–2231 (2023)
19. Schuster, M., Paliwal, K.K.: Bidirectional recurrent neural networks. IEEE Trans. Signal Process. **45**(11), 2673–2681 (1997)
20. Shi, X., Chen, Z., Wang, H., Yeung, D.Y., Wong, W.K., Woo, W.C.: Convolutional lstm network: a machine learning approach for precipitation nowcasting. In: Advances in Neural Information Processing Systems 28 (2015)
21. Shyam, A., Ravichandran, V., Preejith, S., Joseph, J., Sivaprakasam, M.: PPGnet: Deep network for device independent heart rate estimation from photoplethysmogram. In: 2019 41st Annual International Conference of the IEEE Engineering in Medicine and Biology Society (EMBC), pp. 1899–1902. IEEE (2019)
22. Staffini, A., Svensson, T., Chung, U.I., Svensson, A.K.: Heart rate modeling and prediction using autoregressive models and deep learning. Sensors **22**(1), 34 (2021)
23. Sujith, A., Sajja, G.S., Mahalakshmi, V., Nuhmani, S., Prasanalakshmi, B.: Systematic review of smart health monitoring using deep learning and artificial intelligence. Neurosci. Inf. **2**(3), 100028 (2022)
24. Yuchi, M., Jo, J.: Heart rate prediction based on physical activity using feedforwad neural network. In: 2008 International Conference on Convergence and Hybrid Information Technology, pp. 344–350. IEEE (2008)

# Fall Detection Using Intelligent Walking-Aids and Machine Learning Methods

Thanh-Nghi Doan[1,2]([📧]) [iD], Eliane Schroter[3]([📧]), and Thanh-Binh Phan[1,2] [iD]

[1] Faculty of Information Technology, An Giang University, An Giang, Vietnam
{dtnghi,ptbinh}@agu.edu.vn
[2] Vietnam National University, Ho Chi Minh City, Vietnam
[3] Jade-Hochschule, University of Applied Science, Wilhelmshaven, Germany
eliane.schroeter@student.jade-hs.de

**Abstract.** Walking aids are commonly given to older adults to prevent falls, but paradoxically, their use has been identified as a risk factor for falling, which is a prevalent issue among this population, causing serious injuries, disabilities, and even death. This has resulted in a significant increase in public health care and the development of remote health monitoring technology to enhance home care devices. One of the key issues being addressed is the identification of falling incidents, which can aid in the rapid arrival of assistance and prevent additional harm. This paper is to develop intelligent walking aids using machine learning methods and an M5Stack Core2 microcontroller. These aids analyze information from various sensors such as accelerometers and gyroscopes to identify falls. Touch and location sensors are also utilized to determine the device's usage and location in case of an emergency. The collected data is sent to a web server in JSON format via the M5Core2's WiFi module, allowing for a quick response if necessary. The fall detection system has been extensively tested, resulting in 99.62% accurate identification of falls.

**Keywords:** Fall detection · Remote health monitoring · Machine learning · M5Stack Core2 · Intelligent walking-aid

## 1 Introduction

Detecting and preventing falls is crucial, especially for older adults who are more susceptible to falls due to aging-related factors. Home care devices, including fall detection systems, have become essential in ensuring the safety of the elderly and individuals with chronic illnesses who wish to remain at home. This leads to the development of a novel system that incorporates Intelligent walking aids to detect falls and enhance user safety and quality of life. The device provides an added layer of security, allowing users to have confidence in their mobility and engage in activities while ensuring their safety. Over the past years, fall detection and supervision systems have been classified into various categories.

© The Author(s), under exclusive license to Springer Nature Singapore Pte Ltd. 2024
N. Thai-Nghe et al. (Eds.): ISDS 2023, CCIS 1949, pp. 95–109, 2024.
https://doi.org/10.1007/978-981-99-7649-2_8

One of the widely adopted categories is wearable devices, which are worn on the body and equipped with a variety of sensors placed in different areas. Since they are worn all the time, they offer continuous monitoring and tracking of overall activities, making them relatively reliable. The authors in [1] suggested a strong activity recognition method for smart healthcare that involves utilizing body sensors and a deep convolutional neural network (CNN). Various body sensors used in healthcare, such as ECG, magnetometer, accelerometer, and gyroscope sensors, are analyzed to extract significant features from the signal data. The feature extraction process involves Gaussian kernel-based principal component analysis and Z-score normalization. To evaluate the effectiveness of the proposed approach, a publicly available standard dataset is used, and the results are compared to other conventional approaches. The authors in [2] proposed an approach for fall detection that employs three distinct sensors placed at five different locations on the subject's body to gather data for training purposes. The UMAFall dataset is utilized to obtain sensor readings and train models for fall detection. Five different models are trained, each corresponding to one of the five sensor models, and a majority voting classifier is employed to determine the final output. However, for some patients, the constant wearing of these devices depicts an inconvenience or, especially when talking about elderly people, they simply forget to put their intelligent devices on.

In contrast to wearable systems, vision-based systems offer a comfortable and reliable option for fall detection in-home care. The authors in [3] provided a brief overview of vision-based fall detection, outlining recent methods and highlighting their advantages and disadvantages, and touching on possible future research topics. The authors in [4] examined the latest non-intrusive (vision-based) fall detection techniques based on deep learning (DL). They also provide an overview of benchmark datasets for fall detection and explain various metrics used to assess the performance of these systems. The authors in [5] introduced a novel smart camera system for real-time monitoring, recognition, and remote warning of abnormal patient actions. The proposed method is cost-effective and easy to deploy. It does not require ambient sensors and uses regular video camera footage for detection. The system utilizes high-precision human body pose tracking with MediaPipe Pose and employs the Raspberry Pi 4 device and LSTM network for real-time classification and remote monitoring of patient actions. One of the key advantages of these approaches is their ability to differentiate a person's actions without requiring them to wear a device. However, they are limited to a single space and can intrude on a person's privacy. For example, a vision-based device cannot be installed in private areas like the bathroom where falls can also occur. Additionally, some individuals may feel uncomfortable being continuously monitored, despite the data protection measures that are inherent in these devices.

The proposal of using an environmentally based system to address privacy concerns in fall detection. Existing attempts with infrared sensor systems and millimeter-wave radar sensors have limitations but offer the advantage of not being worn and can be installed in private spaces. The paper introduces a intelligent walking aid that combines environmental solutions and wearable devices. By adding sensors to the walking aid, it can detect falls and alert caregivers. The device restores mobility, provides autonomy, increases confidence and safety, and cannot be forgotten. It preserves privacy and can be used anywhere the walking aid is taken. The cost is affordable, and the device focuses on walking activities without the need to differentiate other activities. It primarily aims to improve safety and mobility over long distances. The device can be combined with other fall-detection measures, and further studies can be conducted to gather data on falls with walking aids. The main contributions of this paper consist of:

- Build a comprehensive detection system with a user-friendly interactive interface that can immediately monitor the usage and orientation of the walking aid.
- Create a novel dataset containing over 255 data sets collected from a fall study or real-life scenario.
- Develop a fall detection algorithm for a walking aid that uses touch as input. This algorithm will be implemented on an M5Stack Core2 microcontroller, which will analyze sensor data to detect falls.
- Develop a machine learning model for fall prediction and create a website that is compatible with it. The goal is to interpret the data collected by the M5Core2 through WiFi.

The overview of our proposed system is described in Fig. 1. There are two kinds of walking aids that are frequently used and have been under consideration for this particular situation. These aids are illustrated in Fig. 2. The first aid, shown on the left (a), is a commonly used walking cane. On the right (b) is a walking aid that stands on four feet, making the process of detecting a falling person a bit more complicated.

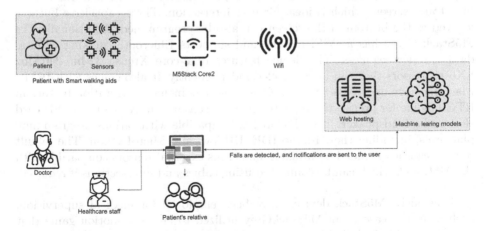

**Fig. 1.** The overview of our proposed system of fall detection.

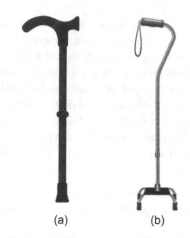

(a)                          (b)

**Fig. 2.** Two common walking aids: (a) Walking cane, (b) Four-legged walking aid

## 2    Materials and Methods

### 2.1    M5Stack Core2

To control the sensors and process data, we utilize an M5Stack Core2, which is equipped with an ESP32 microcontroller (Fig. 3). The device can be charged and connected to other devices with ease through its USB Type-C interface. It comes with an inbuilt lithium battery, featuring a capacity of 390 mAh, and an AXP192 power management chip. The battery level can be checked using the small LED lamp located on the side of the device, which can be turned on or off by pressing the power button located on the side of the device. A reset button is present on the bottom of the device. The M5Stack Core2 features a two-inch touch screen, which is ideal for user interaction. Three capacitive buttons located at the bottom of the screen can also be programmed. Additionally, the M5Stack Core2 has a WiFi module and supports Bluetooth, making wireless data transfer seamless. The device features dual-core Xtensa 32-bit 240 Mhz LX6 processors that can be controlled independently. It also comes with 16 MB Flash memory and 8MB PSRAM. Other features include a vibration motor, an RTC module for precise timing, a 6-axis IMU sensor, a microphone, an SD card slot, and a loudspeaker. The M5Core2 is compatible with various programming platforms, including the Arduino IDE, UIFlow, and MicroPython. The inbuilt sensors can be extended by utilizing numerous external sensors compatible with the M5Core2, which can be connected using cables and intersections if required.

Previously, M5Stack devices have been employed for medical supervision, such as in the case of an M5StackGrey utilized to create a motion game that encourages elderly individuals to remain active and flexible through exercise [6]. The device communicates wirelessly with a computer and features a small compatible vibration motor that can be controlled. In [7], the authors utilized

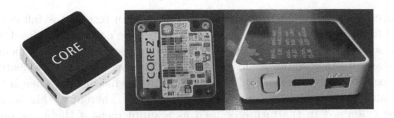

**Fig. 3.** M5Stack Core2 ESP32 IoT Development Kit.

an affordable and easy-to-set-up ESP32-CAM Module by M5, equipped with an OV2640 Wireless Camera Module, for a fall warning system based on visual cues. In a similar vein, the authors in [8] proposed a fall detection system that employs infrared sensors and transfers data wirelessly via WiFi using an M5Stack device. Additionally, the authors in [9] employed the M5Stack Core to create a walking aid for individuals with visual or motor impairments, with the device serving as a sensor data collector and interpreter that sends feedback to a mobile application through vibration alerts when obstacles are present or the path is unclear. The M5Core2 device is particularly relevant to this study due to its built-in gyroscope and accelerometer, which enable the detection of changes in orientation and acceleration and aid in fall identification. Additionally, the device's WiFi module is crucial for transmitting data to a web server. Further sensors capable of identifying touch can be connected to the M5Core2 through cables, as discussed in Sect. 2.2. Lastly, the loudspeaker and touchscreen functionalities are also important to this work, particularly with respect to the alarm tone and interactive interface.

## 2.2  Sensors

The M5Stack Core2's built-in MPU6886, which includes both an accelerometer and a gyroscope, is employed to detect falls. Data can be obtained through either the M5Core2's WiFi module or via Bluetooth. While the WiFi option is a straightforward way to wirelessly transmit data, it is reliant on a stable internet connection and may be subject to local limitations. Alternatively, data can be sent to a phone using Bluetooth, but this approach requires the user to carry the phone at all times and ensure that it is not misplaced. To determine the usage of the walking aid, additional external sensors were considered. Since the sensor unit is fixed to a crutch, it is necessary to use sensors that can detect if the walking aid is currently being used. Various potential sensors were explored, such as heart rate monitors, temperature sensors, and distance measuring sensors. Ultimately, the solution that was chosen is the Adafruit VL6180X, a proximity sensor that can be easily attached to the handle without requiring physical contact. This sensor can measure objects up to a distance of 10 cm and can be installed at the beginning of the handle to detect if a hand is placed on the handle or not. By detecting if an object is placed before it, the sensor can reliably determine if the walking aid is currently being used. This sensor can be integrated with the main code and all its features. In testing the fall detection algorithm, only the acceleration sensor and gyroscope data were utilized to conserve resources. Nevertheless, a reflective sensor or a sensor capable of measuring the distance to the ground could also be considered as an additional feature for fall detection. These sensors could be installed at the handle, but their inclusion would affect the power consumption of the device. Therefore, this possibility may be examined in the future or used as a supplement if the gyroscope and accelerometer data prove to be unreliable over time. Finally, to determine the user's location in case of a falling event, a GPS sensor can be integrated. The M5Stack Atom GPS Sensor is selected for this purpose. With its addition, help can be directed to the right location as soon as possible.

### 2.3   Sending M5Stack Core2 Data to a Web Server

The process of sending M5Stack Core2 data to a web server is described in Algorithm 1. It is written in the programming language C++ with the Arduino IDE and consists of a number of different functions, which together form the final version of the code.

---

**Algorithm 1** Sending M5Core2 data to a web server

---
if Not Found WiFi configuration on Flash then
    Set up WiFi
end if
Connect to WiFi
if Not Found UUID on Flash then
    Create a UUID and write it on Flash.
end if
Get the UUID of M5Core2.
if The proximity sensor detects the usage of the handle then
    The device keeps on taking data from the sensor.
else
    Enter a sleep mode to save energy.
end if
Create a JSON file consisting of sensor data.
Get data acceleration sensor and Gyroscope.
if Fall detected (GyroY<0.8) then Turn on the Alarm.
end if
if Help/Doctor button was pressed then
    Send the JSON file to the Web server immediately.
else
    Turn off the Alarm.
end if
Send the JSON file to the Web server.

---

The fundamental requirement for the fall detection method is a functional WiFi connection to the M5Core2. Without an active internet connection, data transfer is impossible. Therefore, the first step is to verify the availability of an internet connection, preferably one that the device has previously connected to. If the M5Core2 finds the necessary configuration information for the WiFi network in the flash memory, it automatically connects to the network and proceeds to the next stage of the fall detection code. In the absence of any saved WiFi information in the flash memory, the device prompts the user to provide a new WiFi name and password, which are then stored in the flash memory for future use. Once an internet connection is established, the sensor data can be collected and added to the data that will be sent in JSON format, as required.

To ensure that the web hosting can accurately identify the device sending data, it's important to assign a unique identification to each M5Stack Core2. This is achieved by generating a Universally Unique Identifier (UUID) for each device, which is stored in its flash memory. Once generated, the UUID remains the same and is used for all subsequent data transmissions, ensuring the device is correctly paired with its corresponding patient.

The following step is to verify if the device is being used. As previously mentioned, this is accomplished by monitoring the handle using a proximity sensor. If the sensor detects handle usage, the device will continue to gather sensor data. However, if the sensor does not detect usage for an extended period of time, the device will enter sleep mode to save energy and wake up when touch is detected again.

The code also includes an interactive interface with three buttons displayed on the screen. Pressing the "Doctor" or "Help" button will trigger the alarm and send a JSON file to the web hosting to request help immediately. The "Okay" button, on the other hand, will turn off the alarm and send a message to the host indicating that no help is needed if a fall has been detected by the device.

## 3 Results and Discussion

### 3.1 Data Collection

In order to estimate fall risk, several existing methods such as [10–15] have been studied and used to generate fall-related datasets for their research purposes. However, these datasets were not suitable for our system's hardware architecture. Therefore, a new dataset was created for this study to accurately determine the occurrence of falls. The primary tool for detecting falls is the MPU6886 3-axis-acceleration sensor and 3-axis-gyroscope, which is built into the M5Core2. While the accelerometer measures acceleration in all directions, the gyroscope detects the device's orientation and the attached walking aid. Each gyroscope value ranges from 1 to -1, indicating the current position of the crutch. Since the sensor is not attached to the human body but to the walking aid, the methodology for fall detection needs to be reevaluated. To conduct the experimental phase, we need to differentiate between three states in which the crutch can fall together with the user.

The first state that needs to be differentiated is the upright position of the crutch, which occurs when it is not in use when the patient is standing, and when the person is walking. Therefore, smaller impacts on the acceleration and gyroscope data cannot be classified as falls and must be labeled as "normal position." Any actions classified as "normal position" are considered harmless and do not require any action from the system. The upright position is characterized by a y-value of the gyroscope that is close to 1, while the x and z values should be around 0. Since the crutch will bend slightly while the person is walking, small variations are allowed. The forward movement can be observed in the y-value of the acceleration data. While elderly people are generally expected to move more slowly than younger individuals, rapid movements can still occur when they are adjusting their walking aid for the next step. Therefore, it is crucial to monitor both the orientation and acceleration data to determine whether the combination suggests a normal position or a fall. A visual representation of the walking aid in an upright or normal position can be seen in Fig. 4a.

(a) Normal position with the coordinate system.

(b) Falling position according to the coordinate system.

(c) Fallen position according to the coordinate system.

**Fig. 4.** Walking aids position according to the coordinate system

The second state pertains to the falling process, where the crutch loses its upright position and continues to fall for a certain period. Detecting a fall is crucial and significant changes in the gyroscope and acceleration data are key indicators for this state, which is classified as "fall detected". A fall is considered harmful and triggers a message from the device to the web hosting, indicating that assistance might be necessary if no further action is taken. The data gathered during a fall can vary widely, as it can occur in various situations, and the crutch's orientation can differ greatly depending on the fall's direction and speed. Accurately interpreting the combination of sensor data is vital for detecting a fall. During a fall, the acceleration data can rapidly increase to triple digits. The y- and z-values are the most affected when falling to the left or right, while the x- and y-values show the greatest acceleration during a forward or backward fall. At the same time, the crutch's orientation changes from a vertical to a horizontal position. During a fall, the y-value of the gyroscope drops from about 1 to 0, while the x- or z-value changes to 1 or −1, depending on the fall's direction. If the crutch falls to the right, the x-value drops towards −1, while it increases to 1 if it falls to the left. A forward fall is indicated by the z-value climbing up to 1, whereas a rapid decline to −1 indicates a backward fall. Figure 4b shows the walking aid's orientation during a fall, based on the coordinate system.

The final stage is when the crutch is in a fallen position, which indicates that a fall has occurred, and the walking aid is lying still on the floor. This state is labeled as "fallen" and is characterized by little to no variation in the acceleration and gyroscope data. Although it is not considered harmful, it may serve as an indicator of whether the person using the crutch was able to stand up again. The crutch is deemed to have fallen when there is no acceleration detected and the gyroscope values suggest that the crutch is in a horizontal position (y-value close to 0 and one of the other values around 1 or $-1$). A depiction of the fallen walking aid in the coordinate system can be found in Fig. 4c.

In the experimental phase of the study, a significant amount of data was required to be gathered while using the walking aid in different states, including the upright position, the fallen position, and a simulated falling scenario. The collected data was then organized and stored in tables to train the algorithm to accurately distinguish between the three states. To achieve this, various walking patterns were simulated, and data was collected for the "normal position." Moreover, the crutch was made to fall under different conditions and in various directions, and the position of the crutch after each fall was recorded as data.

To collect the first set of falling data, a walking cane was used, and the M5Core2 was fixed on it. A total of over 257 sets of data were collected, divided into three categories: "normal position", "fall detected", and "fallen", and stored in a table. The table includes all the necessary data from the acceleration sensors and gyroscopes, such as the x, y, and z values of the gyroscope and the x, y, and z values of the acceleration data, along with the corresponding label. Several examples of the collected data are shown in Table 1. In order to determine whether a dataset should be labeled as "fallen", "fall detected" or "normal position", the considerations outlined earlier must be taken into account. When the crutch is in an upright or "normal position", the y-value of the gyrosensor will typically be around one. However, during walking, the crutch may be bent, resulting in the y-value dropping to values of around 0.8. Therefore, the label "normal position" is assigned to data where the y-value of the crutch is above 0.8, considering the angle that may be applied while walking.

Should the walking aid fall, the y-value of the gyroscope will drop from what is considered upright into what can be identified as a falling event. Data, where the y-value exceeds 0.8 and a significant acceleration, can be noted is labeled as "fall detected". The walking aid is considered to be falling as long as the acceleration can be identified and the y-value of the gyro-sensor varies between 0.1 and 0.8. When the walking aid falls, the gyroscope's y-value drops from the upright position to indicate a falling event. Data that shows a y-value exceeding 0.8 and a significant acceleration is labeled as "fall detected". The walking aid is considered to be falling as long as the acceleration can be detected and the y-value of the gyro-sensor varies between 0.1 and 0.8. Finally, the label "fallen" is assigned to the data when the walking aid is in a horizontal position and there is minimal acceleration detected, and the y-value of the gyro-sensor is near 0. This combination of sensor readings strongly suggests that the walking aid has fallen and is no longer in an upright position. To use the four-legged walking aid,

**Table 1.** Representative samples of collected falling data with a walking cane.

| No | Label | AccX | AccY | AccZ | GyroX | GyroY | GyroZ |
|----|-------|------|------|------|-------|-------|-------|
| 1 | normal position | 0,7935 | 13,9160 | 8,8501 | −0,0862 | 1,0085 | 0,0388 |
| 2 | normal position | −17,4561 | 39,4287 | 20,9961 | −0,2188 | 1,0361 | 0,0764 |
| 3 | normal position | −1,0986 | 81,4819 | 10,4370 | −0,1743 | 0,9155 | −0,1189 |
| 4 | fall detected | 217,2241 | −10,4370 | 34,6679 | −0,0234 | −0,1667 | −0,3433 |
| 5 | fall detected | 131,8969 | 17,3950 | 13,0005 | 0,0442 | 0,5579 | −0,1936 |
| 6 | fall detected | −141,6016 | 27,4658 | 22,5830 | 0,0718 | 0,5652 | 0,2709 |
| 7 | fallen | 0,2441 | 14,3433 | 11,9019 | −0,9963 | 0,1245 | −0,0508 |
| 8 | fallen | 0,7935 | 14,2212 | 12,207 | −0,0525 | −0,0145 | −0,9446 |
| 9 | fallen | 0,5493 | 12,9395 | 10,7422 | 0,8945 | −0,0452 | −0,3535 |

a second dataset was collected, comprising a total of 255 data sets. These were classified into three categories: "normal position", "fall detected", and "fallen".

## 3.2 Evaluation Metrics

Our performance metrics are precision, recall, and F1-score. Before defining the metrics, we define True Positive (TP), True Negative (TN), False Positive (FP), and False Negative (FN) terms. TP is the number of correctly classified samples by the algorithm. TN is the number of correctly missed samples. FP is the number of wrongly detected samples. FN is the number of wrongly missed samples. The TN is not used in calculating our performance metrics. Precision was the one factor that was used for the main comparison, as it is a good indicator of how reliable a method works on a specific data set. It mainly focuses on the identification of data as true positive or false positive and calculates how accurately a method can predict a falling event of the crutch. It is calculated with Formula 1. The recall metric measures the number of wrongly missed samples by the algorithm. Its value decreases as the number of false negatives increases. Recall is an important evaluation criterion in the applications that necessitate no misses in the classification task. It is computed as the ratio of correctly detected objects to the total number of samples (both correctly classified and missed), the formula of recall is in Formula 2.

$$Precision = \frac{TP}{TP + FP} \tag{1}$$

$$Recall = \frac{TP}{TP + FN} = \frac{TP}{\text{all ground truths}} \tag{2}$$

## 3.3 Applying the Machine Learning Method

In order to analyze the data on the web hosting, a classification problem needs to be solved in order to label the received data correctly to "normal position", "fall detected", and "fallen'. Therefore it needs to be evaluated, what machine learning method is most suitable for the specific problem by cross-validation. The recorded data gets split into training and testing data. Most of the data, depending on the cross-validation method, is used to train the algorithm, whereas the rest of the data is then used to test how reliable the algorithm works on the problem. This is done for every method, to find the best one to use. In this specific case, the training data consisted of 2/3 of the collected data and the testing data consisted of the other 1/3. A 5-folded cross-validation method was used and the process was repeated ten times for each method. In this study, the performance of Support Vector Machine [16], Neural Network [17], and Gradient Boosting [18] models was assessed on a dataset collected from a walking cane.

Using a Cost (C) configuration of 1.00, a Regression loss epsilon ($\epsilon$) of 0.10, a Kernel RBF, a numerical tolerance of 0.0010, and an iteration limit of 100. In this case, our data is six-dimensional, as we have six different types of information that are processed. The Acceleration data in the X, Y, and Z-axis and the data of the gyroscope in the X, Y, and Z-axis. Support Vector Machine finds these Support Vector Classifiers using kernel functions, namely polynomial or radial kernels, and moves data into a higher dimension. The values used for the kernel functions are determined by cross-validation. In order to test SVM on this set of data a radial Kernel was used with a regression loss $\epsilon$ of 0.1. As shown in Fig. 5, SVM achieves an accuracy of 100% for predicting 134 samples of the "fall detected" label, 97.40% for predicting the "normal position" label (75 samples), and 100% for predicting the "fallen" label.

Confusion matrix for SVM (showing number of instances)

| | | Predicted fall detected | fallen | normal position | Σ |
|---|---|---|---|---|---|
| Actual | fall detected | 134 | 0 | 0 | 134 |
| | fallen | 0 | 46 | 0 | 46 |
| | normal position | 2 | 0 | 75 | 77 |
| | Σ | 136 | 46 | 75 | 257 |

**Fig. 5.** Confusion matrix using SVM classifier.

Using the Neural Network machine learning method with a configuration of 100 neurons in the hidden layer, ReLu activation function, Adam optimization solver, regularization of 0.0001, and a maximum number of iterations of 200. In this case, the input nodes are the acceleration and gyroscope data and the output is the classification of the current position of the walking aid. As shown in Fig. 6, It achieves an accuracy of 99.25% for predicting 133 samples of "fall detected" label, 100% for predicting the "normal position" label (77 samples), and 100% for predicting the "fallen" label (46 samples).

Using the Gradient Boosting machine learning method with a configuration of 100 trees, a learning rate of 0.1, a maximum depth limit of 3 for individual trees, and a minimum subset size for the splitting of 2, and using a fraction of

| Confusion matrix for Neural Network (showing number of instances) | | | | | |
|---|---|---|---|---|---|
| | | Predicted | | | |
| | | fall detected | fallen | normal position | Σ |
| Actual | fall detected | 133 | 1 | 0 | 134 |
| | fallen | 0 | 46 | 0 | 46 |
| | normal position | 0 | 0 | 77 | 77 |
| | Σ | 133 | 47 | 77 | 257 |

**Fig. 6.** Confusion matrix using Neural Network classifier.

1.00 of the training instances, As shown in Fig. 7, It achieves an accuracy of 99.25% for predicting 133 samples of "fall detected" label, 98.70% for predicting the "normal position" label (76 samples), and 97.82% for predicting the "fallen" label (45 samples).

| Confusion matrix for Gradient Boosting (showing number of instances) | | | | | |
|---|---|---|---|---|---|
| | | Predicted | | | |
| | | fall detected | fallen | normal position | Σ |
| Actual | fall detected | 133 | 0 | 1 | 134 |
| | fallen | 1 | 45 | 0 | 46 |
| | normal position | 1 | 0 | 76 | 77 |
| | Σ | 135 | 45 | 77 | 257 |

**Fig. 7.** Confusion matrix using Gradient Boosting classifier.

## 3.4 Numerical Results and Discussion

In order to decide, which machine learning method was suited the best for this data set, the confusion matrices were taken into consideration and a ROC analysis was conducted. The default threshold for the ROC analysis was set to a typical 0.5. It shows the rate between True Positives (TP), meaning correctly identified cases of the wanted attribute, and False Positives (FP), falsely classified data as the wanted attribute. As only very few data sets were falsely classified, the graph is drawn very close to one.

The confusion matrices the ROC analysis is based on can be seen in Figs. 5 to 7, which clearly shows how each method performed on the given data set. Figure 5 shows, that the Support Vector Machine falsely predicted two cases as a "fall detected", when in fact they should have been classified as normal positions. As shown in Fig. 6, Neural Network falsely classified a fall as an already fallen walking aid. While Naive Bayes and Gradient boosting show a few more errors with some confusion among all classes (Fig. 7). k-NN performed quite well again, classifying two falls falsely as a normal positions. The overall performance of the methods compared to one another by different parameters can be seen in Table 2. It compares the methods by AUC (Area under the curve), CA (Classification accuracy), F1, Precision, and Recall. AUC compares the curves from the ROC analysis to detect which curve has the greatest area underneath. CA focuses on the proportion of the examples that were classified correctly.

**Table 2.** Comparison between the different machine learning methods in the percentage of correct identifications.

| Machine Learning Method | AUC | Precision | Recall |
|---|---|---|---|
| Support Vector Machine | 99.99 | 99.23 | 99.22 |
| Neural Network | 99.98 | 99.62 | 99.61 |
| Gradient Boosting | 99.34 | 98.84 | 98.83 |

After evaluating the methods using precision as the ultimate metric, it is clear that the Neural Network exhibited the most promising outcomes in detecting a falling event, with a reliability of 99.62%. On the other hand, Naive Bayes had the poorest result, with only 97.71% precision. Overall, the Neural Network model performed the best among all the methods, considering all the given parameters.

Once the data has been modeled and the most appropriate machine-learning method has been chosen, the resulting model can be deployed on the web hosting for further use. To accomplish this, the model must first be saved and then transferred using the pickle library. Once deployed on the web hosting, the model can analyze the data received from the M5 device and aid in identifying instances of falling.

### 3.5   Website and Message Distribution

The data is sent via a WiFi connection to the web host, where the patient's information is stored in a database, allowing the association between the device and the patient. Two scenarios are presented for fall detection: the first involves detecting falls within the device itself, which conserves energy but may have a slight risk of missing falling events during data transmission. The second scenario involves continuously sending and analyzing data using machine learning, providing more precise data collection at the expense of higher power consumption. The website stores various information, including patient data, device data, user group information, and contact lists. As shown in Fig. 8, the patient's data includes personal details, such as name, gender, date of birth, contact information, and address. Device data includes sensor information and a unique identifier for matching devices with patients. The website visually displays the collected data, and there are different sections for patient information, emergency contacts, and doctor information, each with specific privileges and information requirements based on their account types.

**Fig. 8.** Layout of the patient follow-up of the website.

## 4    Conclusion

The paper presented a novel fall detection system using a intelligent walking aid and machine learning methods. The system uses an M5Core2 microcontroller, the Adafruit VL6180X proximity sensor, and the MPU6886 3-axis acceleration and 3-axis gyro sensors to detect falls. An M5Stack Atom GPS sensor can also be used for tracking the current position of the device. The fall detection algorithm and overall system were tested with experiments, and the Neural Network machine learning method performed best, classifying 99.62% of the data correctly. Potential solutions could involve incorporating force sensors on the handle or feet of the walking aid or adding distance sensors to the stick to measure the distance to the legs. Alternatively, machine learning could be utilized to analyze walking patterns and sensor data to detect falls where the walking aid bounces back after impact.

**Acknowledgment.** The authors would like to thank to An Giang University, Jade University, and Vietnam National University in Ho Chi Minh City, Vietnam.

## References

1. Uddin, M.Z., Hassan, M.M.: Activity recognition for cognitive assistance using body sensors data and deep convolutional neural network. IEEE Sens. J. **19**(19), 8413–8419 (2019)
2. Mankodiya, H., et al.: Xai-fall: explainable AI for fall detection on wearable devices using sequence models and XAI techniques. Mathematics (2022)
3. Kong, X.: Vision-based fall detection: methods and challenges. In: 2022 International Conference on Advanced Mechatronic Systems (ICAMechS), pp. 157–160 (2022)

4. Alam, E., Sufian, A., Dutta, P., Leo, M.: Vision-based human fall detection systems using deep learning: a review. Comput. Biol. Med. 105626 (2022)
5. Doan, T.-N.: An efficient patient activity recognition using LSTM network and high-fidelity body pose tracking. Int. J. Adv. Comput. Sci. Appl. 13(8) (2022)
6. Tatsukawa, S., et al.: Development of motion game for elderly based on sensory stimulus presentation. In International Conference on Electrical, Computer and Energy Technologies (ICECET), pp. 1–4 (2021)
7. Sok, C., Saechan, C., Sittipong, C., Kaewmorakot, U., Poolperm, L.: Fall warning machine with image processing via the internet of things. In: 6th International STEM Education Conference (iSTEM-Ed), pp. 1–4 (2021)
8. Tateno, S., Meng, F., Qian, R., Hachiya, Y.: Privacy-preserved fall detection method with three-dimensional convolutional neural network using low-resolution infrared array sensor. Sensors 20(20), 5957 (2020)
9. Grzeskowiak, F., Devigne, L., Pasteau, F., Dutra, G.S.V., Babel, M., Guégan, S.: Swalkit: a generic augmented walker kit to provide haptic feedback navigation assistance to people with both visual and motor impairments. In: International Conference on Rehabilitation Robotics, pp. 1–6 (2022)
10. Yoshida, T., et al.: A data-driven approach for online pre-impact fall detection with wearable devices, pp. 133–147 (2022)
11. Jia, N.: Detecting human falls with a 3-axis digital accelerometer. In: Analog Dialogue (2009)
12. Di, P., et al.: Fall detection and prevention control using walking-aid cane robot. IEEE/ASME Trans. Mechatron. 21(2), 625–637 (2016)
13. Malasinghe, L., Ramzan, N., Dahal, K.: Remote patient monitoring: a comprehensive study. J. Ambient. Intell. Humaniz. Comput. 10, 01 (2019)
14. Pan, D., Liu, H., Dongming, Q., Zhang, Z.: Human falling detection algorithm based on multisensor data fusion with svm. Mob. Inf. Syst. 1–9(10), 2020 (2020)
15. Ramirez, H., Velastin, S.A., Meza, I., Fabregas, E., Makris, D., Farias, G.: Fall detection and activity recognition using human skeleton features. IEEE Access 9, 33532–33542 (2021)
16. Cortes, C., Vapnik, V.: Support-vector networks. Mach. Learn. 20(3), 273–297 (1995)
17. McCulloch, W.S., Pitts, W.: A logical calculus of the ideas immanent in nervous activity. Bullet. Math. Biophys. 5(4), 115–133 (1943)
18. Friedman, J.H. Greedy function approximation: a gradient boosting machine. Annal. Statist. 1189–1232 (2001)

# A Cloud-Based Intelligent Virtual Assistant for Adolescents

Zakia Afrin, Dewan Md. Farid$^{(\boxtimes)}$, and Khondaker Abdullah Al Mamun

Department of Computer Science and Engineering, United International University
United City, Madani Avenue, Badda, Dhaka 1212, Bangladesh
{mamun,dewanfarid}@cse.uiu.ac.bd
https://cse.uiu.ac.bd/profiles/dewanfarid/

**Abstract.** Adolescence is a transitional period that occurs between puberty and adulthood. To make a healthy transition into adulthood, adolescents need to have easy access to healthcare and education and a friendly environment both at home and in society. Adolescent-friendly services are not a familiar concept in Bangladesh as they are treated as a healthy group of people. They often feel depressed as they have no available resources to solve the issues that they face in their daily life. In this paper, we present a prototype of Bondhu (), a cloud-based intelligent virtual assistant for Bangladeshi adolescents. The cloud service stores all users' data that data considered as a knowledge base for future responses to the users. The knowledge base will improve by analyzing global users' responses. It acts as a virtual friend to deal with mental health challenges and guides them to express their negative emotions, thereby easing anxiety.

**Keywords:** Adolescents · Mental health · Virtual assistant

## 1 Introduction

Adolescence is a challenging phase of life defined by abrupt changes and elevated emotional and social issues. Typically, it spans the ages of 10 to 18 years old [47]. In reality, adolescents are persons in their second decade of life who are confronted with internal problems or circumstances that can immediately affect their mental health [20]. Adolescents make up 23% of the population in Bangladesh, where there are 36 million of them [4]. They represent a prospective resource whose potential can either be squandered or fostered positively. Mental health issues in adolescents who do not receive treatment are associated with poor academic performance, poverty, substance abuse, harmful behaviors, criminality, poor sexual and reproductive health, self-mutilation, and inadequate personal hygiene [39]. Consequently, adolescent mental health disorders pose substantial threats to global public health. With the aid of preventative measures, the progression of adolescent risk factors can be avoided. Detection at an early stage can mitigate the severity of growth and development risk factors.

N. Thai-Nghe et al. (Eds.): ISDS 2023, CCIS 1949, pp. 110–124, 2024.
https://doi.org/10.1007/978-981-99-7649-2_9

In a developing country like Bangladesh, adolescents do not have adequate access to sexual and reproductive health information [8]. They face numerous mental and physical problems as a result of living in a society where reproductive and sexual health issues are taboo. There is a dearth of online forums for adolescents where they can interact with each other and health professionals to find solutions to their problems [4]. People in Bangladesh experience inadequate mental healthcare due to a lack of mental health facilities, a shortage of qualified mental health professionals, financial issues, and deficient health policies [27]. In addition, many adolescents avoid mental health services because they prefer to handle their problems on their own and are unsure of the efficacy of professional services [22]. The government of Bangladesh adopted adolescent-friendly health services (AFHS) to provide health care for adolescents [5]. It operates in thirteen of the sixty-four districts. However, adolescents and their families are hesitant to seek AFHS services due to social stigma and reluctance. There are also safety and privacy concerns associated with the use of these services. These challenges should motivate the development of an intelligent system that aids adolescents in overcoming mental health issues.

Mental distress is characterized by symptoms of depression and anxiety [35]. Intelligent and interactive virtual assistants, also known as conversational agents or chatbots, may assist troubled adolescents in coping with their emotional distress. These assistants have drawn considerable interest in a variety of fields and may produce positive social effects by nurturing the autonomy, effectiveness, and paradoxical social connection of their users. Virtual medical assistants can enhance healthcare services by reducing costs and increasing accessibility [43]. Chatbots for mental health provide assistance to individuals who may avoid treatment due to stigma, cost, or distress associated with in-person therapy [46]. A chatbot is designed to assist adolescents with academic, bullying, emotional, romantic, and sex-related issues [22]. It was simpler to be transparent to a chatbot than to a human as chatbots offer anonymity and confidentiality. Virtual assistants for mental health have been designed for clinical [45] and non-clinical [25] adult patient groups. Several studies that have evaluated mental health chatbots for young people have found that they are helpful in identifying and alleviating symptoms related to anxiety, stress, and depression [25,29]. The mental health of chatbot users improves, and they perceive the services to be beneficial and trustworthy, according to a study [23]. Conversational agents with a focus on mental health are able to identify individuals experiencing psychological distress and assist in alleviating this distress [38]. Also, these agents play a vital part in providing empathy and harmony [9].

Chatbots can enhance mental health care by giving information, resources, and reminders for appointments and self-care. Chatbots cannot express empathy, understand complexity, or provide nuanced responses like mental health specialists. We aim to develop Bondhu (), an intelligent virtual assistant geared towards teenage groups that would be used in both urban and rural areas of Bangladesh. The majority of users are college and university students, while only a few are high school students. Due to the absence of personal smartphones, the majority of school-aged children in Bangladesh share their parents' devices [30]. According to a survey, adolescents in Bangladesh use mobile phones at a rate of over 90%, and at least half of them are smartphones [1]. As a virtual friend, it can interact with adolescents to provide fundamental knowledge. It helps anxious adolescents gradually express their negative emotions by comforting them and listening to them. It also helps adolescents by providing them with encouraging ideas and solutions to their problems. While chatbots and other AI tools can bring some advantages to the field of mental health, they should not be used as a replacement for qualified medical personnel. Our proposed model facilitates expert psychological counseling services to its users. We gather data from 1008 adolescents to assess an adolescent's stress level and provide support in accordance with that level. We assess the perceptions of users by developing a prototype of Bondhu (), a virtual assistant in Bangla language. We collected responses from 57 university freshmen at Northern University Bangladesh in order to determine how users perceive the virtual assistant's appearance and functionality. Our study shows that most of the participants admired the prototype's design and were eager to use the assistant.

The remaining portion of the paper is organized as follows. Section 2 investigates related works and discusses the current state of adolescent health and related technologies in Bangladesh. In Sect. 3, we provide an overview of the Bondhu () framework that we have developed. Additionally, we collect data to identify anxiety and conduct a survey on our prototype. Finally, we conclude the paper with discussing several potential future scopes for our research in Sects. 4 and 5.

## 2  Related Works

The rapid expansion of the Internet has caused an increase in Internet-delivered health promotion initiatives [40]. Internet-delivered health promotion programs may be especially well-suited to target the current generation of adolescents because this generation grew up on the Internet and is more receptive to the new opportunities this medium offers than the adults of today [33,42]. However, in order to reach this generation, these activities must be adapted to the demands and Internet usage of teenagers [15,17]. In the Netherlands, for instance, almost 90% of teens use instant messaging services online (e.g., Windows Live Messenger [WLM], formerly known as MSN Messenger, being the most popular instant messaging tool in the Netherlands) [34]. Currently, chatting robots or chatbots can also be added to the friend lists.

A chatbot is an artificially intelligent chat agent that includes emotional communication like human, such as by allowing users to ask questions (i.e., queries) and, in response, provide meaningful responses to those questions [18]. The chatbot parses the content of the user's input and links it to a message database containing possible responses that are matched to the user's input. Since the advent of "ELIZA" in 1966, chatbots have existed. "ELIZA" is a dialogue system that resembles a conversation with a Rogerian therapist [41]. The chatbot technology has been utilized in a therapeutic setting for smoking cessation, for instance

[26]. Moreover, as a tool for public planning procedures [10]. This exploratory project will focus on a chatbot for WLM that responds to inquiries from Dutch-speaking teens about sex, drugs, and alcohol. In addition to topic-related content, the chatbot can chat informally (e.g., exchanging greetings). The chatbot seeks to target adolescents as an alternative to and supplement to their other personalized information offerings (i.e. information lines). This chatbot is compatible with the inclination of adolescents to look for this information online [44]. This attitude could be explained by the difficulty parents have in conveying these topics [11], particularly sex-related topics [19,48], which is represented in their tendency to inquire (instead of providing answers to the questions asked by their adolescent children). This may be harmful in and of itself, leading to adolescents' risky conduct [24], which is widespread among Dutch teenagers [28]. Consequently, a demand-driven chatbot that allows teens to pose queries could be a suitable alternative. Offers scripted drug information through WLM. The user's input is limited to numerals, comparable to an Interactive Voice Response system for the telephone. According to our knowledge, Bzz is the first chatbot in the field of health promotion that allows for open interaction and intends to answer queries regarding sex, drugs, and alcohol posed by teens. Nevertheless, according to a recent systematic review, Internet-delivered health promotion initiatives are frequently moderately utilized by adolescents, and it is recommended to track and report objective use measures (e.g., server registrations) because there are few data on actual engagement [16]. Consequently, the first critical step is to research whether and how the chatbot is utilized, as well as how adolescents perceive it. This study is required to evaluate the potential of chatbots in the field of health promotion and to provide recommendations for future research and practice based on evidence.

In order to acquire a deeper understanding of the potential of chatbots in the field of health promotion, a comparison conducted with two other channels that adolescents regularly utilize to find answers to sexual, drug, and alcohol-related inquiries. Including the following: (1) the telephone information lines already provided by the same national health promotion entities that created the chatbot's material. (2) Search engines (such as Google) in general, as they are frequently used to find information about sexual health [13,14]. In this exploratory phase, an effect evaluation would be incorrect because the chatbot is demand-driven (i.e., by the inquiries of adolescents) and does not intend to immediately expand knowledge in general or influence determinants or behavior (however, it does aim to do so in the long-term). This study focuses on adolescents employment and evaluation of chatbots, particularly in comparison to information lines and search engines.

Mayalogy Limited is an online health and well-being company based in Dhaka, Bangladesh that provides on-demand information [2], that also provides an anonymous messaging service, whereby users can pose their health, psychosocial, and legal questions. Maya Apa Plus, a premium that provides answers of subscribers' questions in 10 min. Shornokishore Network Foundation (SKNF) [36] established secondary school-based clubs, health education and classroom-

based training, peer education at the community for non-school going adolescents, community mobilization for addressing barriers, advocacy for supporting
policy. Dnet provides solutions and implements development projects for the
improvement of the lives of marginalized women, children, and youth [32]. "Digital Platform for Adolescent" is an android-based application with a backend
panel for adolescent club members to connect, share knowledge and have access
to information [3]. "Adolescent Club (APC) Bangladesh" [12] is an android application to prevent child marriage, teenage pregnancies, micronutrient deficiencies,
risky sexual behavior etc. Aponjon Koishor [6] is a novel mobile app in Bangla for
adolescents that helps create awareness among adolescent boys and girls about
physical and mental changes that they go through as well as about adolescent
reproductive health. Pathfinder International works to empower adolescents and
youth to make their own decisions about their bodies and their futures by changing community norms related to gender, early marriage, and childbearing [37].

## 3  Methodology

The Bondhu () is a virtual assistant for young people in Bangladesh that is hosted
on the cloud. People primarily between the ages of 12 and 20 use this system.
An adolescent can use our app right away if they have a smartphone with an
internet connection. The virtual assistant facilitates an interactive question-and-
answer session. It offers multiple-choice questions to the user and analyses their
responses to determine how they are feeling. It stores all of the user's data and
quickly processes solutions for all issues. To get help from the virtual assistant, a
user must install the Bondhu () application in his smartphone. For the first time,
he will need to enter his phone number throughout the registration process. If he
is a new user, the virtual assistant will ask him to learn about his mental state.
The assistant will respond to him based on this data, earning his trust in the
process. If the user has previously entered data into the app, the assistant will
continue the conversation from where it left off. The assistant offers guidance if it
detects a critical problem. The virtual assistant also offers expert psychological
counseling to its users through the MAYA health service [2]. If the user requires
immediate assistance, he may dial 999 (National Emergency Service). The course
of action is shown with a flow diagram in Fig. 1.

Figure 2 depicts the detailed architectural design of the proposed virtual assistant. In the initial question-and-answer session, the user will be asked a few questions with multiple choices. The bot service works here to figure out what the
next question should be based on the answer to the last question. The bot service
determines the user's mental condition after asking a few questions. The user
is then asked to express his concerns/feelings by text message using the in-app
messaging platform. The dialog engine sends the text data to the backend NLP
module. This NLP tool pre-processes the raw text data and extracts emotional
or sentimental keywords. Then, we submit this content to a pre-trained model [7]
that will categorize the text's emotional state. In order to re-train our decision
tree, we map this categorization response with previously obtained question-
answers. If both the decision tree and text classification answers are found to be

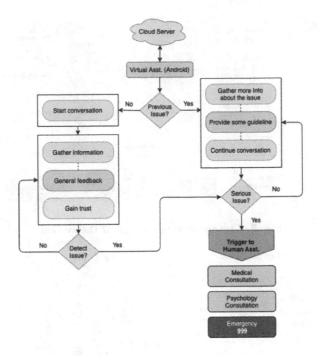

**Fig. 1.** Bondhu () framework.

important, the action service sends the text data and question-answer pairs to an expert through the MAYA service [2]. An expert will provide feedback which will be informed to the user when available. We also tag the answer so that we can offer it to other users who will encounter similar issues in the future.

The server is developed in Golang, and the user's data is stored in a MySQL database. We used the Google Cloud Platform to run the server. The prototype is developed for Android phones, which connect to a server in the cloud.

### 3.1 Dataset Collection

We have collected data on 1008 students from various Bangladeshi schools, colleges, and universities. Participants' ages ranged between 14 and 20 years old. All interviews were carried out online using a Google form between June and August 2021. The Bengali version of the Anxiety Scale developed by Deeba and Begum [21] for the Bangladeshi population was used for the survey. This scale is used in clinical psychology surveys and was developed specifically for Bangladeshis. There were 41% male respondents and 59% female respondents to the questionnaire. On the anxiety scale, there are 36 questions. Figure 3 illustrates the gender and age of the participants of the interview.

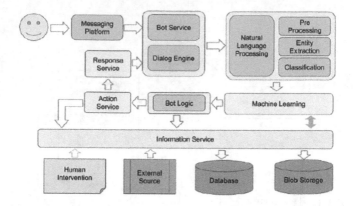

**Fig. 2.** The architecture design of the system

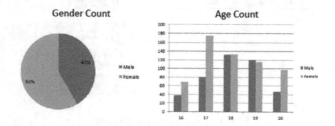

**Fig. 3.** Gender and Age count of the participants

## 3.2   Data Analysis

A total of 1,008 adolescents responded to the 36 anxiety-related questions. There are five options for each question, and participants must select one. The following are the options for answers:

(Not at all; Very little; Roughly; More; Much more)

Let's look at an example:
Question: (Are you short of breath?)
Answer: (Very little)

These answer choices have points from 0 to 4 accordingly. The sum of all points from each answer determines the level of anxiety of an individual adolescent. This process is formulated in a published article [21]. If the total point of an adolescent is greater than 53, that individual is considered to have anxiety. Based on the answers from 1008 adolescents, this process finds out that 244 individuals are having anxiety. Some important observations are given below:

As the answer to question, 297 individuals choose it as (Much more).

As the answer to question, 213 individuals choose it as (Much more).

It has been found that 36 percent of adolescents who fear death are suffering from anxiety.

## 3.3  Decision Model

This questionnaire contains 36 questions to analyze the anxiety level of an individual. If we present adolescents with a lengthy list of questions, the system will not be youth-friendly. Because they will lose motivation to respond to all questions. Therefore, the optimal strategy is to design a model with fewer queries based on their answer options. We use J48 classifier as our decision tree for this exercise. The 36 questions are symbolized from Q1 to Q36 and the answer choices are denoted as A, B, C, D, E sequentially. The complete dataset results with 98.41% accuracy. We have found out that this training model is overfitted. So we run this model 12-fold and create a decent decision tree with up to 8 layers and get 91.58% accuracy. If we train our model with more data, we can still increase the accuracy by limiting the number of levels below 8.

## 3.4  User Interface Design

User Experience (UX) refers to the overall experience related to the perception (emotion and thought), reaction, and behavior that a user feels and thinks through his or her direct or indirect use of a system, product, content, or service [31]. Recent studies show that user interface/user experience (UI/UX) is an important part of smartphone application design. An interface, similar to a human face, has a direct advantage in attracting users. A well-designed interface can guide and lead users to complete an operation. And rationally designed interfaces can give users a sense of accomplishment. Figure 4 displays some prototype screenshot examples. Our application is specifically designed in the Bengali language to cater to Bengali-speaking users. Upon login, users are greeted by a feature-rich home page. Here, they can access a mental health assessment tool, connect with mental health experts, and access a wealth of valuable tips and information on mental health issues.

## 3.5  Survey on Prototype

The purpose of this survey is to comprehensively evaluate the effectiveness of Bondhu () Application. The fundamental measure is knowledge acquisition. We conducted an opinion poll of 57 Northern University Bangladesh students to gather feedback on the application. In August 2022, the survey was performed online using Google forms. Each responder took about two minutes on average to complete the survey. We obtained two demographic variables (gender and age) from the survey participants. There were 59.6% men and 40.4% women among the 57 participants. Figure 4 depicts the gender distribution of the participants. The participants ranged in age from 17 to 20 years. The gender of the participant's age groups is represented in Fig. 5.

**Fig. 4.** Snapshots of the prototype

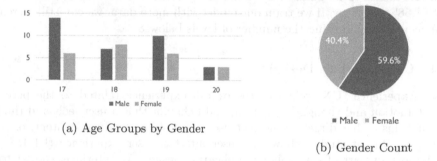

(a) Age Groups by Gender

(b) Gender Count

**Fig. 5.** The age and gender distribution of survey participants

Before conducting the survey, we engaged the respondents in a thorough discussion of the features and benefits of the Bondhu () application. Then, we presented the application's prototype to the participants. Then, the survey respondents were asked five questions to assess their understanding of the apps and to obtain suggestions for improving our system's design. The first question asked, "To what extent do they agree or disagree that the app's features are easy to use?" Each of the 57 respondents provided an answer. As displayed in Fig. 6 below, 27 respondents strongly agreed with the statement, 8 agreed, 16 agreed but had no opinion, and 6 disagreed.

The participants were questioned regarding the level of satisfaction they had with the user interface design and functionality of the application. There were 57 participants total, and 28 of those persons were satisfied, 23 were only moderately satisfied, and the remaining participants were not satisfied. This is illustrated in Fig. 7. A greater number of respondents (89.4 percent) said they were impressed with the application.

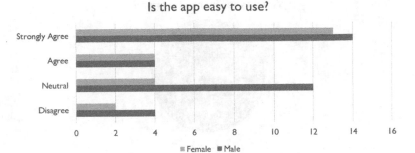

Fig. 6. Parception of how ease the app feature

**Fig. 7.** The participants' degree of satisfaction

Participants were requested to add any additional features of their preference. The majority of them offer no new suggestions. The significant findings we
received are:

1. Voice Recognition
2. Push message after conversation
3. Conversation transcript

The following poll question was, "Will they suggest our app to their friends
and family?" As seen in Fig. 8, 75.4% of individuals agreed to suggest in the
future, while the remaining 24.6% disagreed.

Finally, as shown in Fig. 9, individuals were asked how frequently they will
use our app. The conclusion indicated that 22 individuals will utilize this at
least once each week. 19 will utilize a couple of times per month, 11 will use it
everyday, while 5 will use it less than once per month.

Will they recommend in future?

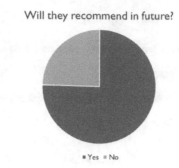

■Yes ■No

**Fig. 8.** The proportion who will recommend the application in the future

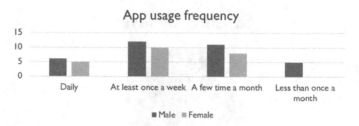

**Fig. 9.** Perception of the participants' app usage frequency

## 4  Discussion

Our research presents a virtual assistant to assist adolescents in dealing with mental health difficulties. We seek to comprehend user views and identify the robust design potential and challenges of a virtual assistant. We believe this research will be useful in various health-related projects. We strongly believe that the virtual assistant can provide valuable assistance to the adolescent community. Virtual assistants protect adolescent privacy and serve as trustworthy partners. It has the potential to raise reproductive health awareness and reduce mental stress in adolescents, paving the way for a better and healthier future.

According to our knowledge, today there is still a paucity of virtual assistant-based interventions for well-being promotion and life skills training among adolescents. In addition, there are insufficient data and guidelines for developing the user experience of educational virtual assistants in this intervention domain. However, the usage of the Internet and digital solutions by youth is rising globally, creating cost-effective prospects for the delivery of self-help educational programs to this group. Also, recent research supports our findings, demonstrating that digital therapies based on a virtual assistant can be highly engaging, promote well-being, and reduce stress in adolescents.

The virtual assistant intervention that is designed and evaluated in this study was rather easy and fast to implement and was well-received by adolescents. The initial rounds of our participatory design process incorporated the comments and recommendations of a total of 23 adolescents; The findings lead to the future development of virtual coaching systems for the training of young-sters, Solutions that have been assessed to be credible and attractive for this user group. The deployment of educational resources and dialogue-based interaction with the virtual assistant proved to be highly fascinating and beneficial for self-reflection on the difficult scenarios given, rendering virtual coaching sessions a highly promising digital platform for experiential forms of learning in adoles-cents. Our designed virtual coaching experience could be effectively integrated into existing school mental health assessment programs; this coaching experience might also be considered as a method for providing mental health education and training outside of the classroom.

This research is restricted by the small sample size and the fact that this study was undertaken at only a few institutions in Bangladesh. It impacts the applicability of our results to other countries and circumstances. However, men-tal health promotion programs for adolescents provided by a variety of delivery platforms, including digital media, are strongly recommended by public health policies in many countries and by the World Health Organization. These recom-mendations strengthen the global applicability of our work.

## 5    Conclusion

In conclusion, the potential for smart healthcare is immense. For individual users, smart virtual assistants can promote improved self-management of mental health. Timely and adequate medical services can be offered when required and the essence of medical services will be more personalized. A competent virtual assistant can improve a person's experience. For research institutions, a virtual assistant can reduce the cost and duration of research and increase the overall efficiency of research. However, there are still some problems in the develop-ment process. The solution to these problems depends not only on technological progress, but also on the joint efforts of users, doctors, health institutions, and technology companies. This study demonstrated the design and formative evalu-ation of an intelligent and interactive virtual assistant intervention for assessing adolescents' mental health and supporting them in overcoming mental health-related problems. This study is based on a positive technology approach and was well embraced by adolescents. More study is needed to conduct a more thorough examination of this intervention's effectiveness in improving skills and resilience. In the future, we intend to incorporate the virtual assistant's limitations by analyzing our findings and design issues and delivering the service on a broad scale.

**Acknowledgements.** We appreciate the support for this research received from the Innovation Fund (G. O. No: 373) from Information & Communication Technology Division, Government of the People's Republic of Bangladesh.

## References

1. Survey: Over 90% adolescents use mobile phones in bangladesh. https://www.dhakatribune.com/bangladesh/238379/survey-over-90

2. (May 2023). https://en.wikipedia.org/wiki/Mayalogy
3. Adian, K.M.: Business development activities at riseup labs (2021)
4. Ainul, S., Bajracharya, A., Reichenbach, L., Gilles, K.: Adolescents in Bangladesh: a situation analysis of programmatic approaches to sexual and reproductive health education and services (2017)
5. Ainul, S., Ehsan, I., Tanjeen, T., Reichenbach, L.: Adolescent friendly health corners (AFHCS) in selected government health facilities in Bangladesh: an early qualitative assessment (2017)
6. Ananya Raihan, C., et al.: Aponjon (2020)
7. Azmin, S., Dhar, K.: Emotion detection from Bangla text corpus using Naive Bayes classifier. In: 2019 4th International Conference on Electrical Information and Communication Technology (EICT), pp. 1–5. IEEE (2019)
8. Barkat, A., Majid, M.: Adolescent reproductive health in Bangladesh: Status, policies, programs and issues. Policy Project Report, USAID Asia/Near East Bureau (2003)
9. Bickmore, T., Gruber, A., Picard, R.: Establishing the computer-patient working alliance in automated health behavior change interventions. Patient Educ. Couns. **59**(1), 21–30 (2005)
10. Boden, C., Fischer, J., Herbig, K., Spierling, U.: CitizenTalk: application of chatbot infotainment to e-democracy. In: Göbel, S., Malkewitz, R., Iurgel, I. (eds.) TIDSE 2006. LNCS, vol. 4326, pp. 370–381. Springer, Heidelberg (2006). https://doi.org/10.1007/11944577_37
11. Boone, T.L., Lefkowitz, E.S.: Mother-adolescent health communication: are all conversations created equally? J. Youth Adolesc. **36**, 1038–1047 (2007)
12. Calpona, E.: The impact of COVID-19 on the child protection system in Bangladesh. Ph.D. thesis, Queen Margaret University, Edinburgh (2023)
13. Cline, R.J., Haynes, K.M.: Consumer health information seeking on the internet: the state of the art. Health Educ. Res. **16**(6), 671–692 (2001)
14. Crutzen, R., de Nooijer, J., Brouwer, W., Oenema, A., Brug, J., de Vries, N.K.: Qualitative assessment of adolescents' views about improving exposure to internet-delivered interventions. Health Educ. **108**(2), 105–116 (2008)
15. Crutzen, R., de Nooijer, J., Brouwer, W., Oenema, A., Brug, J., de Vries, N.K.: A conceptual framework for understanding and improving adolescents' exposure to internet-delivered interventions. Health Promot. Int. **24**(3), 277–284 (2009)
16. Crutzen, R., de Nooijer, J., Brouwer, W., Oenema, A., Brug, J., de Vries, N.K.: Strategies to facilitate exposure to internet-delivered health behavior change interventions aimed at adolescents or young adults: a systematic review. Health Educ. Behavior **38**(1), 49–62 (2011)
17. Crutzen, R., de Nooijer, J., de Vries, N.: How to reach a target group with internet-delivered interventions? Eur. Health Psychol. **10**(4), 77–79 (2008)
18. Crutzen, R., Peters, G.J.Y., Portugal, S.D., Fisser, E.M., Grolleman, J.J.: An artificially intelligent chat agent that answers adolescents' questions related to sex, drugs, and alcohol: an exploratory study. J. Adolesc. Health **48**(5), 514–519 (2011)
19. Daddis, C., Randolph, D.: Dating and disclosure: adolescent management of information regarding romantic involvement. J. Adolesc. **33**(2), 309–320 (2010)
20. De Sanctis, V., et al.: A practical approach to adolescent health care: a brief overview. Rivista Italiana di Medicina dell'Adolescenza-Volume **12**(1), 1–10 (2014)
21. Deeba, F., Begum, R.: Development of an anxiety scale for Bangladeshi population. Bangladesh Psychol. Stud. **14**, 39–54 (2004)

22. Dosovitsky, G., Bunge, E.: Development of a chatbot for depression: adolescent perceptions and recommendations. Child Adolesc. Mental Health **28**(1), 124–127 (2023)
23. Fitzpatrick, K.K., Darcy, A., Vierhile, M.: Delivering cognitive behavior therapy to young adults with symptoms of depression and anxiety using a fully automated conversational agent (woebot): a randomized controlled trial. JMIR Mental Health **4**(2), e7785 (2017)
24. Fitzsimons, G.J., Moore, S.G.: Should we ask our children about sex, drugs and rock & roll? potentially harmful effects of asking questions about risky behaviors. J. Consum. Psychol. **18**(2), 82–95 (2008)
25. Fulmer, R., Joerin, A., Gentile, B., Lakerink, L., Rauws, M., et al.: Using psychological artificial intelligence (tess) to relieve symptoms of depression and anxiety: randomized controlled trial. JMIR Mental Health **5**(4), e9782 (2018)
26. Grolleman, J., van Dijk, B., Nijholt, A., van Emst, A.: Break the habit! designing an e-therapy intervention using a virtual coach in aid of smoking cessation. In: IJsselsteijn, W.A., de Kort, Y.A.W., Midden, C., Eggen, B., van den Hoven, E. (eds.) PERSUASIVE 2006. LNCS, vol. 3962, pp. 133–141. Springer, Heidelberg (2006). https://doi.org/10.1007/11755494_19
27. Hasan, M.T., et al.: The current state of mental healthcare in Bangladesh: part 1-an updated country profile. BJPsych Int. **18**(4), 78–82 (2021)
28. Hibell, B.: The 2007 espad report. Substance use among students in **35**, 1–408 (2009)
29. Huang, J., et al.: *TeenChat*: a chatterbot system for sensing and releasing adolescents' stress. In: Yin, X., Ho, K., Zeng, D., Aickelin, U., Zhou, R., Wang, H. (eds.) HIS 2015. LNCS, vol. 9085, pp. 133–145. Springer, Cham (2015). https://doi.org/10.1007/978-3-319-19156-0_14
30. Kabir, A.H., Akter, F.: Parental involvement in the secondary schools in Bangladesh: challenges and a way forward. Int. J. Whole Sch. **10**(2), 1–18 (2014)
31. Kim, S.J., Cho, D.E.: Technology trends for UX/UI of smart contents. Rev. Korea Contents Assoc.iation **14**(1), 29–33 (2016)
32. Laizu, Z., Armarego, J., Sudweeks, F.: The role of ICT in women's empowerment in rural Bangladesh (2010)
33. Leung, L.: Impacts of net-generation attributes, seductive properties of the internet, and gratifications-obtained on internet use. Telematics Inform. **20**(2), 107–129 (2003)
34. van de Mheen, D.: Wat doen jongeren op internet en hoe verslavend is dit?
35. Mirowsky, J., Ross, C.E.: Measurement for a human science. J. Health Soc. Beh. 152–170 (2002)
36. Mridha, M.K., et al.: Investing in adolescent girls' nutrition in Bangladesh: situation analysis of trends and ways forward (2019)
37. Parvez, J., Islam, A., Woodard, J.: Mobile financial services in Bangladesh. USAID, mSTAR and fhi360 (2015)
38. Philip, P., et al.: Virtual human as a new diagnostic tool, a proof of concept study in the field of major depressive disorders. Sci. Rep. **7**(1), 42656 (2017)
39. Pinto, A.C.S., Luna, I.T., Sivla, A.D.A., Pinheiro, P.N.D.C., Braga, V.A.B., Souza, Â.M.A.: Risk factors associated with mental health issues in adolescents: a integrative review. Revista da Escola de Enfermagem da USP **48**, 555–564 (2014)
40. Portnoy, D.B., Scott-Sheldon, L.A., Johnson, B.T., Carey, M.P.: Computer-delivered interventions for health promotion and behavioral risk reduction: a meta-analysis of 75 randomized controlled trials, 1988–2007. Prev. Med. **47**(1), 3–16 (2008)

41. Pruijt, H.: Social interaction with computers: an interpretation of weizenbaum's eliza and her heritage. Soc. Sci. Comput. Rev. **24**(4), 516–523 (2006)
42. Roberts, D.F., Foehr, U.G.: Trends in media use. Fut. Children 11–37 (2008)
43. Sharma, P.: Chatbots in medical research: advantages and limitations of artificial intelligence-enabled writing with a focus on chatgpt as an author. Clin. Nuclear Med. 10–1097 (2022)
44. Suzuki, L.K., Calzo, J.P.: The search for peer advice in cyberspace: an examination of online teen bulletin boards about health and sexuality. J. Appl. Dev. Psychol. **25**(6), 685–698 (2004)
45. Tielman, M.L., Neerincx, M.A., Bidarra, R., Kybartas, B., Brinkman, W.P.: A therapy system for post-traumatic stress disorder using a virtual agent and virtual storytelling to reconstruct traumatic memories. J. Med. Syst. **41**, 1–10 (2017)
46. Vaidyam, A.N., Wisniewski, H., Halamka, J.D., Kashavan, M.S., Torous, J.B.: Chatbots and conversational agents in mental health: a review of the psychiatric landscape. Can. J. Psychiatry **64**(7), 456–464 (2019)
47. Vilhjalmsson, R., Thorlindsson, T.: Factors related to physical activity: a study of adolescents. Soc. Sci. Med. **47**(5), 665–675 (1998)
48. Wilson, E.K., Dalberth, B.T., Koo, H.P., Gard, J.C.: Parents' perspectives on talking to preteenage children about sex. Perspect. Sex. Reprod. Health **42**(1), 56–63 (2010)

# Appling Digital Transformation in Intelligent Production Planning of Vietnam's Garment Industry

Huu Dang Quoc[✉] [iD]

Thuong Mai University, 79 Ho Tung Mau, Cau Giay, Hanoi, Vietnam
huudq@tmu.edu.vn

**Abstract.** Currently, the 4.0 industrial revolution is taking place across the globe, and digital transformation is considered a revolution to change how businesses operate and business models fundamentally. In Vietnam, the industrial garment manufacturing sector accounts for 12−16% of the country's total export turnover, accounting for 5.2% of the global market share. This paper presents the digital transformation in industrial garment production using IoT devices to collect production data on sewing lines and methods to digitize the sewing line data industry. After digitizing the sewing line data, the paper proposes an intelligent planning method to improve production efficiency and reduce the time to execute garment contracts. The planning method is based on applying the Real-RCPSP problem combined with the Cuckoo Search algorithm to find a feasible schedule for assigning the resources executing tasks. The experiment results show that the garment contract execution time decreased from 5.97% to 33.99%, depending on the dataset.

**Keywords:** Digital transformation · industrial garment production management · project scheduling problem · optimization algorithm

## 1 Introduction

Digital transformation in businesses is taking place firmly, supporting businesses to optimize their production and business processes, improve efficiency, increase profits, and also allow businesses to facilitate negotiations with partners and get more contracts. In industrial garment production, having management information systems (MIS) to monitor production becomes mandatory for big partners to sign economic contracts with enterprises. This is accelerating the digital transformation process in industrial garment production enterprises recently. Many studies have shown the factors affecting the digital transformation of enterprises; with industrial garment enterprises, two factors directly affecting production efficiency are intelligent production planning based on available resources to execute contracts that manufacture and apply the IoT devices for real-time performance monitoring. These factors encourage industrial garment enterprises to transform digitally and develop and exploit information technology applications in their production and business activities.

© The Author(s), under exclusive license to Springer Nature Singapore Pte Ltd. 2024
N. Thai-Nghe et al. (Eds.): ISDS 2023, CCIS 1949, pp. 125–137, 2024.
https://doi.org/10.1007/978-981-99-7649-2_10

The application of information technology in production planning assists transport and transportation operators in maximizing resources and improving the efficiency of production and business activities. To do this work, data on product manufacturing processes, production lines, resources,... Need to be digitized, and optimization problems applied to find a suitable plan for the manufacturing sector. However, data digitization faces many difficulties due to the need for complete descriptions of production data and no effective digital transformation methods, so it can not apply computational models and methods to automatic production planning. Therefore, most production planning is still conducted by experience or manually, leading to many limitations in specific quantitative calculations and allocating labor, resources, and facilities in the field product manufacturing process.

This paper will present digital transformation in enterprises that support monitoring and tracking the efficiency of production work by unit, each production segment over time, and data digitization methods line industrial garment production. After that, it proposes an intelligent production planning method applying the resource-constrained project scheduling problem Real-RCPSP [1, 2]. The Real-RCPSP problem extends the multi-skill resource-constrained project scheduling problem [3, 4], which is applied in a practical production environment. This problem adds a new constraint that presents the execution time of tasks depending on the power of the execution resources.

The following content of the paper includes the following main parts: Sect. 2 presents the research related to digital transformation and the Real-RCPSP problem; Sect. 3 shows the method of integrating IoT devices in industrial garment production and how to digitize data of industrial product production lines; Sect. 4 gives the scheduling problem used in scheduling and coordinating production activities based on the Real-RCPSP limited resource project scheduling problem. Sect. 5 shows the experiment results of the Real-RCPSP problem [1, 2] on actual data collected from TNG company. Section 6 presents conclusions and directions for further research.

## 2   Related Works

Digital transformation in enterprises and many manufacturing sectors has garnered significant attention and research interest aiming to enhance production efficiency. The researchers have identified factors that influence the success of digital transformation in enterprises. Specifically, Swen and Reinhard [5] highlighted three critical factors affecting the digital transformation process, namely the adoption of new technologies, information technology, and communication in operations, along with the digital competence of leaders. These three key factors are pivotal in enabling successful digital transformation within businesses. Reis et al. [6] classified the significance of digital transformation into three categories: technology, organization, and society. Technology entails using emerging technologies, emphasizing the importance of integrating IoT devices in manufacturing. Vogelsang et al. [7] identified three groups of factors that influence the effectiveness of digital transformation, including organizational aspects (management, employees, data, customers), the business environment (corporate culture, industry characteristics, operational domains), and technology (information technology infrastructure, adoption of new technologies, and information security).

Many researchers have emphasized the role of information technology infrastructure and application strategies that significantly affect the success of enterprises' digital transformation. Osmundsen et al. [8] indicated that one of the eight factors affecting the digital transformation of enterprises is information technology infrastructure; in addition to the readiness for digital transformation, human resources is an essential factor strongly promoted in our research. Marzenna et al. [9] emphasized the impact of information technology on enterprise argumentation, and Muhammad and Anton [10] also analyzed three factors of the digital transformation process affected by the change. New technology changes include adaptability, resource addition, and innovation ability that positively impact the development of digital transformation.

According to the authors' research [11], digital transformation agents in enterprises usually include 5 main factors: leaders, human resources, IT infrastructure, the budget, and services in digital transformation. In particular, the IT infrastructure factor plays a vital role in the success of the digital transformation process of enterprises.

To make intelligent production planning, the authors in the paper [1] stated the project implementation accounting scheduler with limited resources Real-RCPSP, this problem exhibits a highly relevant characteristic in the industrial garment manufacturing sector, as it incorporates constraints on project resource allocation and the scheduling of tasks within a project (in the context of garment production, this refers to the execution of garment contracts). The problem's constraints indicate that higher-skilled resources (workers) can complete tasks in shorter durations or achieve higher quality. This paper will employ the Real-RCPSP and actual production data to coordinate production planning tasks.

## 3    Integrating IoT Devices and Digitizing Industrial Sewing Data

Digital transformation uses digital technologies to change business models, creating new opportunities, revenue, and value [12]. Digital transformation in enterprises is understood as changing from a traditional model to a digital business by applying new technologies such as big data, the Internet of Things (IoT), and cloud computing to change the way of controlling, managing, leadership, work processes, and corporate culture. It is about rethinking how organizations bring together people, data, and processes to create better new value. Digital transformation is the only consulting restructuring in distributing data, processes, and people that creates new value. In industrial garment production, the digital transformation process is associated with the equipment and integration of IoT devices so that managers can real-time monitoring production progress and plan innovative production to improve work efficiency.

### 3.1    IoT Devices Integration

One of the factors to improve production efficiency is that managers need to grasp the production progress, which is continuously updated, thereby providing timely management solutions. With the development of technology in the current industrial sewing industry, industrial sewing machines can integrate with IoT devices to collect product information about the production process and the sewing machine's working process. Over time, from which to evaluate the detailed working efficiency of each worker, each machine can also monitor the performance of investment equipment. IoT devices attached to sewing machines have been exploited by some research, production, and development units for sewing lines at enterprises, one of the commonly used devices is the Brother device [13]. The operation procedure of the device is carried out through the steps shown in Fig. 1.

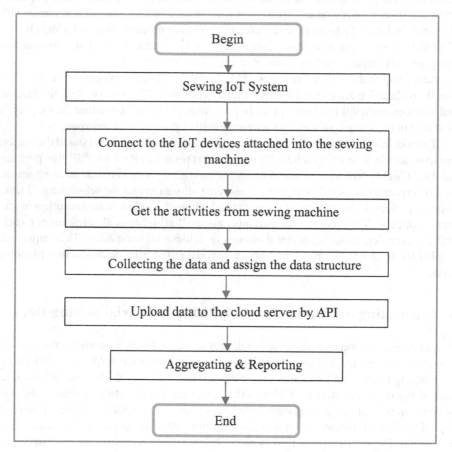

**Fig. 1.** Steps to collect data from Brother devices

The collected data will be sent to the IoT device provider's cloud server center. The enterprises will monitor via the web application of the IoT device provider system or management information system (MIS) of the companies to synthesize reports according to the enterprise's administrative requirements. Figure 2 tracks the performance of the 21 lines in a garment company on MIS.

The MIS integrates with IoT devices' data by connecting to the data system stored on the IoT device provider's Cloud server. Connecting and retrieving data from Cloud Server is done through the API through the steps shown in Fig. 3 below.

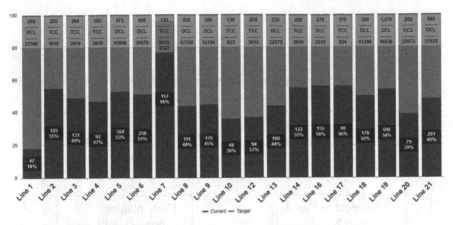

**Fig. 2.** Productivity chart of a company with 21 sewing lines

## 3.2 Production Line Data

The production lines usually observe the sequence of stages from input materials at the beginning and create products at the last stage. At each stage, input materials and middle products at different outputs will be used. To complete a production stage, it is necessary to use different resources such as equipment, machinery, raw materials, and labor. In particular, machinery, equipment, and labor resources are often limited to a certain number, and these resources also have different capacities.

Production data typically includes the parameters described in Table 1 below.

## 3.3 Sewing Line Data

In industrial garments, to complete a contract (with many products of the same type), the company will organize sewing lines to carry out the production stages of the product. Each sewing line has many workers with different skill levels; Sewing stages in a sewing line are arranged in the order of priority of the production process.

*Description of sewing line data*

For each garment contract, which will be performed on the sewing lines, the manager will make a production plan based on the data: contract, labor, labor qualification, and other related resources. Constraints for planning production on the sewing line are described in Table 2 below.

Based on the characteristics of sewing line data, towards the application of automatic production planning, sewing line data can digitize according to the following rules:

- Each garment contract is one project
- Each stage of the product is a task
- The execution time of a stage is the time to execute a task
- Workers have skill levels from 1 to 7; the level of workers will correspond to the skill level in the Real-RCPSP problem model.
- Each worker is a renewable resource; the resource will have a specific skill level
- The priority of stage is the priority of tasks.

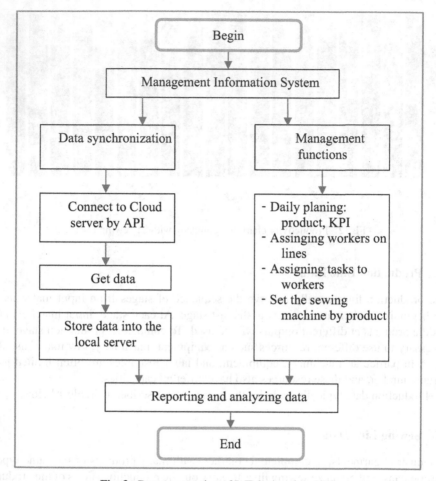

**Fig. 3.** Data aggregation of IoT devices into MIS

*Example 1:*

Digitize data on TNG's garment contracts [19] is shown in Table 3 below.

These contracts can be organized by 04 sewing lines with the corresponding number of workers on each line as 37,39,47,41.

Applying the rule of digitizing sewing line data with contract data in Table 3, we get a standard dataset suitable for inputting the Real-RCPSP problem, as shown in Table 4 below.

In Table 4, the column *"Performance Time (PT)"* is the total actual contract execution time in hours.

**Table 1.** Production data parameters

| No | Content |
|----|---------|
| 1 | Products to be produced |
| 2 | Number of products to be produced |
| 3 | Number of products in production stage |
| 4 | Order of execution of stages |
| 5 | Resources needed for each stage |
| 6 | Skills (Capability) of Resources |
| 7 | Resource type:<br>- Resource Consumption (use one time)<br>- Renewable resources (machinery, labor, etc.) |
| 8 | Capability resource requirements for each stage |
| 9 | The execution time of each state of work corresponds to different capacities of the resource |
| 10 | The costs |

**Table 2.** Constraints for planning on the horizon

| No | Constraint |
|----|-----------|
| 1 | Each contract will order one product type |
| 2 | Each type of product under the contract will have a different number of stages |
| 3 | Each stage will require the worker with the specific level of skill |
| 4 | A worker with a higher level of kill can perform better that lower rank |
| 5 | The stages have a priority relationship with each other, the previous stage is not finish, the later stage can not be start |
| 6 | Based on the contract, the company may assign workers to execute |
| 7 | The company setups many lines (team) to run the contract |

**Table 3.** Garment contracts

| No | Contract | Product Type | Product number | Number of task |
|----|----------|-------------|----------------|----------------|
| 1 | WE1190/1698402 Liner Buy Mar 14-F19 | T-shirt | 33,693 | 71 |
| 2 | FM4013/ 1536181 buy 11/11-F19 | swimming trunks | 83,340 | 137 |

**Table 4.** TNG dataset

| Datasets | Tasks | Resources | Precedence Constraints | Number of skills | Performance time (PT) |
|----------|-------|-----------|------------------------|------------------|-----------------------|
| TNG1 | 71 | 37 | 1026 | 6 | 409 |
| TNG2 | 71 | 39 | 1026 | 6 | 325 |
| TNG3 | 71 | 41 | 1026 | 6 | 296 |
| TNG4 | 71 | 45 | 1026 | 6 | 392 |
| TNG5 | 137 | 37 | 1894 | 6 | 1174 |
| TNG6 | 137 | 39 | 1894 | 6 | 1052 |
| TNG7 | 137 | 41 | 1894 | 6 | 871 |
| TNG8 | 137 | 45 | 1894 | 6 | 996 |

## 4  The Real-RCPSP Problem

The scheduling of projects, economic contracts, or actual production processes is always constrained by requirements such as completion time or resources used. The RCPSP (Resource Constraint Project Scheduling Problem) [14, 15] is a problem that aims to solve the situation schedule for projects with limited resources which is proven to be an NP-Hard class, so it can not find the optimal solution in the polynomial time. By applying the approximate methods to the RCPSP problem, we can find reasonable scheduling solutions that reduce project implementation time and cost. In reality, RCPSP can be applied in many areas of life, such as the economy, military, transportation, and garment industry,... A sub-class extending from the RCSPP problem is the Real-RCPSP problem, added two new constraints as follows:

- A resource can have many different skills; each task will require the execution resource to satisfy the skill requirements at the specific ability.
- A resource whose skill matches the task's requirements and has a higher skill level will be able to perform the task better.

With the addition of 2 new constraints, the Real-RCPSP problem has high applicability in planning and operating production.

The Real-RCPSP problem can be conceptually formulated based on the notations in Table 5.

The Real-RCPSP problem could be state as follow:

$$f(P) \rightarrow min \tag{1}$$

where:

$$f(P) = \max_{W_i \in W} \{E_i\} - \min_{W_k \in W} \{B_k\} \tag{2}$$

**Table 5.** The notations

| Symbol | Description |
|---|---|
| $C_i$ | The set of tasks need to be completed before task i can be executed |
| $S$ | The set of all resource's skills $S^i$: the subset of skills owned by the resource $i$, $S^i \subseteq S$; |
| $S_i$ | The skill $i$; |
| $t_j$ | The duration of task $j$ |
| $L$ | The resources used to execute tasks of the project |
| $L^k$ | The subset of the resources which can be performed task $k$; $L^k \subseteq L$ |
| $L_i$ | The resource i |
| $W$ | The tasks of the project need to do |
| $W^k$ | The subset of task which can be executed by the resource $k$, $W^k \supseteq W$ |
| $W_i$ | The task $i$ |
| $r^i$ | The subset of the skill required by task $i$. A resource has the same skill and skill level equal to or greater than the requirement that can be performed |
| $B_k, E_k$ | The begin time and end time of the task $k$ |
| $A_{u,v}{}^t$ | The variable to identify the resource $v$ is running task $u$ at time $t$; 1: yes, 0: no; |
| $h_i$ | The skill level $i$; |
| $g_i$ | Type of skill $i$; |
| $m$ | Makespan of the schedule |
| $P$ | The feasible solution |
| $P_{all}$ | The set of all solution |
| $f(P)$: | The function to calculate the makespan of $P$ solution |
| $n$ | Task number |
| $z$ | Resource number |

Subject to the following constraints:

- 
$$S^k \neq \emptyset \ \forall L_k \in L \tag{3}$$

- 
$$t_{jk} \geq 0 \ \forall W_j \in W, \ \forall L_k \in L \tag{4}$$

- 
$$E_j \geq 0 \ \forall W_j \in W \tag{5}$$

- 
$$E_i \leq E_j - t_j \ \forall W_j \in W, \ j \neq 1, W_i \in C_j \tag{6}$$

$$\forall \, W_i \in W^k \exists S_q \in S^k : g_{S_q} = g_{r_i} \text{ and } h_{S_q} \geq h_{r_i} \tag{7}$$

$$\forall L_k \in L, \forall q \in m : \sum_{i=1}^{n} A_{i,k}^q \leq 1 \tag{8}$$

$$\forall W_j \in W \exists ! q \in [0, m], !L_k \in L : A_{j,k}^q = 1; \text{ where } A_{j,k}^q \in \{0; 1\} \tag{9}$$

$$t_{ik} \leq t_{il} vo' ih_k \leq h_l \forall (r^k, r^l) \in \left\{ S^k \times S^l \right\} \tag{10}$$

In the Real-RCPSP problem, each task has additional skill (skill) requirements of the resource needed to perform, each resource is also divided into different skill levels.

## 5  Appling the Real-RCPSP Problem in Industrial Sewing Production Planning

To plan production coordination in each sewing line, we apply the Real-RCPSP problem with the digitized sewing line datasets shown in Table 5.

Real-RCPSP is a problem of class NP-Hard, which can not be solved in polynomial time, so it is necessary to use some evolutionary algorithms to find the best schedule for each dataset. In this paper, the author applies CS (Cuckoo Search) algorithm [16–18] to find a schedule. The result can be used to make the production coordination plan by assigning the resources to perform the stages of industrial garment production.

The application of the Real-RCPSP problem to the industrial garment production process allows the leaders can build an automatic production plan, without using traditional, manual methods, according to experience. The production plan is automatically generated based on the input datasets (the digitized orders) for the Real-RCPSP problem combined with the CS algorithm. The calculations are made in detail based on constraints of the labor requirements, execution time, and number of products,...

*Experimental parameters*

The experiment to find a solution using the Real-RPCSP problem and the CS algorithm is performed with the following parameters:

- Dataset: 08 datasets presented in Table 4
- Population size $N_p$: 70
- Number of generations, $N_g$: 60,000
- Number of test runs: 20.

Actual environment: Microsoft Visual Studio 2019, C#

*Experimental Results*

The test run results can be shown in Table 6 below.

**Table 6.** Experiment results

| Datasets | TNG PT | CS | Deviating | % |
|----------|--------|-----|-----------|--------|
| TNG1 | 409 | 270 | 139 | 33.99% |
| TNG2 | 325 | 247 | 78 | 24.00% |
| TNG3 | 296 | 236 | 60 | 20.27% |
| TNG4 | 392 | 260 | 132 | 33.67% |
| TNG5 | 1174 | 947 | 227 | 19.34% |
| TNG6 | 1052 | 953 | 99 | 9.41% |
| TNG7 | 871 | 819 | 52 | 5.97% |
| TNG8 | 996 | 861 | 135 | 13.55% |

In Table 5, column TNG PT is the actual contract performance time (in hours) of TNG company, column CS is the planned execution time (schedule) set by the application of the Real-RCPSP problem model combined with the solution by the CS algorithm.

Table 6 shows that, when applying automatic scheduling by evolutionary methods, it will bring better contract completion time than current practice from 5.97% to 33.99%. The difference in execution time can be done by a tool as shown in the expression in Fig. 4.

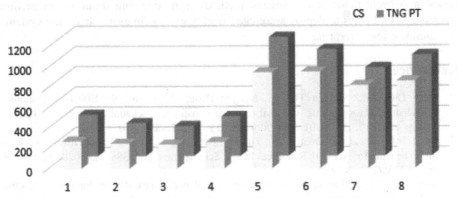

**Fig. 4.** Comparison of execution time between CS and TNG PT algorithms

The conducting result shows that it is possible to automatically calculate the workflow coordination and the assignment of stages in the industrial sewing line. The industrial sewing data is suitable for the model of the Real-RCPSP problem because the data and process characteristics are pretty matched. Deploying an automatic production planning model permits optimizing resources for executing garment contracts, thereby reducing contract execution time and improving profits for garment companies. The method of automated production planning based on applying Real-RCPSP and approximation algorithms is an intelligent solution for production in the recent. This also contributes to increasing the proportion of information technology applications in production automation, an inevitable trend in the coming period.

## 6 Conclusion

Digital transformation in industrial garment enterprises in the current period is necessary to help businesses improve the efficiency of production and business activities. Two essential elements of digital transformation in garment enterprises are the application of IoT devices to monitor production activities and automatic production coordination planning to reduce the time to execute garment contracts. For automated production planning, businesses need to digitize production processes. The paper presented the essential parts of data to implement digitization, proposed the process of digitizing production line data, and specifically applied it in digitizing the sewing line data of TNG company. In order to implement automatic production coordination scheduling, the paper stated the Real-RCPSP problem and proposed a CS algorithm. Next, it experimented on the TNG dataset; the results show that the planning has effectively reduced contract execution time from 5.97% to 33.99% compared to actual time. Therefore, applying these plans will improve enterprises' profits and business efficiency, thereby enhancing the competitiveness and ability of enterprises to integrate into the international environment.

In the coming time, the author will continue to explore the Real-RCPSP problem associated with practical implementation in some production lines and business enterprises in different fields. It will suggest methods to digitize data about the production process and find ways to deploy automated workflow coordination scheduling systems in manufacturing enterprises.

## References

1. Dang Quoc, H., Nguyen The, L., Nguyen Doan, C., Xiong, N.: Effective evolutionary algorithm for solving the real-resource constrained scheduling problem. J. Adv. Transp. 8897710, 11 (2020). https://doi.org/10.1155/2020/8897710
2. Quoc, H.D., The, L.N., Doan, C.N., Thanh, T.P., Xiong, N.N.: Intelligent Differential Evolution Scheme for Network Resources in IoT. Sci. Program. 8860384, 12 (2020). https://doi.org/10.1155/2020/8860384
3. Myszkowski, P.B., Laszczyk, M.: Investigation of benchmark dataset for many-objective multi-skill resource constrained project scheduling problem. Appl. Soft Comput. 127, 109253 (2022)
4. Myszkowski, P.B., Laszczyk, M., Nikulin, I., Skowro, M.: IMOPSE: a library for bicriteria optimization in multi-skill resource-constrained project scheduling problem. Soft Comput. J. 23, 32397 (2019)
5. Nadkarni, S., Prügl, R.: Digital transformation: a review, synthesis and opportunities for future research. Manag. Rev. Q. 71, 233–341 (2020)
6. Reis, J., Amorim, M., Melão, N., Matos, P.: Digital transformation: a literature review and guidelines for future research. In: Rocha, Á., Adeli, H., Reis, L.P., Costanzo, S. (eds.) Trends and Advances in Information Systems and Technologies. WorldCIST'18 2018. Advances in Intelligent Systems and Computing, vol. 745. Springer, Cham (2018). https://doi.org/10.1007/978-3-319-77703-0_41

7. Vogelsang, K., Liere-Netheler, K., Packmohr, S., Hoppe, U.: Success factors for fostering a digital transformation in manufacturing companies. J. Enterp. Transform. **8**(1–2), 121–142 (2019). https://doi.org/10.1080/19488289.2019.1578839
8. Osmundsen, K., Iden, J., Bygstad, B.: Digital transformation: drivers, success factors, and implications. In: MCIS 2018 Proceedings (2018)
9. Cichosz, M., Wallenburg, C.M., Knemeyer, A.M.: Digital transformation at logistics service providers: barriers, success factors and leading practices. Int. J. Logist. Manag. **31**(2), 209–238 (2020). https://doi.org/10.1108/IJLM-08-2019-0229
10. Taufani, M., Widjaja, A.W.: Digital transformation for enhancing LSP (Logistic Service Provider) performance. In: Proceeding of the International Conference on Family Business and Entrepreneurship, vol. 2. no. 1 (2022)
11. Le Viet, H., Dang Quoc, H.: The factors affecting digital transformation in Vietnam logistics enterprises. Electronics 12 (1825). https://doi.org/10.3390/electronics12081825
12. Ribeiro-Navarrete, S., Botella-Carrubi, D., Palacios-Marqués, D., Orero-Blat, M.: The effect of digitalization on business performance: an applied study of KIBS. J. Bus. Res. **126**, 319–326 (2021).https://doi.org/10.1016/j.jbusres.2020.12.065
13. Brother devices for textile industry. https://www.brother-usa.com/industries/textiles
14. Klein, R.: Scheduling of Resource Constrained project, Springer Science Business Media NewYork, Kluwer Academic Publisher (2000). ISBN 978-1-4613-7093-2
15. Blazewicz, J., Lenstra, J.K., Kan, A.R.: Scheduling subject to resource constraints: classification and complexity. Discrete Appl. Math. **5**, 11–24 (1983)
16. Yang, X.S., Deb, S.: Cuckoo search via Lévy flights. In: Proceedings of World Congress on Nature & Biologically Inspired Computing (NaBIC 2009), India. IEEE Publications, USA, pp. 210–214 (2009). https://doi.org/10.1109/NABIC.2009.5393690
17. Yang, X.S.: Nature-Inspired Metaheuristic Algorithms, Luniver Press (2010). ISBN-13: 978-1-905986-28-6
18. Solihin, M.I., Zanil, M.F.: Performance comparison of Cuckoo search and differential evolution algorithm for constrained optimization. In: International Engineering Research and Innovation Symposium (IRIS), vol. 160, no. 1, pp. 1–7 (2016).https://doi.org/10.1088/1757-899X/160/1/012108
19. TNG Investment and Trading Joint Stock Company, 434/1 Bac Kan street - Thai Nguyen city, Viet Nam; Website http://www.tng.vn

# Blockchain-Based Platform for IoT Sensor Data Management

An Cong Tran[1]([✉]), Tran Minh Tai[1], Phan Lam Nhut Huy[1],
and Ngoc Huynh Pham[2]

[1] Can Tho University, Can Tho, Vietnam
tcan@cit.ctu.edu.vn
[2] FPT University, FPT Polytechnic, Can Tho, Vietnam
ngocph11@fpt.edu.vn

**Abstract.** The advancement of information technology has significantly improved environmental monitoring by enabling the use of automated devices that measure environmental impact indicators and transmit data through various communication protocols. The water environment monitoring platform is specifically designed to support efficient device management and long-term data storage. The platform targets two types of users: system administrators and end users. End users, consisting of device owners and users with shared permissions to view devices' data, can make identity requests to store data, manage created devices, monitor and track data, and share permission to view any sensors' data on the device with other users. Meanwhile, system administrators verify device creation requests and renew a device's access token. Our platform leverages several technologies, including Blockchain Hyperledger Fabric to store device data, Firebase to store device information and manage users, Nodejs to create server-side APIs that communicate with the Blockchain network, and ReactJS and Reactnative to develop client-side interfaces. The effectiveness of the platform has been evaluated through a series of experiments, and the results show reasonable throughput. These technologies have enabled effective and efficient environmental monitoring and data management in modern agriculture and aquaculture.

**Keywords:** IoT Sensor Data Management · Blockchain · Hyperledger Fabric · Caliper

## 1 Introduction

The Internet of Things (IoT) is an evolution of data communication that enables direct, persistent, and automated device-to-device communication. It can be considered a global network that allows communication between human-to-human, human-to-things, and things-to-things. This is a networking paradigm where interconnected smart objects continuously generate and transmit data over the Internet. It is based on various standards and enabling technologies with different sensing, connectivity, storage, computational, and other capabilities. Numerous IoT applications exist in various domains such as agriculture, aquaculture,

N. Thai-Nghe et al. (Eds.): ISDS 2023, CCIS 1949, pp. 138–152, 2024.
https://doi.org/10.1007/978-981-99-7649-2_11

health, logistics, retail, smart cities, wastewater management, etc., as surveyed in [1,2]. For example, many IoT applications have been successfully implemented for intelligent traffic systems, smart cities, control of logistics chains, industrial automation, environment monitoring, etc.

For example, IoT is used in various ways in logistics to improve the productivity of supply chains and, therefore, the profitability of companies [3,4]. In this area, IoT devices can accurately register entries and exits of goods. Volume and weight sensors in shelves can also help to out if the good being stored in them is the right one. Alternatively, intelligent labels can let us find the goods and their exact locations in the warehouse. In cargo, remote monitoring sensors for temperature and humidity ensure the handling of goods meets the required care, etc. In smart cities, IoT creates an intelligent network of connected objects and machines that transmit data using wireless technology and the cloud [5]. The data received in a real-time manner can be analyzed to help municipalities, enterprises, and citizens make better decisions that improve quality of life. Communities can improve energy distribution, streamline trash collection, decrease traffic congestion, and even improve air quality with help from the IoT [6,7]. For instance, intelligent garbage cans automatically send data to waste management companies and schedule pick-up as needed versus on a pre-planned schedule [8]. Connected traffic lights receive data from sensors and cars, adjusting soft cadence and timing to respond to real-time traffic, reducing road congestion [9], etc.

In agriculture and aquaculture, IoT is helping farmers to change the way they work through precision farming, especially in the current situation where the climate is changing, the water supply is limited, and fossil fuels are dwindling [10]. One of the most important applications of IoT is to help farmers monitor and manage their farms remotely and automatically through sensor networks and actuators. IoT sensing networks can collect various environmental parameters such as temperature, solved oxygen level, pH, etc. The collected data from IoT sensors can be used to control the actuators and provide farming automation. In addition, the data collected in a long-term observation can be used to optimize and modify the environment and predict future situations, e.g., the water need of crops in the future [1,2,11].

The wide use of IoT leads to large-scale or massive IoT data that provides invaluable information. In addition, it is a challenge to ensure data integrity for cloud-based IoT applications because of the dynamic of the IoT data. For example, in an intelligent transportation system, supporting that we have 1 million GPS vehicle devices, which record the GPS ID, company ID, vehicle ID, GPS time, GPS longitude, GPS latitude, and some other related data every 30 s, each of them is 50 bytes. The total size of the data in one day will be $10^6 \times 50B \times 2 \times 60 \times \times 24 \approx 144\,\text{GB}$. A report on the IoT management market of Mordor Intelligence[1], a fully revenue-funded organization from its founding in 2014 with more than 1.700 partner enterprises across 20 industries, shows that the IoT data management market is expected to grow at a CAGR (Compounded

---

[1] https://www.mordorintelligence.com/.

Annual Growth Rate) of 16.58% during the period 2021–2026. According to Cisco, IoT is expected to generate about 507.5 zettabytes (1 zettabyte = 1 trillion gigabytes) of data by 2019.

There are several approaches to storing and managing IoT data. A very early approach is to use relational database management systems to store and query IoT data [12]. As the data is stored in RDBMS tables, the concurrency is not supported due to the RDBMS lock mechanism. Thus, this solution performance could be better, especially compared to the noSQL-based approach [13–15]. In [14], the authors propose a noSQL-based solution called IOTMDB which better supports storing massive and heterogeneous IoT data. Besides the storage strategy, this solution also cared for sharing the data based on ontology and syntaxes for IoT queries.

The available frameworks of data integrity verification with public audibility cannot avoid the Third Party Auditors (TPAs). However, in a dynamic environment, such as the IoT, the reliability of the TPA-based frameworks could be more satisfactory. Under a blockchain-based framework for Data Integrity Service framework, a more reliable data integrity verification can be provided for both the Data Owners and the Data Consumers without relying on any Third Party Auditor (TPA). Cloud users typically have no control over the cloud storage servers being used, which means there are risks to Data Confidentiality, Data Integrity, and Data Availability. One way to provide trustworthiness in IoT data is through a distributed service trusted by all its participants that guarantees that the data remains immutable. If all participants have the data and the means to verify that the data have not been tampered with since the first definition, trustworthiness can be achieved. Moreover, having a system that guarantees data reliability would allow governments to share and securely transfer information with citizens. Aside from the heterogeneity and integration challenges present in the IoT, the trustworthiness of its data is also an essential issue to **bear in mind**. One way to provide trustworthiness in IoT data is through a distributed service trusted by all its participants that guarantees that the data remains immutable. If all participants have the data and the means to verify that the data have not been tampered with since the first definition, trustworthiness can be achieved. Data immutability becomes a key challenge in many areas where regulations require detailed traceability of assets during their life cycle. Data sources can be identified anytime and remain immutable, increasing their security.

One potential research gap in the blockchain-based IoT device-sharing system is the need to emphasize privacy and security concerns more. While blockchain technology is often touted as being highly secure and decentralized, there are still potential vulnerabilities that malicious actors could exploit. For example, if an IoT device owner shares their device with others on the blockchain-based platform, they may unintentionally expose sensitive data to third parties. Additionally, there is the risk that blockchain-based IoT device-sharing systems could be used for illegal activities, such as money laundering or terrorism financing. Therefore, future research could focus on developing more robust privacy and security measures for blockchain-based IoT device-sharing systems. This could include exploring methods for ensuring that user data is securely stored and transmitted and examining how blockchain technology can be used to prevent fraud and other types of malicious behavior on the platform. Researchers can promote greater trust and adoption of blockchain-based IoT device-sharing systems by addressing these issues.

The contributions of this study can be highlighted as follows:

- First, we have proposed a model that combines blockchain and traditional databases to store IoT data to ensure information security in sharing environments.
- Second, we have proposed a database structure based on Firebase to manage sharing information.
- We have performed experiments to evaluate the proposed blockchain with various test cases and deployed it into an application with graphs illustrating sensor data.

The paper is organized as follows. The next section (Sect. 2) will be relevant studies. We will present our proposed architecture and experimental results for IoT data sharing in Sects. 3, 4, respectively. Finally, we summarize closing remarks in Sect. 5.

## 2   Related Work

Blockchain-based systems have garnered significant attention in recent years, and various works have been carried out to explore their potential in different domains. For instance, Ojha et al. [1] employed smart contracts licensing the technology to ensure flexible access control of multi-sharing, which provided increased security to the system of Blockchain-Based Privacy-Preserving and Rewarding Private Data Sharing for IoT. In another work, authors in [16] utilized smart contracts to filter fabricated information and to employ effective voting and consensus mechanisms that prevented unauthenticated participants from sharing garbage information. Similarly, Hasan et al. [17] incorporated blockchain technology with Software-Defined Wireless Body Area Networks (SDWBANs) to facilitate secure data sharing. In addition, they designed and integrated a wise contract-based fine-grained access control policy, which ensured that only data owners had complete control over the health data.

Moreover, Yang et al. [18] proposed EdgeShare, a data-sharing framework for the Industrial Internet of Things. In this framework, IoT data-sharing activities were recorded in a blockchain for secured transaction logging and auditing. In another study, the authors in [19] introduced a blockchain-based data-sharing approach in the post-quantum era. They utilized three blockchain networks: Hyperledger Fabric, Ethereum, and Quorum. They compared the parallelization performance of Toom-Cook's and Karatsuba's computation methods, selecting NTRU as their quantum-resistant security algorithm. Furthermore, Abubaker et al. [20] proposed integrating IoT and blockchain to monetize IoT data and provide trustful data trading.

Deng et al. [21] introduced a Secure and Efficient Access Control Scheme for Shared IoT Devices over Blockchain. They implemented a new lightweight authentication protocol with instruction data encrypted by a temporary session key negotiated between the user and the IoT device, enabled nontamperable transactions, and prevented central corruption and single point of failure. Their system included three functions: achieving the prepayment of users and settlement for the service contributor, participating in a verification step during the critical negotiation to prevent malicious behavior from users or devices, and recording the workload of the gateway. In [22], the authors used the consortium blockchain to establish a trustworthy environment. A Role-Based Access Control (RBAC) model is then deployed using our proposed multi-signature protocol and intelligent contract methods. Another work in [23] proposed CloudChain where nodes communicated synchronously by direct memory accesses with the shared-memory model with the Remote Direct Memory Access technology, based on which we propose a shared-memory consensus algorithm to ensure persistence and liveness, the two crucial blockchain security properties countering Byzantine nodes.

Additionally, Xiong et al. [24] provided a coordinated recognition scheme over multiple IoT devices, which expanded the visible coverage and exchanged detection information, overcoming the limitations of single devices, such as shelter, target omission, line of sight, blur video, and foggy weather. Attkan et al. [25] presented a comprehensive quality study for researchers on authentication and session keys, integrating IoT with blockchain and AI-based authentication in cybersecurity. They discussed the traditional key generation and distribution systems, perks, and limitations of IoT attacks. Finally, Auer et al. [26] presented a high-level architecture for a blockchain-IoT-based platform for promoting shared mobility, combining car-sharing and car-leasing. They developed a prototype from the OEM's point of view by developing a blockchain-IoT-based platform streamlining car-sharing and leasing processes by considering primary stakeholders. While describing access control mechanisms, Bagga et al. [27] elaborated on each use case, such as smart home, smart grid, healthcare, and intelligent agriculture.

## 3    Blockchain-Based Platform for IoT Data Sharing

The water environment monitoring platform is aimed at all users, including individuals, organizations, businesses, etc. The platform allows authorized users to register and manage monitoring devices and sensors' data of the devices. In addition, the system provides data storage on the Hyperledger Fabric blockchain network to help users ensure the immutability, transparency, and accuracy of data. Users can retrieve information on their devices in real-time and share access rights to data on their devices with other system users. We use Blockchain technology with Hyperledger Fabric, a distributed storage system on a ledger, building chain code with JavaScript language.

## 3.1  System Architecture

The platform is built from critical components: Hyperledger Fabric network implemented with chain codes for users to interact with the Ledger, Nodejs server based on ExpressJS Framework, and Firebase database service to store and manage authentication. The ultimate user experience is a ReactJS and React Native library for building user interfaces on web and mobile apps running on Android devices.

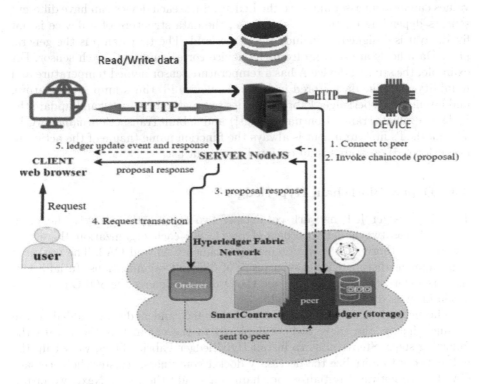

**Fig. 1.** The proposed architecture

The build system consists of the following components: Server Nodejs, Hyperledger Fabric network, Firebase, and user interface (Web and mobile app), as shown in Fig. 1. Users can access the system via a Web browser or mobile application to request user authentication by calling the corresponding API to the Nodejs server. The server will interact with the FireStore cloud database to perform authentication for a real user. Once the user has been authenticated, they can perform their functions, such as viewing device data, adding devices, etc. When the user needs to read the device's data, it will send a request to the Nodejs server; then, the server will interact with the Hyperledger Fabric network to authenticate the device's identity on the network and return the data to the user. When sending data to the network, monitoring devices must also go through the Nodejs server to authenticate the device with a token signed by the device identifier information in the Hyperledger Fabric network; when the authentication is successful, the data will be updated in the network.

## 3.2  Design the Chain Code (Smart Contract)

We define the structure of the data of the chain code with the sensor values sent from the device along with the device code and timestamp information. These values can form an asset and save the Ledger. Since each device can have different sensors depending on the user's purpose, the data structure of a device is not fixed, so it is designed to include the device's Id. The timestamp is the general part. Data fields are sent up from the device corresponding to each sensor. For example, the sending device A has a temperature sensor named temperature and humidity, then the data stored includes the device Id, timestamp, temperature, and humidity. Hyperledger Fabric provides two methods to query and update the Ledger, evaluateTransaction(name, args]) and submitTransaction(name, args]). The method's first argument is always the function name (name of the action to be performed), and args is an optional argument list.

## 3.3  Deploy Blockchain Hyperledger

The Hyperledger IoT network consists of two organizations, org1, and org2, and each has two nodes, peer0, and peer1. For each organization there is a certificate authority for each organization named CA1 and CA2. The org1 and org2 domains distinguish between peers of different organizations. In addition, there is an Orderer organization, and the Orderer node uses the SOLO consensus mechanism as the Orderer method.

The network is set up, the channel is created, and nodes are added to the channel; finally, the chain code is deployed to the channel as Fig. 2 with the following steps. Step 1 aims to install Hyperledger Fabric. Then, we set up the IoT network to initialize the necessary docker containers, register the necessary CAs for the network, initialize the channel and add the peers. Next, we install the chain code on the peer nodes in the channel. Then, we approve and commit the chain code to the channel.

## 3.4  External Database

To increase flexibility in data access and reduce the cost of storing and accessing data, some data is stored in an external database. This system uses the Firebase Authentication service provided by Firebase to manage the user logins and Firestore cloud service, which acts as a NoSQL database and supports real-time synchronization to store some data of the system such as sensor information, device information, and user information.

The system uses the Firebase Authentication service provided by Firebase that allows users who already have a Google account to log in to the system.

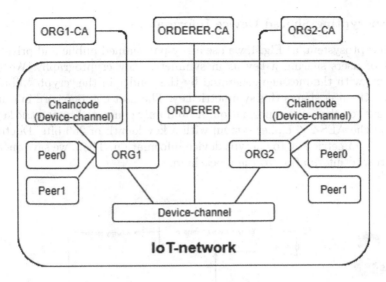

**Fig. 2.** IoT-network architecture.

Once the user is logged into the system, Firebase generates a unique UID, and the information related to the Google account is saved to Firebase. All user activities can be managed based on this information (Fig. 3).

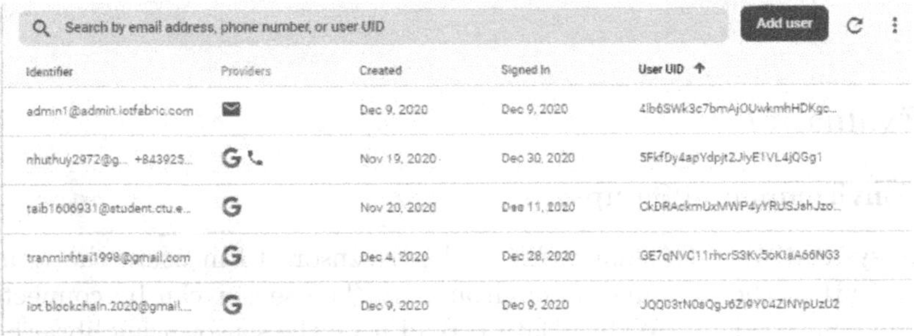

| Identifier | Providers | Created | Signed In | User UID ↑ |
|---|---|---|---|---|
| admin1@admin.iotfabric.com | ✉ | Dec 9, 2020 | Dec 9, 2020 | 4lb6SWk3c7bmAjOUwkmhHDKgc... |
| nhuthuy2972@g... +843925... | G ☎ | Nov 19, 2020 | Dec 30, 2020 | 5FtdDy4apYdpjt2JiyE1VL4jQGg1 |
| taib1606931@student.ctu.e... | G | Nov 20, 2020 | Dec 11, 2020 | CkDRAckmUxMWP4yYRUSJahJzo... |
| tranminhtai1998@gmail.com | G | Dec 4, 2020 | Dec 28, 2020 | GE7qNVC11rhcrS3Kv5oKlsA66NG3 |
| iot.blockchain.2020@gmail... | G | Dec 9, 2020 | Dec 9, 2020 | JQQ03tN0sQgJ6Zi9Y04ZiNYpUzU2 |

**Fig. 3.** System external (Firebase) database structure.

The database uses the Firestore cloud service provided by Firebase, which acts as a NoSQL database and supports real-time to help synchronize data on the Client easily. The database consists of 4 Collections (equivalent to one table in SQL), including devices: a collection containing device information, bcAccounts containing device identifiers in the blockchain network, fieldRef: a collection containing a user's shared sensor; and adminUser: contains administrative user information.

## 3.5   Encrypting Shared Device Information

In this cryptosystem, as Fig. 4, we use a key pair named public and private keys instead of users sharing a key as in symmetric key cryptography. We send a plain text with the message generated by the sender in the "crypto" library. In addition, the user shares this symmetric key. The user enters record information (plain text) on the device information. A randomly generated symmetric key (Ks) based on the AES-256 cipher system with a key length of 256 bits. Doctor Bob uses this symmetric key to encrypt device information. Then, we save encrypted device record information to the blockchain.

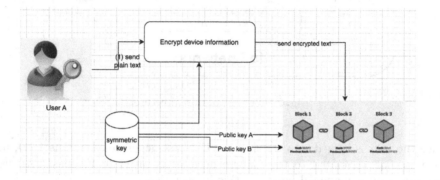

**Fig. 4.** An illustration of information on users in Firebase.

## 4   Evaluation

### 4.1   Environmental Setups

In this system, we use temperature and pH sensors to measure the temperature and pH of the surrounding environment. The sensor can be connected to Waspmote Smart device Water. Data is read from the sensor using libraries provided by the device manufacturer. In addition, we then use HTTP protocol to send this payload, including timestamp value, device token, device id, and measured values. This payload is sent to the NodeJS server acting as a client on the network, where the Client generates a transaction proposal from the received data and sends it to peers in the Blockchain network. This transaction is only accepted if at least one peer has signed it from each organization. Once the transaction is signed and sent, it is validated for signatures. The data is added to the Ledger if the transaction is valid. First, the Smart Water device reads values from the sensor. Next, the data is timestamped and then sent to the NodeJS server. The Nodejs application then processes this payload and generates a transaction proposal. The proposed transaction is signed by at least one peer from each institution. This signed transaction proposal is then submitted to the Blockchain network. Inside the Blockchain network, the chain code and the appropriate functions are called. The chain code then makes updates to the Ledger and returns a response. Then a response is generated by the chain code and returned to the application.

Farmers often use various IoT devices in precision agriculture to monitor soil moisture, temperature, and other environmental factors. To connect these devices, a Waspmote can be programmed to communicate with each device using specific protocols or drivers. For example, a Waspmote could be configured to communicate with a soil moisture sensor using the ZigBee wireless communication protocol. Once the Waspmote is connected to the sensor, it can collect data on soil moisture levels and send this information to a central database for analysis. This data can then be used to optimize irrigation schedules and improve crop yields. In addition, the Waspmote can be programmed to connect to multiple pH sensors and collect data from each sensor. The Waspmote can then combine this data into a single dataset that can be sent to a central monitoring system for water quality analysis.

On the server side, we have deployed the Hyperledger Fabric blockchain network on the Ubuntu operating system. In addition, we build and use smart contracts to interact with data stored on the Ledger. We have implemented applications with NodeJS with ExpressJS Framework combined with information stored on Firebase. The Server side can manage functions affecting the blockchain network through Hyperledger Fabric SDK for Node.js, and query data can be returned to the Client according to the requests the Client has sent to the server. On the client side, such as the Web side, we can use browsers such as Google Chrome and Firefox designed with the ReactJS framework. For the mobile application side, we build with React Native framework for Android devices. In addition, the system must be run on the Ubuntu operating system, and the Hyperledger Fabric network and chain code must be installed. Another issue that needs to be ensured is the Internet connection to use libraries and Firebase.

## 4.2  Experimental Results

**Network Performance with the Blockchain-Based System.** Hyperledger Caliper is a framework for evaluating blockchain network performance (as illustrated in Fig. 5), allowing users to test different blockchain solutions with customers' use cases and get a set of performance test results. Caliper creates a workload based on a tested system and continuously monitors its response. Finally, it can generate a report based on the observed responses. The speed at which transactions are entered into the blockchain system is a factor in performance tests. Caliper allows customization of controller specifications for users to experiment with a custom mechanism. Fixed Rate is the controller that supports sending the number of transactions at a pre-specified fixed time with the unit of Transactions Per Second (TPS). Fixed Load denotes the controller automatically adjusting support to the device configuration. In addition, it helps the evaluation process to be manageable, not to assess the maximum load capacity of the blockchain network accurately.

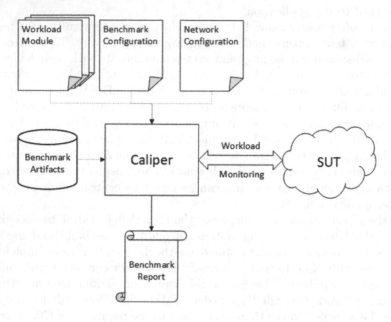

**Fig. 5.** Testing network configuration

**Table 1.** Evaluation throughput of the proposed blockchain. 'Sent' and 'Received' are measured in TPS

| Test cases | Success | Fail | Sent | Max delay | Min delay | Mean delay | Received |
|---|---|---|---|---|---|---|---|
| 10asset100tps | 6008 | 0 | 100.1 | 0.14 | 0.01 | 0.04 | 100 |
| 10asset200tps | 12008 | 0 | 200.1 | 0.16 | 0.01 | 0.02 | 200 |
| 10asset500tps | 24417 | 5518 | 498.7 | 19.92 | 0.03 | 10.73 | 466.9 |
| 10asset1000tps | 21276 | 9096 | 506.1 | 21.51 | 0.77 | 13.97 | 472.2 |
| 100asset100tps | 6008 | 0 | 100.1 | 0.17 | 0.01 | 0.04 | 100.1 |
| 100asset200tps | 12008 | 0 | 200.1 | 0.11 | 0.01 | 0.01 | 200 |
| 100asset500tps | 18616 | 9493 | 468.1 | 25.84 | 0.07 | 15.05 | 431.7 |
| 100asset1000tps | 17205 | 13247 | 507.4 | 29.79 | 2.08 | 17.53 | 470.6 |

Hyperledger Caliper is leveraged to test with the following cases: + 100, 200, 500, 100 TPS on the server equipped with a CPU: I7 6700HQ, a RAM of 16GB running on Docker environment version 20.10.1 with eight workers in 60 s.

Table 1 reveals the results of test cases. In 4 test cases at less than 200 transactions/sec, the highest latency was only about 0.17 s. All transactions executed are successful. For the case of 10asset500tps with 500 transactions/second, the highest latency is 19.92 s, and the average latency is about 10.73. The number of successful transactions, 24417, accounts for about 81.5% of the total transactions. In the case of 10asset1000tps with a setting of 1000 transactions/second, the highest latency is 21.51 s, and the average latency is 13.97 s. Sending throughput reached 506.1 compared to 1000 transactions/sec at the set level. The number of successful transactions is 21276, accounting for 70% of the total transactions. For the 100asset500tps test at 500 transactions/sec, the transactions started showing the highest latency at 25.84 s, with the average latency at 15.05 s. Sending traffic reached 468.1 compared to the set level of 500 transactions/sec. In this case, the number of failed transactions is 9493, which accounts for about 33.7% of the total transactions. In the case of 100asset1000tps testing at 1000 transactions/second, the latency appeared at a high of 29.79, and the average latency was also at 17.53. Save volume only reached 507 compared to 1000 at the set level. In this case, the number of failed transactions is 13247, accounting for 43% of the total transactions.

**Table 2.** Results with Fixed Load controller

| Test case | Success | Fail | Sent | Max delay | Min delay(s) | Mean delay(s) | Received |
|---|---|---|---|---|---|---|---|
| fixedLoad | 15067 | 0 | 251.1 | 0.17 | 0.01 | 0.04 | 249.9 |
| Fixed Rate | 21248 | 0 | 354.1 | 3.21 | 0.01 | 0.29 | 354 |

We also obtain similar performance, as shown in Table 2 using Fixed Load and Fixed rate controllers with the same parameters. Caliper automatically adjusts sending traffic at 251.1; latency is low. This shows that the load capacity can be higher than the recommended level of the controller. Based on the results of 70% successful transactions at 1000 transactions/sec setup with a sending throughput of about 506.1 transactions/sec, set at 354 (70% success of 506.1) transactions/sec to try. experience. In this case, the latency is already stable. All successful transactions show that the load capacity at 354 transactions/sec is acceptable.

**Table 3.** Computational resource usage

| Blockchain node | Max % CPU | Mean % CPU | Max % RAM | Mean % RAM |
|---|---|---|---|---|
| Organization 1 | 80.73 | 12.04 | 85.9 | 71.2 |
| Organization 2 | 72.97 | 11.65 | 82.2 | 69.8 |
| Peer 1 | 155.55 | 27.16 | 382 | 375 |
| Peer 2 | 137.25 | 27.18 | 373 | 366 |
| Orderer | 3.36 | 0.93 | 130 | 125 |

The results show the load capacity of the Blockchain network, as exhibited in Table 3, to evaluate, improve, and upgrade the system when necessary.

**Deployment in an Application.** We designed and implemented some fundamental functions. This section illustrates a real-time device viewing function in Fig. 6. This function can view the function interface that shows device information, real-time data of the device, data graph of sensors, and a metric function button to view data over time certain.

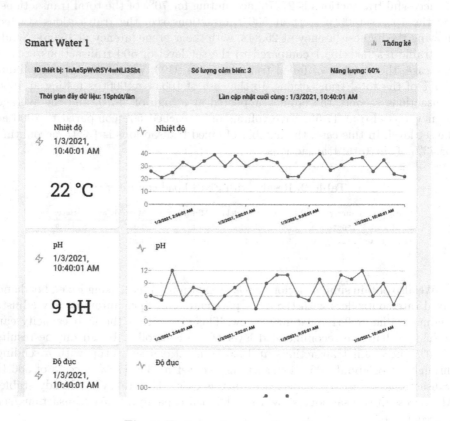

**Fig. 6.** Testing network configuration

## 5    Conclusion

This study proposes a platform that enables efficient equipment management, long-term data storage, monitoring, and user sharing. The platform employs various technologies such as Blockchain Hyperledger Fabric, Firebase, Nodejs, ReactJS, and Reactnative to achieve this. The platform has been evaluated through various experiments, and the results show a reasonable throughput. Furthermore, these technologies have enabled effective and efficient environmental monitoring and data management, essential for modern agriculture.

IoT device-sharing platforms could be an area of future work, as these platforms would allow for easier sharing and management of IoT devices among different users. In the future, these platforms could include features such as user authentication, device tracking, and access control and could be accessed through a centralized dashboard or app. Additionally, these platforms could facilitate sharing of data generated by IoT devices, enabling users to collaborate and analyze data more effectively. However, developing such platforms would require addressing data privacy and security issues and ensuring compatibility with a wide range of IoT devices for further research.

# References

1. Ojha, T., Misra, S., Raghuwanshi, N.S.: Wireless sensor networks for agriculture: the state-of-the-art in practice and future challenges. Comput. Electron. Agric. **118**, 66–84 (2015)
2. Talavera, J.M., et al.: Review of IoT applications in agro-industrial and environmental fields. Comput. Electron. Agric. **142**, 283–297 (2017)
3. Tu, M.: An exploratory study of Internet of Things (IoT) adoption intention in logistics and supply chain management. Int. J. Logist. Manag. (2018)
4. Hopkins, J., Hawking, P.: Big data analytics and IoT in logistics: a case study. Int. J. Logist. Manag. (2018)
5. Rajab, H., Cinkelr, T.: IoT based smart cities. In: 2018 International Symposium on Networks, Computers and Communications (ISNCC), pp. 1–4. IEEE (2018)
6. Kim, T.H., Ramos, C., Mohammed, S.: Smart city and IoT (2017)
7. Hammi, B., Khatoun, R., Zeadally, S., Fayad, A., Khoukhi, L.: IoT technologies for smart cities. IET Networks **7**(1), 1–13 (2017)
8. Putra, R.H., Kusuma, F.T., Damayanti, T.N., Ramadan, D.N.: IoT: smart garbage monitoring using android and real time database. Telkomnika **17**(3), 1483–1491 (2019)
9. Frank, A., Al Aamri, Y.S.K., Zayegh, A.: IoT based smart traffic density control using image processing. In: 2019 4th MEC International Conference on Big Data and Smart City (ICBDSC), pp. 1–4. IEEE (2019)
10. Ahmed, N., De, D., Hussain, I.: Internet of Things (IoT) for smart precision agriculture and farming in rural areas. IEEE Internet Things J. **5**(6), 4890–4899 (2018)
11. Muangprathub, J., Boonnam, N., Kajornkasirat, S., Lekbangpong, N., Wanichsombat, A., Nillaor, P.: IoT and agriculture data analysis for smart farm. Comput. Electron. Agric. **156**, 467–474 (2019)
12. Ding, Z., Gao, X.: A database cluster system framework for managing massive sensor sampling data in the internet of things. Chin. J. Comput. **35**(6), 1175–1191 (2012)
13. Cruz, F., Gomes, P., Oliveira, R., Pereira, J.: Assessing NOSQL databases for telecom applications. In: 2011 IEEE 13th Conference on Commerce and Enterprise Computing, pp. 267–270. IEEE (2011)
14. Li, T., Liu, Y., Tian, Y., Shen, S., Mao, W.: A storage solution for massive IoT data based on NOSQL. In: 2012 IEEE International Conference on Green Computing and Communications, pp. 50–57. IEEE (2012)
15. Győrödi, C., Győrödi, R., Pecherle, G., Olah, A.: A comparative study: MONGODB vs. MYSQL. In: 2015 13th International Conference on Engineering of Modern Electric Systems (EMES), pp. 1–6. IEEE (2015)

16. Liu, Y., et al.: A blockchain-based decentralized, fair and authenticated information sharing scheme in zero trust internet-of-things. IEEE Trans. Comput. **72**(2), 501–512 (2023)

17. Hasan, K., Chowdhury, M.J.M., Biswas, K., Ahmed, K., Islam, M.S., Usman, M.: A blockchain-based secure data-sharing framework for software defined wireless body area networks. Comput. Networks **211**, 109004 (2022). https://doi.org/10.1016/j.comnet.2022.109004

18. Yang, L., Zou, W., Wang, J., Tang, Z.: EdgeShare: a blockchain-based edge data-sharing framework for industrial internet of things. Neurocomputing **485**, 219–232 (2022). https://doi.org/10.1016/j.neucom.2021.01.147

19. Zeydan, E., Turk, Y., Ozturk, S.B., Mutlu, H., Dundar, A.A.: Post-quantum blockchain-based data sharing for IoT service providers. IEEE Internet Things Mag. **5**(2), 96–101 (2022)

20. Abubaker, Z., Khan, A.U., Almogren, A., Abbas, S., Javaid, A., Radwan, A., Javaid, N.: Trustful data trading through monetizing IoT data using BlockChain based review system. Concurr. Comput. Pract. Exp. **34**(5) (2021). https://doi.org/10.1002/cpe.6739

21. Deng, Y., Wang, S., Zhang, Q., Wang, J.: A secure and efficient access control scheme for shared IoT devices over blockchain. Mob. Inf. Syst. 2022, 1–21 (2022). https://doi.org/10.1155/2022/4496486

22. Gai, K., She, Y., Zhu, L., Choo, K.K.R., Wan, Z.: A blockchain-based access control scheme for zero trust cross-organizational data sharing. ACM Trans. Internet Technol. (2022). https://doi.org/10.1145/3511899

23. Xu, M., Liu, S., Yu, D., Cheng, X., Guo, S., Yu, J.: Cloudchain: a cloud blockchain using shared memory consensus and RDMA. IEEE Trans. Comput. **71**(12), 3242–3253 (2022)

24. Xiong, F., Xu, C., Ren, W., Zheng, R., Gong, P., Ren, Y.: A blockchain-based edge collaborative detection scheme for construction internet of things. Autom. Const. **134**, 104066 (2022). https://doi.org/10.1016/j.autcon.2021.104066

25. Attkan, A., Ranga, V.: Cyber-physical security for IoT networks: a comprehensive review on traditional, blockchain and artificial intelligence based key-security. Complex Intell. Syst. **8**(4), 3559–3591 (2022). https://doi.org/10.1007/s40747-022-00667-z

26. Auer, S., Nagler, S., Mazumdar, S., Mukkamala, R.R.: Towards blockchain-IoT based shared mobility: car-sharing and leasing as a case study. J. Network Comput. Appl. **200**, 103316 (2022). https://doi.org/10.1016/j.jnca.2021.103316

27. Bagga, P., Das, A.K., Chamola, V., Guizani, M.: Blockchain-envisioned access control for internet of things applications: a comprehensive survey and future directions. Telecommun. Syst. **81**(1), 125–173 (2022). https://doi.org/10.1007/s11235-022-00938-7

# Deep Learning and Natural Language Processing

# Bangla News Classification Employing Deep Learning

Abu Sayem Md. Siam, Md. Mehedi Hasan, Md. Mushfikur Talukdar,
Md. Yeasir Arafat, Sayed Hossain Jobayer, and Dewan Md. Farid[✉]

Department of Computer Science and Engineering, United International University,
United City, Madani Avenue, Badda, Dhaka 1212, Bangladesh
dewanfarid@cse.uiu.ac.bd
https://cse.uiu.ac.bd/profiles/dewanfarid/

**Abstract.** In this paper, we have introduced a deep learning model for
the classification of Bangla news articles as it is an important task for
maintaining and managing Bangla news articles. The proposed model
combines a hybrid Recurrent Neural Network (RNN) comprising Bi-
directional Long-Short Term Memory (LSTM) and Bi-directional Gated
Recurrent Unit (GRU) algorithms to improve classification accuracy. We
tested the performance of proposed classifier with traditional machine
learning techniques e.g. naïve Bayes and decision tree induction. The
experimental results indicate that the deep learning models achieved
almost 90% accuracy. This study provides valuable insights into the performance of various machine learning and deep learning techniques for
Bangla news classification and highlights the potential of deep learning
for this task.

**Keywords:** Bangla News Classification · Deep Learning · Recurrent
Neural Network

## 1 Introduction

In recent years, the rapid growth of digital media and the widespread availability
of online news sources have led to an information overload for individuals seeking
reliable and relevant news content. This surge in news articles has presented a
significant challenge for readers in terms of identifying and accessing the information that is most important to them. As a result, there has been an increasing need to develop effective solutions for news classification and organization.
Bangla, the official language of Bangladesh, is one of the most widely spoken
languages in the world. With a significant number of Bangla speakers accessing
digital media, there is a pressing demand for a reliable and efficient method to
classify Bangla news articles automatically. The classification of Bangla news
articles can greatly reduce the time and effort required for readers to find articles that align with their interests. News classification refers to the process of
categorizing or labelling news articles based on their content or topic. Traditional news classification methods have primarily focused on English-language

N. Thai-Nghe et al. (Eds.): ISDS 2023, CCIS 1949, pp. 155–169, 2024.
https://doi.org/10.1007/978-981-99-7649-2_12

articles [1] However, the classification of Bangla news articles poses unique challenges due to the linguistic characteristics of the language. One of the main challenges of Bangla news classification is the lack of sufficient labeled training data. Machine learning algorithms rely on large amounts of labeled data to learn patterns and make accurate predictions. Unfortunately, the availability of labeled training data for Bangla news articles is limited, making it difficult to develop robust classification models specifically for Bangla. Furthermore, the linguistic complexity of Bangla poses another challenge for news classification. The language is rich in vocabulary and has complex sentence structures that may differ from those in English. As a result, traditional classification models trained on English news articles may not perform well when applied to Bangla texts.

To address these challenges, researchers have started exploring different approaches to Bangla news classification. Some studies have focused on using transfer learning techniques [2], where pre-trained models trained on other languages are fine-tuned on Bangla news articles. Others have experimented with rule-based methods that make use of Bangla-specific linguistic characteristics. [3]. Random forests is an ensemble learning method for classification, regression and other tasks that operate by constructing a multitude of decision trees at training time and outputting the class that is the mode of the classes or mean prediction of the individual trees [4]. Random Forest can be applied to text classification tasks and it has shown good performance in handling high-dimensional data and avoiding overfitting [5]. Convolutional Neural Networks (CNNs) are widely used in computer vision tasks, but it has also been used successfully in Natural Language Processing tasks. CNNs can capture local patterns in textual data by using convolutional filters and pooling operations. They have shown promising results in sentence classification and document classification tasks [6].

Recurrent Neural Networks (RNNs) particularly Long Short-Term Memory (LSTM) and Gated Recurrent Unit (GRU) architectures, have been used for sequential data processing, including text classification. RNNs can capture the contextual information and dependencies between words in a text by utilising recurrent connections. They have been effective in sentiment analysis, text summarisation, and other text classification tasks [7]. Transformer models, such as the famous BERT (Bidirectional Encoder Representations from Transformers), have revolutionised natural language processing tasks. These models leverage self-attention mechanisms to capture the relationships between words in a text and have achieved state-of-the-art performance in various text classification tasks [8]. Research on these algorithms involves, experimenting with different architectures, hyper-parameter tuning, and incorporating various techniques like pre-training on large corpora, transfer learning, and fine-tuning. The goal is to improve the accuracy, efficiency, and interpretability of the models for text classification in different languages, including Bangla. Moreover researchers have also explored techniques for handling imbalanced datasets, incorporating domain-specific knowledge, and developing hybrid models that combine different algorithms to leverage their strengths. Additionally, feature engineering, such as word embeddings, character-level representations, and attention mechanisms, has been investigated to enhance the representation and understanding of textual data.

In this paper, we present a detailed study of the application and comparison of different machine learning models for Bangla news classification. Our main objective is to evaluate the accuracy and performance of several models, including Decision Tree, Naive Bayes, Neural Network and Recurrent Neural Network (Bi-directional LSTM and Bi-directional GRU). To conduct our experiments and evaluations, the dataset for this study was collected from Kaggle [9]. The dataset consists of 378471 news articles from the online newspaper Prothom Alo published between 2013 to 2019. The articles were classified into eight categories: Bangladesh, sports, international, entertainment, economy, opinion, technology and lifestyle. The collected dataset provides valuable insights into the news trends and preferences of the Bangladeshi audience. The use of this dataset allows us to comprehensively analyze and compare the performance of the various models on a wide range of news articles. We aim to determine which model achieves the highest accuracy in classifying the news articles into their respective categories. Our first model, the Decision Tree, is known for its simplicity and interoperability. Next, we employ the Naive Bayes algorithm, a probabilistic classifier based on Bayes' theorem. By training the Naive Bayes model on our Bangla news dataset, we can evaluate its performance and compare it with the previous Decision Tree model. In our study, we also investigate the effectiveness of Neural Networks for Bangla news classification. Neural Networks are powerful models capable of learning complex patterns in data by utilizing multiple layers of interconnected neurons. By designing and training a Neural Network model with our dataset, we are able to assess its accuracy in effectively categorizing Bangla news articles. Furthermore, we explore the potential of Recurrent Neural Networks (RNNs) for Bangla news classification. Specifically, we utilize a combination of Bi-directional LSTM layers, which are known for their ability to capture long-term dependencies in sequential data. By employing Bi-directional LSTM layers in our Neural Network model, we aim to enhance its performance and accuracy in classifying Bangla news articles. In addition to the Bi-directional LSTM layers, we also investigate the effectiveness of incorporating Bi-directional GRU (Gated Recurrent Unit) layers. GRUs have a simpler structure compared to LSTMs, but they still offer valuable learning capabilities. By combining both LSTM and GRU layers in our model, we aim to improve the representation power and performance of the classifier. Throughout our experiments, we meticulously evaluate the accuracy and performance of each model using various metrics such as precision, recall, and F1 score. Additionally, we conduct in-depth comparative analyses to determine which model outperforms the others in terms of accuracy and effectiveness. The findings of our experiments are discussed and analyzed in detail in the subsequent sections of this paper. In this paper, we hope to contribute to the development of accurate and efficient solutions for Bangla news classification, which can enhance the browsing experience for Bangla news readers and alleviate the information overload they often face.

The remaining part of the paper is organised as follows Sect. 2 contains the prior works for solving the same problems followed by our proposed method in Sect. 3 presents the research methodology which describes proposed methods. Section 3 holds the experiment details and outcomes obtained from the study. Finally, this work is concluded and provides future direction in Sect. 5.

## 2    Related Works

One of the earliest problems in natural language processing (NLP) is text classification. Over time, the scope of application areas has broadened and the difficulty of dealing with new areas has increased and several approaches to Bangla text classifications, such as using transformers, deep convolution neural networks, and recurrent nerual networks [10]. Maisha and Masum (2021) proposed a method for NLP. To extract features and train the dataset, machine learning (ML) classifiers are used, together with Ciountvectoriser, transformer (TF-IDF), and other composite process techniques of the "Pipeline" class. They split the students into two classes, one positive class with 1700 papers and the other negative class with 1800 papers. In order to analyse the best classifier for the given model, the model typically uses Multinomial Naïve Bayes (MNB), K-Nearest Neighbour (K-NN), Random Forest (RF), (C4.5) Decision Tree (DT), Logistic Regression (LR), and Linear Support Vector Machine (LSVM). RF has beat other classifiers across all six by 99% accuracy. Even though LSVM's output has the lowest accuracy of 80%, it is still regarded as good [11]. They employed foundational models like Naive Bayes, Logistic Regression, Random Forest, Linear SVM, and CNN as well as deep learning models like BiLSTM. They discovered that the Support Vector Machine used as the base model, and CNN, used for deep learning, produced the best results for their data-set. The retrieved text cannot be processed directly by machine learning algorithms, hence Monika Vhumika and Dhruv(2021) used pre-processing and feature engineering techniques including count vectorizer, TFIDF, and word2vec. The model accepts Hindi news headlines from a variety of predefined categories, and after prepossessing, the corpus size with unique terms was 54,44,997 words. Multinomial Naive Bayes with a count vectorizer has (85.47%) accuracy [12]. Rifat and Imran (2022) seeks to determine the sentiment expressed in Bangla newspaper headlines based on an attribute. This effect could be neutral, negative, or both. Aspect-based sentiment analysis is the technique used. Bernoulli Naive Bayes fared better on this model when compared to Logistic Regression (LR), SGD Classifier, SVM, Random Forest, MLP Classifier, and Voting Classifier among all the algorithms, with an F1 score of 70.75 [13].

A new model provided by Rahman and Mithila (2021) to classify Bangla news by applying Decision Tree, supervised machine learning algorithm. Mainly Two approaches have been engendered for purpose and DT makes best results with 95% accuracy and 95% Fl-Score [14]. Das and Sammi (2023) provide a method using six most popular algorithms and they got the highest accuracy for the decision tree classifier which is about 93.72% accuracy. They got this accuracy by using a dataset of 2727 bers of data which is collected from different sources and also got better results for f1-score and confusion matrix [15]. Nusrat Khan et al. (2022) research for (NLP) using several machine and deep learning algorithms. Those Algorithms are Decision Tree, Random Forest, LSTM, BiLSTM, BERT used for eight distinct categories. Dataset is comprising 6293 training and 1574 testing samples which are collected from Prothom-Alo and kaggle from 2019 to 2017 and divided into six predetermined crimes. As a result, automatic labeling data shows accuracy not greater than 90.15%. But the Decision Tree algorithm shows 91.83% accuracy which is higher than others [16]. Categorising news manually requires a lot of time and work. So, text Manually classifying news requires a lot of time and work. Text categorisation is therefore essential in the modern era, leading Main two authors Salehin and Alam (2021) to apply five machine learning classifiers and two neural networks to classify Bangla news articles from a variety of categories, including politics, sports, business, entertainment, and so on. As a result Neural Networks provide best results with 0.87% f1-score [17].

Yeasmin et al. (2021) research based on Bangla news articles. To classify Bangla news articles into two distinct datasets and five classes using data gathered from Prothom Alo and Kaggle from papers published in 2013 to 2019. These classes are technology, sports, international, economic, and entertainment. The corpus of this dataset contains 1425 documents. The results show Multi-layer Neural networks provide better performance with 94.99% accuracy [18]. Data was gathered from Facebook pages, blogs, newspapers, etc. by the main authors, Mahmud et al. (2022). And sort the nearly 2272 bits of data into the three categories of good, negative, and neutral. With 99.43% accuracy, CNN outperformed the competition [19]. Recurrent Neural Networks were used by Rahman and Chakraborty (2021) to describe a model, and they achieved 98.33% accuracy. Using a dataset of 40,000 divided in 12 categories of data which provide 16,889,327 words in total and 413,542 words are unique. They use 30% data as testing on this model [20]. Two deep learning models for categorization Bangla text were proposed by Mostaq Ahmed, Partha Chakraborty, and Tanupriya Choudhury in their article "A New Method for Classified Bangla Text Documents" published in 2012. A dataset with 10 sections, 16,889,327 words overall, and 13,810 sentences overall has been picked for this model. After that, include training data into the model's architecture design. As a result the accuracy and F1-score 97.72% for the deep learning recurrent neural network attention layer and recurrent neural network contain 86.56% accuracy [21].

# 3    Methodology

## 3.1    Decision Tree

A decision tree classifier is a popular machine learning algorithm used for solving both classification and regression problems offering a fast and useful solution for classifying instances in large datasets with a huge quantity of variables. It is a tree-structured alike classifier, where internal nodes represent the features of a dataset, branches represent the decision rules and each leaf node represents the outcome [22]. The decision tree algorithm works by recursively splitting the data into subsets based on the most significant feature at each node of the tree. This process continues until all or the majority of records have been classified under specific class labels. The objective of a decision tree classifier is to construct a tree that can accurately categorise new data points using the training data. Decision trees are explainable and can handle both numerical and categorical data, but may not generalise well to unseen data.

$$Gini(D) = 1 - \sum_{i=1}^{K} (p_i)^2 \qquad (1)$$

$$Gini(D, A) = \frac{|D1|}{|D|} \cdot Gini(D1) + \frac{|D2|}{|D|} \cdot Gini(D2) \qquad (2)$$

In Eq. 2, the Gini index of D given that partitioning by A is the weighted sum of the Gini indices of D1 and D2, where the weights are the relative sizes of D1 and D2.

In this study, the "gini" criterion was used which minimizes the Gini impurity and min_samples_split was set as 2. Additionally, nodes were expanded until all leaves contained less than min_samples_split samples.

## 3.2 Naïve Bayes

Naïve Bayes is a probabilistic ML algorithm which is used for classification tasks. It considers that all the features are conditionally independent where the class is given and simplifies the probabilistic calculation. It works based on Bayes' theorem (Eq. 3) for calculating the posterior probability of each class given the input features and then selects the class with the highest probability. Naive Bayes classifiers are highly scalable and can handle large datasets with high dimensional feature spaces efficiently. They are particularly useful when dealing with text classification and spam filtering tasks [23]. In this experiment, we have used Multinomial Naïve Bayes(MNB) classifier. In MNB the data is multinomial distributed basically depending upon the frequency of each word in every text or document. It's primarily used in document classification problems like classifying a particular document into categories such as Sport, Politics, Economy, Education, etc.

$$P(y|X) = \frac{P(X_1, X_2, ..., X_n|y) \cdot P(y)}{P(X_1, X_2, ..., X_n)} \qquad (3)$$

Here in Eq. 3, P(y|X) Probability of a data point being in class y given features X. P(X1, X2, ..., Xn|y) Likelihood of observing features X1, X2, ..., Xn if the class is y. p(y) Prior probability of class y. P(X1, X2, ..., Xn) Probability of observing features X1, X2, ..., Xn without considering a specific class, used for normalization.

In this study, the Additive smoothing parameter was 1.0 and the prior probabilities were estimated from the training data.

### 3.3    Neural Network

Dense Neural Network is a type of artificial neural network architecture where every neuron in a given layer is connected to every neuron in the subsequent layer. It is Identified by densely connected layers where information flows in a Forward propagation manner from the input layer to the output layer. Each neuron applies a weighted sum of inputs followed by an activation function to produce an output. Dense Neural Networks are effective for capturing complex patterns in data but can be prone to overfitting and require careful tuning of hyperparameters (Fig. 1).

**Fig. 1.** Neural Network Model Architecture.

In Eq. 4 $X_i$ represents the i-th element of the input vector X, $W_i$ represents the i-th element of the weight vector W and b represents bais term.

$$y = f\left(\sum_{i=1}^{n} w_i \cdot x_i + b\right) \qquad (4)$$

In this model, we have used three hidden layers which contains 256,128 and 64 neurons respectively. In the hidden layer we have use "relu" as an activation function and the output layer contains 8 neurons with a softmax activation function. Adam was used as an optimizer with a learning rate of 0.001 and categorical_crossentropy was used as a loss function.

### 3.4    Proposed Hybrid RNN

A hybrid Recurrent Neural Network (RNN) comprising Bi-directional Long-Short Term Memory (LSTM) and Bi-directional Gated Recurrent Unit (GRU) layers is a powerful architecture commonly used for sequential data modelling tasks. This model is particularly useful when dealing with time series, language translation, text classification and speech recognition applications. The model architecture comprises the embedding input layer, the Bi-directional LSTM layer, the Bi-directional GRU layer, a flatten layer, a dense layer and the output layer. The Bi-directional LSTM combs through the sequential data across both timelines, while the Bi-directional GRU identifies short-term dependencies within the input data. The flatten or pooling layer is then used to simplify and reduce the number of parameters in the model before the output layer. The model is trained through iterative epochs and the optimal number of epochs is determined through the validation accuracy (Fig. 2).

**Fig. 2.** Proposed Hybrid RNN Model Architecture.

The model architecture begins with a Sequential layer, followed by a Flatten layer to simplify the data. Next, a Dense layer comprising 128 units employs the ReLU activation function, optimizing feature extraction. Concluding the model, another Dense layer with 8 units and a softmax activation function ensures proper value normalization.

# 4   Experimental Results

## 4.1   Experimental Setup

The experimental setup involved conducting experiments on Google Colab. The programming language used in the experiment was Python v3.11.4, along with four libraries - Numpy, Pandas and SKlearn beside we use popular machine learning and deep learning framework TensorFlow v2.13.0. For training purposes, e took a batch size of 128 for 20 epochs.

## 4.2   Data-Set Collection and Prepossessing

**Data Collection.** In this study, we have collected data from different news sources such as 'Prothom Alo', 'Kaler Kontho' and 'Bangladesh Protidin'. The news articles were classified into eight categories: Bangladesh, sports, international, entertainment, economy, opinion, technology and lifestyle.

**Data Preprocessing.** To train our model, we have used a dataset from Kaggle, a popular platform for machine learning datasets. We chose twelve categories, including Bangladesh, sports, international, entertainment, economy, opinion, technology, and lifestyle to train our model. The dataset contained 378,471 news articles, from which we randomly undersampled 40,000 news articles to avoid overfitting and ensure each category had an equal representation of 5,000 articles, which helped the model learn the distribution of the data more accurately.

For our evaluations, we collected news article data from various online news portals. Our dataset included a diverse range of articles encompassing eight categories - Bangladesh, sports, international, entertainment, economy, opinion, technology, and lifestyle - with a total of 800 examples across all categories.

1. **Drop unnecessary column:** In this dataset, there are some unnecessary columns like author, published date, modification date, tag, comment count, URL which do not carry significant meaning. So we drop those columns from the whole dataset.

2. **Punctuation, special character removal:** Punctuation like (; : " , ? !, etc.) and special character like @, , \$, %, , (, *, ), etc. that is not important for classification are removed from the whole Dataset.

3. **Stop word removal:** Stop words are common words that are often removed from text data before any analysis or processing because those words do not carry significant meaning. To enhance the accuracy of text classification, it is necessary to remove stop words. In this regard, we have collected Bangla stop words and removed those words from our dataset. Additionally, there are several single letter words that hold little relevance to the text classification task. Hence, we have removed these single-letter words in the prepossessing phase. As a result, we could eliminate the redundant words and improve the accuracy of our classification model.

4. **Feature Selection and Extraction:** We have used TF-IDF model for converting string features into numerical features for performing the mathematical operation [24]. TF-IDF stands for Term Frequency-Inverse Document Frequency. It is a numerical statistic that reflects how important a word is to a document in a collection or corpus of documents. Term Frequency (TF) measures how frequently a term (word) appears in a document. Inverse Document Frequency (IDF) measures how important a term is to the entire collection or corpus of documents. Term frequency(TF) and Inverse Document Frequency (IDF) can be defined as:

$$TF = \frac{No.\,of\,time\,word\,appear\,in\,the\,doc.}{Total\,no.\,of\,word\,in\,the\,documents}$$

$$IDF = \log \frac{No.\,of\,Documents}{No\,of\,Docs\,in\,which\,the\,word\,appears}$$

TF-IDF is a multiplication between TF and IDF value. It reduces the importance of the common word that is used in a different document. And only take important words that are used in classification. For this data-set, the best result is found by considering maximum document frequency of 0.1, which means the words that are on the 10% document or more than that. Here, 43279 most frequent words are used as a feature vector.

5. **Categorical encoding:** techniques include label encoding and one-hot encoding. Label encoding assigns a unique integer to each label based on alphabetical ordering, while one-hot encoding represents each category as a one-hot vector, where only one bit is hot or true at a time. Table III provides an example of a one-hot encoding for a dataset with two categories. In machine learning algorithms, label encoding is typically used to encode categories, while one-hot encoding is more commonly employed for multi-layer dense neural networks.

**Splitting Data-Set into Training and Testing Set.** After successfully balancing the dataset, the dataset split into train and test datasets. The dataset is divided into 4:1. That means 80% of data from the datasets are used for training, and the rest 20% is considered as testing (Fig. 3).

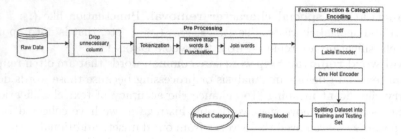

**Fig. 3.** Overview of whole process.

## 4.3   Results and Analysis

See Table 1.

**Table 1.** Model evaluation analysis.

| Model | Accuracy | Precision | Recall | F1-Score |
|---|---|---|---|---|
| Decision Tree | 0.68 | 0.68 | 0.68 | 0.68 |
| Naïve Bayes | 0.84 | 0.85 | 0.84 | 0.84 |
| Dense Neural Network | 0.90 | 0.90 | 0.90 | 0.90 |
| Hybrid Recurrent Neural Networks | 0.89 | 0.89 | 0.89 | 0.89 |

**Evaluation Matrices.** In this section, the performances of the model are analyzed on different machine learning algorithms and neural network for the test dataset. To measure the performance of the model we have used different evaluation matrices such as accuracy, precision, recall, F1 score and confusion matrix. Accuracy calculates the percentage of correctly predicted instances. it is calculated by the total number of correctly predicted samples divided by the total number of samples.

$$Accuracy = \frac{TP + TN}{TP + TN + FP + FN}$$

And precision calculates the true positive predictions and overall positive predictions, recall measures the true positive predictions out of all actual positive instances.

$$Precision = \frac{TP}{TP + FP}$$

$$Recall = \frac{TP}{TP + FN}$$

The F1 score is a popular metric to use for classification models as it provides robust results for both balanced and imbalanced datasets, unlike accuracy It is defined as the harmonic mean of the model's precision and recall

$$F1 - Score = \frac{2 * (\, Precision \, * \, Recall \,),}{Precision \, + \, Recall}$$

Finally, A confusion matrix is a table used to evaluate the performance of a classification model, by comparing the predicted labels of the model with the actual labels of the data. It can be misleading to rely only on classification accuracy, particularly in situations where there are not an equal number of observations in each class. The matrix presents the quantities of true positives, false positives, true negatives, and false negatives that correspond to the model's predictions, giving a thorough overview.

**Decision Tree Model.** This Decision Tree model also gives different accuracy rates from the different 8 categories and the average accuracy of the model is 68%, along with Sports having the highest accuracy (78%), and Lifestyle having the lowest (59%). A confusion matrix is shown below (Fig. 4), which may demonstrate which class achieves better outcomes.

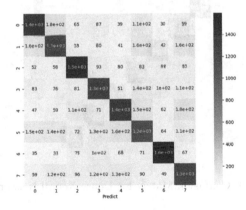

**Fig. 4.** Confusion Matrix for Decision Tree Model.

**Naïve Bayes Classifier.** The accuracy rate of the Naïve Bayes model is (84%). Opinion has the lowest accuracy rate in the model, while the Sports category has a higher accuracy rate. A confusion matrix is shown below (Fig. 5), which may demonstrate which class achieves better outcomes.

**Confusion Matrix for Dense Neural Network.** It was found that the Dense Neural Network (DNN) achieved the highest accuracy among the various models

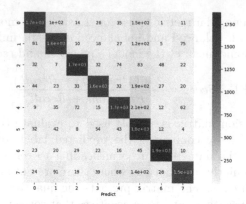

**Fig. 5.** Confusion Matrix for Naïve Bayes Classifier

tested, reaching an impressive accuracy rate of (91.71%). The confusion matrix is shown below (Fig. 6) which can show which class has a better result.

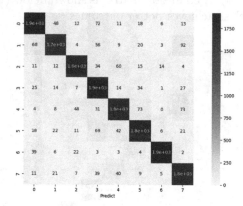

**Fig. 6.** Confusion Matrix for Dense Neural Network

**Hybrid RNN.** It was found that the Hybrid Recurrent Neural Network (HRNN) achieved less accuracy than the Neural Network Model, reaching an impressive accuracy rate of (89.0%). We got height precession for economy class. The precession was (94%) for this class. The confusion matrix is shown below (Fig. 7) which can show which class has a better result.

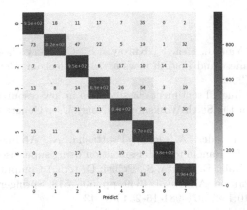

**Fig. 7.** Confusion Matrix for Dense Neural Network

Here Dense Neural network performs the HRNN because it is more complex then Nense Neural network. Moreover, have more parameters and can be sensitive to overfitting. However, It's possible that the ANN architecture used in this experiment was more suitable or better tuned for the specific dataset.

## 5  Conclusion

In conclusion, the Newspaper Classification model presented in this paper is a promising solution for automatically categorizing news articles into different categories using NLP techniques. Through experimentation and evaluation, we have gained valuable insights into the performance and capabilities of these different methods. The model was evaluated using multiple machine learning algorithms and the results demonstrated that Naïve Bayes Classifier performs better than the other machine learning algorithms in terms of accuracy and precision and Neural Network performs better than all other machine learning models. This model has the potential to be a valuable tool for simplifying the categorization of news articles, enabling users to quickly and efficiently access relevant information. However, the field of text classification is evolving day by day. Also, there are still many opportunities for improvement and additional research. Future research may involve refining existing algorithms, addressing imbalanced datasets, and exploring hybrid models to leverage the strengths of different approaches. Overall, this work contributes to the advancement of Bangla news classification and provides a solid groundwork for future study in this field. With the continued efforts and advancements in this field, we can expect improved accuracy and efficiency in automatically categorizing news articles and individualized news recommendations for Bangla-speaking audience

# References

1. Bracewell, D.B., Yan, J., Ren, F., Kuroiwa, S.: Category classification and topic discovery of Japanese and English news articles. Electron. Notes Theor. Comput. Sci. **225**, 51–65 (2009)
2. Sazzed, S.: Cross-lingual sentiment classification in low-resource Bengali language. In: Proceedings of the Sixth Workshop on Noisy User-Generated Text (W-NUT), pp. 50–60 (2020)
3. Kowsher, M., Tahabilder, A., Jahan Prottasha, N., Abdur-Rakib, M., Moyez Uddin, M., Saha, P.: Bangla topic classification using supervised learning. In: Das, A.K., Nayak, J., Naik, B., Dutta, S., Pelusi, D. (eds.) Computational Intelligence in Pattern Recognition. AISC, vol. 1349, pp. 505–518. Springer, Singapore (2022). https://doi.org/10.1007/978-981-16-2543-5_43
4. Ho, T.K.: The random subspace method for constructing decision forests. IEEE Trans. Pattern Anal. Mach. Intell. **20**(8), 832–844 (1998)
5. Ho, T.K.: Random decision forests. In: Proceedings of 3rd International Conference on Document Analysis and Recognition, vol. 1, pp. 278–282. IEEE (1995)
6. Amin, R., Sworna, N.S., Hossain, N.: Multiclass classification for Bangla news tags with parallel CNN using word level data augmentation. In: IEEE Region 10 Symposium (TENSYMP), pp. 174–177. IEEE (2020)
7. Al Imran, A., Wahid, Z., Ahmed, T.: BNnet: a deep neural network for the identification of satire and fake Bangla news. In: Chellappan, S., Choo, K.-K.R., Phan, N.H. (eds.) CSoNet 2020. LNCS, vol. 12575, pp. 464–475. Springer, Cham (2020). https://doi.org/10.1007/978-3-030-66046-8_38
8. Aurpa, T.T., Sadik, R., Ahmed, M.S.: Abusive Bangla comments detection on Facebook using transformer-based deep learning models. Soc. Netw. Anal. Mining **12**(1), 24 (2022)
9. Nazi, Z.A.: Bangla newspaper dataset (2020). https://www.kaggle.com/datasets/furcifer/bangla-newspaper-dataset
10. Chowdhury, P., Eumi, E.M., Sarkar, O., Ahamed, M.F.: Bangla news classification using GloVe vectorization, LSTM, and CNN. In: Arefin, M.S., Kaiser, M.S., Bandyopadhyay, A., Ahad, M.A.R., Ray, K. (eds.) Proceedings of the International Conference on Big Data, IoT, and Machine Learning. LNDECT, vol. 95, pp. 723–731. Springer, Singapore (2022). https://doi.org/10.1007/978-981-16-6636-0_54
11. Maisha, S.J., Nafisa, N., Masum, A.K.M.: Supervised machine learning algorithms for sentiment analysis of Bangla newspaper. Int. J. Innov. Comput. **11**(2), 15–23 (2021)
12. Disayiram, N., Rupasingha, R.A.H.M.: A comparative study of classifying English news articles using machine learning algorithms. In: Trends in Electrical, Electronics, Computer Engineering Conference (TEECCON), pp. 50–55. IEEE (2022)
13. Wahid, Z., Imran, A.A., Rifat, M.R.I.: Bnnetxtreme: an enhanced methodology for Bangla fake news detection online. In: Dinh, T.N., Li, M. (eds.) CSoNet 2022. LNCS, vol. 13831, pp. 157–166. Springer, Cham (2022). https://doi.org/10.1007/978-3-031-26303-3_14
14. Rahman, S., Mithila, S.K., Akther, A., Alam, K.M.: An empirical study of machine learning-based Bangla news classification methods. In: 12th International Conference on Computing Communication and Networking Technologies (ICCCNT), pp. 1–6. IEEE (2021)

15. Das, R.K., Sammi, S.S., Kobra, K., Ajmain, M.R., Khushbu, S.A., Noori, S.R.H.: Analysis of Bangla transformation of sentences using machine learning. In: Troiano, L., Vaccaro, A., Kesswani, N., Díaz Rodriguez, I., Brigui, I., Pastor-Escuredo, D. (eds.) ICDLAIR 2022. LNNS, vol. 670, pp. 36–52. Springer, Cham (2023). https://doi.org/10.1007/978-3-031-30396-8_4

16. Khan, N., Islam, M.S., Chowdhury, F., Siham, A.S., Sakib, N.: Bengali crime news classification based on newspaper headlines using (NLP). In: 25th International Conference on Computer and Information Technology (ICCIT), pp. 194–199. IEEE (2022)

17. Salehin, K., Alam, M.K., Nabi, M.A., Ahmed, F., Ashraf, F.B.: A comparative study of different text classification approaches for Bangla news classification. In: 24th International Conference on Computer and Information Technology (ICCIT), pp. 1–6. IEEE (2021)

18. Yeasmin, S., Kuri, R., Rana, A.M.H., Uddin, A., Pathan, A.S.U., Riaz, H.: Multi-category Bangla news classification using machine learning classifiers and multi-layer dense neural network. Int. J. Adv. Comput. Sci. Appl. 12(5) (2021)

19. Mahmud, M.S., Islam, M.T., Bonny, A.J., Shorna, R.K., Omi, J.H., Rahman, M.S.: Deep learning based sentiment analysis from Bangla text using glove word embedding along with convolutional neural network. In: 13th International Conference on Computing Communication and Networking Technologies (ICCCNT), pp. 1–6. IEEE (2022)

20. Rahman, S., Chakraborty, P.: Bangla document classification using deep recurrent neural network with BiLSTM. In: Prateek, M., Singh, T.P., Choudhury, T., Pandey, H.M., Gia Nhu, N. (eds.) Proceedings of International Conference on Machine Intelligence and Data Science Applications. AIS, pp. 507–519. Springer, Singapore (2021). https://doi.org/10.1007/978-981-33-4087-9_43

21. Ahmed, M., Chakraborty, P., Choudhury, T.: Bangla document categorization using deep RNN model with attention mechanism. In: Tavares, J.M.R.S., Dutta, P., Dutta, S., Samanta, D. (eds.) Cyber Intelligence and Information Retrieval. LNNS, vol. 291, pp. 137–147. Springer, Singapore (2022). https://doi.org/10.1007/978-981-16-4284-5_13

22. Qorib, M., Oladunni, T., Denis, M., Ososanya, E., Cotae, P.: Covid-19 vaccine hesitancy: text mining, sentiment analysis and machine learning on covid-19 vaccination twitter dataset. Expert Syst. Appl. 212, 118715 (2023)

23. Sreedhar, L., Kavya, B., Kiran, H.S., Bhaskar, C.V.: Email spam detection using machine learning algorithms. Network 52(4) (2023)

24. Sunagar, P., Kanavalli, A., Shetty, N.D.: Feature extraction and selection techniques for text classification: a survey. Int. J. Adv. Res. Eng. Technol. 11(12), 2871–2881 (2020)

# Bangla Social Media Cyberbullying Detection Using Deep Learning

Anika Tasnim Rodela[1], Huu-Hoa Nguyen[2], Dewan Md. Farid[3(✉)],
and Mohammad Nurul Huda[3]

[1] Department of Information and Communication Technology, Bangladesh
University of Professionals, Mirpur Cantonment, Dhaka, Bangladesh
[2] College of Information and Communication Technology, Can Tho University,
3/2 Street, Ninh Kieu District, Can Tho City, Vietnam
[3] Department of Computer Science and Engineering, United International
University, United City, Madani Avenue, Badda, Dhaka 1212, Bangladesh
{anika,dewanfarid,mnh}@cse.uiu.ac.bd, nhhoa@ctu.edu.vn
https://cse.uiu.ac.bd/profiles/dewanfarid/

**Abstract.** The growth of social media over the past decade has been
nothing short of phenomenal. An exponential increase has been seen on
platforms like Facebook, Twitter, Instagram, LinkedIn, and YouTube,
in their user base, accumulating billions of active users worldwide. Technology advancements, the ubiquitous use of smartphones, and the innate
human desire for connection don't always contribute constructively; they
might additionally end up in spreading violence in the form of bullying
of others which is known as cyberbullying. As a result of cyberbullying, the victims will frequently experience anxiety, depression, and other
mental diseases which can even result in suicide. Therefore the extensive
need of detecting and controlling cyberbullying motivated us to automate
this process. In this paper, we have introduced an approach to detect
cyberbullying from social media data by using deep learning models. We
have used Long Short Term Memory (LSTM), Gated Recurrent Unit
(GRU), and Convolutional Neural Network in the proposed hybrid approach along with word embedding technique called fastText. We achieved
an accuracy of 91.63% with the proposed model by using a publicly available dataset containing 16,073 samples which outperformed all the state
of the art models.

**Keywords:** Cyberbullying · Deep Learning · Embedding

## 1 Introduction

In the modern age of technology, cyberbullying, a widespread form of online
harassment, has emerged as an alarming issue. It describes the practice of using
social media platforms or any electronic media to harass, bully, or threaten others. Cyberbullying can have serious negative effects on the victims, including
psychological harm, depression, anxiety, sometimes leading them to self-harm,

N. Thai-Nghe et al. (Eds.): ISDS 2023, CCIS 1949, pp. 170–184, 2024.
https://doi.org/10.1007/978-981-99-7649-2_13

and even catastrophic outcomes like suicide. From 1997 people were using social media and now there are 4.26 billion users worldwide, which is more than half of the total world population [16]. Cyberbullying is continuously being proliferated with the growth of social media, necessitating the development of efficient detection and prevention techniques. According to a report in [31], 73% of students stated that they were bullied at least once in their lifetime and about 44% faced the same in the last 30 days. These concerning statistics highlight the pressing need for proactive measures to stop cyberbullying and defend vulnerable individuals.

While cyberbullying exists in all languages, there are a lot of bullying occurrences in the native languages as well. Bangla or Bengali is the sixth most spoken language in the world and currently 234 million native speakers are using it [21]. Bangladesh is a developing country and almost all the people of the country speak Bangla as their first language. With an annual growth rate of 10.1% in 2021–2022, Bangladesh had 50.3 million active social media users on a monthly basis as mentioned in [1]. Especially since the COVID 19 pandemic people are even more indulged in social media and this leverages different bullying situations or occurrences. Hence it has also become mandatory to introduce proper techniques to detect such cyberbullying occurrences in the native language like Bangla. It somewhat requires more strenuous efforts to detect cyberbullying in a native language such as Bangla rather than international languages such as English because the preprocessing, stop word removal, using proper stemmer etc. are different and more challenging for Bangla for the lack of proper resources. According to Kumar and Sachdeva [20] the linguistic challenges are also contributed by the cultural diversity, using hash-tags on trending topics which is often region specific, unconventional use of typographical resources including capital letters, punctuation, and emojis, as well as the accessibility of keyboards in one's native language. All this information points to the fact that detection of such harassment or bullying in the native language is demanding and at the same time more challenging.

Automatic detection and classification of cyberbullying from social media data is a task that requires both natural language understanding and generic text classification [20]. Recently deep learning (DL) models are being considered by researchers as an attempt to perform automated text classification and natural language processing. In particular, recurrent neural networks (RNNs) such as long short-term memory (LSTM) networks have been utilised to analyse the sequential connections or patterns and temporal dynamics in social media posts. To identify instances of cyberbullying, these models can effectively understand the proper context, linguistic sequences and sentiment. Convolutional neural networks (CNNs) have been used as well to extract useful information from text by taking into account the relationships and underlying structure found within sentences and phrases. The accuracy of cyberbullying detection systems has been improved by combining these deep learning models with approaches like word embeddings and attention mechanisms.

Machine learning and deep learning are being used frequently in classification tasks such as cyberbullying. There are several works on this topic, such as, toxic comments detection using supervised learning [6], abusive Bangla comments detection by using transformer-based models [5], cyberbullying detection from deep neural networks [3], hate speech detection [2], Classification Benchmarks based on Multichannel Convolutional-LSTM Network [19], Multi Labeled Bengali Toxic Comments Classification [7] etc. In spite of existing a good number of research works, most of them lack in some areas. Most of the research works lacked in introducing datasets with proper amount of data and the datasets being unbalanced. Moreover, some of them used very complex models which degraded the efficiency of the system. Addressing the issues, we have conducted our work on a large dataset of 16073 samples and we proposed a model which is comparatively easily and lightly implemented and resulted in much higher accuracy.

The rapid recognition of cyberbullying instances is essential in order to lessen its adverse effects. However, it is challenging and time-consuming to manually monitor and identify cyberbullying instances from the large volume of social media data. Hence, the development of automated processes utilising modern technology like deep learning has emerged as a possible remedy. In this regard we have performed classification of cyberbullying on a dataset containing 16073 instances where 8488 instances were cyberbullying and the other 7585 were non cyberbullying instances. We have used a hybrid approach of LSTM-GRU-CNN and corporate with three different embeddings: (1) fastText, (2) GloVe, and (3) Word2Vec. The proposed model achieved an accuracy of 91.63% which outperformned all the other models from our study and the recent studies.

The following parts of the paper go as follows: Sect. 2 represents the Literature Review stating and elaborating a comparative study on all the state of the art models. Section 3 discusses about the utilisation of deep learning, the dataset's description and our proposed model in details. Section 4 is the summary of the experimental results and Sect. 5 concludes the whole study.

## 2   Literature Review

There are several research works conducted related to this work of cyberbullying detection and classification. Various research works are focused on various factors such as, monolingual or multilingual content, binary or multi-class classification and machine learning or deep learning or hybrid models.

### 2.1   Monolingual and Multilingual

Based on the types of lingual contents, we can divide the research works into two categories such as, monolingual and multilingual. There have been many studies conducted on monolingual content of natural language processing such as, English language for hate speech or cyberbullying detection or any text classification tasks [17,23,25,32]. Other than English, Ptaszynski et al. conducted cyberbullying detection in Polish language [28], Gordeev [13] reported verbal aggression detection in Russian and English, Ibrohim et al. [18] and Pratiwi et al. [27] conducted hate speech and abusive language detection from Indonesian tweets. For multilingual contents, there are diverse existing research works as well. Haider et al. [14,15] introduced a multilingual cyberbullying detection system using machine learning and deep learning techniques and they performed the validation on Facebook and Twitter data in Arabic language. Another multilingual cyberbullying detection system was proposed by Pawar et al. [26] in two Indian languages namely: Hindi and Marathi. Arreerard et al. [4] proposed a machine learning model for classification of defamatory Facebook comments in the Thai language. Kumar and Sachdeva [20] developed a similar system to detect cyberbullying in Hindi-English code-mixing data using deep neural networks and transfer learning for cyberbullying detection.

## 2.2 Binary-Multilabel and Machine Learning-Deep Learning Classification

All the researchers in this domain formulated cyberbullying detection as a text classification problem. Based on the type of classification, in most papers we have found mostly binary classification while some incorporated multi-class classification as well. Chakraborty and Seddiqui [9] detected threat and abusive language in Bangla by using both machine learning and deep learning models while formulating the problem as a binary classification problem. They achieved an accuracy of 78% when using SVM with linear kernel. Malik et al. [24]performed toxic speech detection using machine learning and deep learning models along with word embedding techniques such as, BERT and fasttext. For machine learning models they used Logistic Regression, Support Vector Machine, Decision Tree, Random Forest, XGBoost and for deep learning models they used Convolutional Neural Network (CNN), Multi Layer Perceptron, Long Short Term Memory (LSTM) They achieved 82% accuracy for binary classification using CNN. To categorise Bengali comments as toxic and non-toxic, Banik et al. [6] employed both supervised machine learning models such Naive Bayes, Support Vector Machine, Logistic Regression, and deep learning models like Convolutional Neural Network, Long Short Term Memory. For binary classification, Convolutional Neural Networks had the highest accuracy (95.30%). They utilised word2vec as a word embedding technique.

Das et al. [10] to detect Bengali hate speech employed Convolutional Neural Networks with Long Short Term Memory (CNN-LSTM), Convolutional Neural Networks with Gated Recurrent Unit (CNN-GRU), and attention-based Convolutional Neural Network while achieving 77% accuracy. To detect cyberbullying Ahmed et al. [3] used Convolutional Neural Network with Long Short Term Memory (CNN-LSTM) and gained 87.91% accuracy for binary classification and 85% accuracy for multi-class classification. Belal et al. [7] categorised Bangla toxic comments using a deep learning based pipeline. They used BERT as the word embedding technique and performed binary classification using Long

Short Term Memory (LSTM) achieving 89.42% accuracy. For multi-label classification, the authors used a combination of Convolutional Neural Network and Bi-directional Long Short Term Memory (CNNBiLSTM) with attention mechanism and achieved 78.92% accuracy. Using Multichannel Convolutional-Long Short Term Memory (MConv-LSTM) with BangFastText, Word2Vec, and GloVe for word embedding, Karim et al. [19] identified Bengali hate speech. Their model received a 92.30% F1-score using BangFastText embedding.

Based on the models used, we can divide the research works into three different categories such as, models based on machine learning algorithms, deep learning algorithms and a hybrid model where simultaneously multiple approaches are used. The authors in [8,9,29,30] used machine learning based approaches to detect cyberbullying or hate comments. Deep learning algorithms such as Convolutional Neural Networks (CNNs), Long Short Term Memory Networks (LSTMs), Recurrent Neural Networks (RNNs) have been used by many researchers [7,9,24]. Many researchers are also exploring hybrid models to get better performance in this text classification problem. The authors in [6,7,12,19] used Convolutional Neural Networks with Long Short Term Memory (CNN-LSTM), Convolutional Neural Networks with Gated Recurrent Unit (CNN-GRU), Convolutional Neural Network and Bi-directional Long Short Term Memory (CNNBiLSTM) with attention mechanism to build their models for this classification problem of cyberbullying detection.

## 3 Learning with Deep Learning

Deep learning has transformed artificial intelligence and data analysis through delivering capabilities that were unattainable for comprehending and analysing complex data. Deep learning algorithms have demonstrated outstanding results in applications such as image and speech recognition, natural language processing, and predictive analytics by applying artificial neural networks with multiple layers. Deep learning techniques enable machines to learn from enormous datasets and make intelligent choices by mimicking the complex structure and activity of the human brain. These algorithms utilise the strength of neural networks, which are constructed from interconnected "neurons" or nodes arranged in layers. Deep learning algorithms can derive increasingly abstract representations of data through multiple layers, enabling advanced and deeper understanding.

Most recently Natural language processing (NLP) tasks such as text classification activities have been significantly influenced by deep learning, evolving the field of study and resulting in advancements in numerous areas of language understanding, development, and analysis. Recurrent neural networks (RNNs) including Long Short-Term Memory (LSTM) and Gated Recurrent Unit (GRU), have emerged as prevalent deep learning models in natural language processing. These architectural frameworks are made to manage sequential data and recognise dependencies in text composed of natural language. For text categorisation problems, deep learning models like Convolutional Neural Networks (CNNs), Recurrent Neural Networks (RNNs) (including LSTM and GRU), and

Transformer-based models have been extensively used currently. Deep learning based word embedding techniques are largely used as well. Deep learning models often leverage word embeddings like Word2Vec or GloVe to represent words as dense vectors in a continuous space. The model can learn accurate representations of words due to these embeddings, which recognise semantic connections as well as contextual information. For sequential data analysis, Recurrent Neural Network (RNN) is widely used. Long Short-Term Memory (LSTM) and Gated Recurrent Units (GRU) are the variants of RNN.

## 3.1 Long Short Term Memory (LSTM)

In sequential data to understand the context and long term dependencies LSTM [11] is largely applied. It addresses the vanishing gradient problem which allows it to learn better of long term dependencies in sequential data. Three gating mechanisms-the input gate, forget gate, and output gate-as well as memory cells are introduced by LSTM. By controlling the information flow, these gates allow the model to selectively recall or forget previous knowledge according to the present context. Due to this feature, LSTM performs particularly well in tasks involving sequential data, including speech recognition, sentiment analysis, and machine translation.

## 3.2 Gated Recurrent Unit (GRU)

GRU addresses some limitations of the traditional RNNs as well such as vanishing gradients and failure to capture long term dependencies. GRU follows gating mechanisms to control the flow of information- Update gate and reset gate. Update gate is for combining new information and determining to which context the previous hidden state should be updated. Reset gate is to control forgetting or resetting the previous hidden state. GRU keeps track of a hidden state that serves as the network's memory. Using the gating mechanisms, the current input and the prior hidden state are combined to update the hidden state at each time step. It collects essential information from previous time steps and encodes it into the hidden state that is currently in effect.

## 3.3 Convolutional Neural Network (CNN)

CNN is a deep learning architecture that is typically employed for signal and image processing tasks. Through convolutional and pooling layers, it is supposed to automatically generate hierarchical representations of data. CNNs excelled at tasks including image classification, object identification, and image segmentation because they use convolutional filters to identify local patterns and spatial dependencies. The spatial dimensions are reduced with the help of the pooling layers, while the key features are kept intact. By considering the input as a one-dimensional signal, CNNs have been further developed to handle sequential data, such as text categorisation and speech recognition.

### 3.4   Proposed Method

In this research, we have used Bangla social media text for cyberbullying detection. Therefore, following the most common and standard convention, we have pre-processed the raw text. The dataset of almost 16000 instances was used for this purpose. The properly annotated and labeled dataset was set for preprocessing and for this purpose at first, tokenization was employed to extract words from the texts. This step was completed by leveraging Natural Language Toolkit tokeniser. The next step was to remove all the punctuations from the text. Then we used a stemmer to convert the tokens to the basic form of each word by using Bangla Natural Language Processing Toolkit (BNLTK) stemmer. From the stem or the base words, it is convenient to convert each sentence into a numeric form. We have used a pool of 746 Bangla words as stop words which added some values to these texts. In the next procedure, all the texts were ensured to be of the same length which is 140 characters and this was our input sequence. At this point word embedding techniques such as GloVe, fastText and Word2Vec are used in order to convert the sentences into numeric vectors. These word embedding techniques work as vectorisers. For our Bangla dataset we used fastText which is trained by Wikipedia data. We have compared the performance with GloVe and Word2Vec as well (Fig. 1).

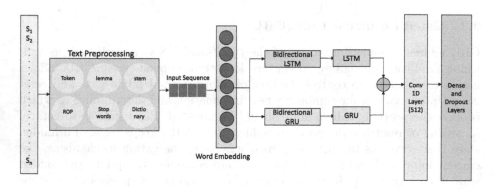

**Fig. 1.** Proposed model for cyberbullying detection.

In natural language processing (NLP), eliminating stop words and punctuation can alter the meaning of sentences or words, which can affect the classification tasks. Stop words like "and", "the", "is", etc. are often removed since they are common and don't have a specific or significant meaning, whereas punctuation can give significant clues about the structure and meaning of a sentence. To hold up the meaning of the sentences even after preprocessing we used lemmatization or stemming. Stemming can be used to reduce words to their root or base forms as opposed to fully eliminating them. This helps in retaining some of the semantic meaning while still reducing dimensionality. For example, "eating" and "ate" can be reduced to "eat". Again, we have used custom stop word lists; these lists are made to keep domain-specific or contextually appropriate stop words rather than eliminating all stop words. Word2Vec, GloVe, and other pre-trained word embeddings, are trained on large corpora and capture the semantic connections between words. Even when stop words and punctuation are removed, they can understand the complex meanings of words and their context. Moreover, as a result of using hybrid Approaches (combining traditional NLP techniques with neural network models), the problem can also be solved. At first texts are preprocessed by removing stop words and punctuation and then using a neural network model is used to learn to capture the context and meaning from the remaining words. Finally, we also assessed the effects of the preprocessing decisions. It's important to compare the performance of the model with and without these preprocessing procedures because sometimes eliminating stop words and punctuation may be deteriorating to the task at hand (Fig. 2).

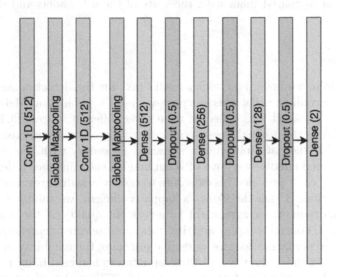

**Fig. 2.** CNN, Dense and Dropout Layers from methodological framework.

In our model, we have used a bidirectional LSTM layer and a bidirectional GRU layer of 200 neurons. Combining these two parallel layers, the output is fed into a conv1D layer which has 512 neurons. The combination of LSTM, GRU and CNN is used for capturing long term dependencies and extracting insightful features. After the CNN layer max pooling is used, another layer of Conv1D and maxpooling were embedded. Finally three dense layers having 512, 256 and 128 neurons were used where the dropout is 50%. ReLU activation function was used in every layer except for the last or the output layer. For the output layer Softmax activation function was used. A dense layer with two neurons was the output layer. After training for 30 epochs and setting the learning rate as 0.001 the best accuracy was achieved although at 3 epoch we achieved the highest accuracy after fine tuning our model. This proves the efficiency and lightness of our model in terms of both time and memory and other resources.

To train the model we selected 13000 instances among 16000 in the dataset by splitting the dataset in 80:20 ratio. For the test set the rest of the instances (3000 instances) were selected. The accuracy, sensitivity and positivity along with the correctly predicted values (both positive and negative) were calculated in order to evaluate the performance of the model. A heat map was used to show the distribution of the predicted results. The proposed model is compared to other state of the art models and a performance comparison is shown.

## 4    Experimental Results

We have evaluated the performance of the proposed model in the standard metrics as well as compared them with the state of the art models and embedding techniques.

### 4.1    Dataset

In this research, we have used social media text for the detection and classification of cyberbullying. For this purpose we used a publicly available dataset of Belal et al. [7] which was collected from three different sources [3,19,22] and was relabelled as the original ones were neither fully correct nor consistent. The authors labeled the data following a standard procedure with the guidance of three professional annotators. Two of them labeled each instance independently and the third expert provided his expertise to resolve any disagreements through proper discussions. Using the Cohen's kappa coefficient, we validated and evaluated the inter-annotator agreement to assess the quality of the annotations. By choosing 30 control samples and 100 randomly selected sentences that had already been expertly labeled, we were able to ensure the reliability of the annotators. Based on the evaluation, all three annotators had credibility ratings that were higher than 80%. The dataset contains 16, 073 social media texts in total of which 7, 585 are classified as non-cyberbullying and 8, 488 are classified as cyberbullying. Table 1 shows the distribution of cyberbullying instances by class.

### 4.2    Performance Comparison with Different Embeddings

Table 2 presents metrics and performance measures for the proposed models with three embedding techniques: fastText and Word2Vec and GloVe. These models were evaluated using our dataset, and their performance was assessed based on various metrics. The first metric, accuracy, measures the overall correctness of the models' predictions. In this case, fastText achieved an accuracy of 91.63%, higher than Word2Vec's accuracy of 90.42% and GloVe's 90.67%. These values indicate that all models were able to make accurate predictions on the given dataset. FastText achieved a sensitivity of 0.9423, much higher than Word2Vec's sensitivity of 0.9146 and GloVe's 0.9170. This suggests that fastText was slightly

**Table 1.** Various properties of dataset.

| Properties | Values |
|---|---|
| Total instances | 16073 |
| Cyberbullying text | 8488 |
| Non cyberbullying text | 7585 |
| Training instances | 13000(0:13000) |
| Testing instances | 3000(0:3000) |
| Number of words before preprocessing | 34340 |
| Number of words after preprocessing | 32009 |

better at correctly identifying positive instances. Specificity, or the true negative rate, measures the proportion of actual negative instances correctly identified by the models. The models had similar specificity values, with fastText at 0.8873, Word2Vec at 0.8926 and GloVe at 0.8952. These values indicate that the models were effective in correctly identifying negative instances. The table also provides two additional metrics, $P_0$ and $P_1$. It appears that the scores associated with the positive predictive value is ($P_1$) and negative predictive valus is ($P_0$). These findings provide insights into the models' performance and can inform the reader about the effectiveness of these approaches in the given research context.

**Table 2.** Performance evaluation with different embeddings.

| Metrics | Proposed Model(fastText) | Word2Vec | GloVe |
|---|---|---|---|
| Accuracy | **0.9163** | 0.9042 | 0.9067 |
| True Positive | **1600** | 1553 | 1557 |
| False Positive | 171 | 163 | 159 |
| False Negative | 98 | 145 | 141 |
| True Negative | 1346 | 1354 | **1358** |
| Sensitivity | **0.9423** | 0.9146 | 0.9170 |
| Specificity | 0.8873 | 0.8926 | **0.8952** |
| $P_0$ | **0.9321** | 0.9033 | 0.9059 |
| $P_1$ | 0.9034 | 0.9050 | **0.9073** |

**Table 3.** Performance evaluation with models.

| Model | Accuracy | Precision | Recall | F1-score |
|---|---|---|---|---|
| LSTM+GRU+CNN+FastText(Proposed Model) | **91.63%** | **0.9170** | **0.9163** | **0.9162** |
| LSTM + BERT embedding | 89.42% | 0.89 | 0.89 | 0.89 |
| MConv-LSTM + BERT embedding 8 | 87.87% | 0.88 | 0.88 | 0.88 |
| Bangla BERT fine-tune | 88.57% | 0.89 | 0.88 | 0.89 |

### 4.3    Performance Comparison with Different Related Models

The proposed model, which combines LSTM, GRU, and CNN with fastText embedding achieved the highest accuracy of 91.63% as mentioned in Table 3. The results of the three baseline methods have been obtained by rerunning those models on the mentioned dataset. We achieved the same results as mentioned in the paper publishing the original dataset. It demonstrated strong precision (0.9170), recall (0.9163), and F1-score (0.9162), indicating its effectiveness in accurately classifying text and outperformed the related models for the used dataset by Belal et al. [7]. The LSTM + BERT embedding model used by the author achieved an accuracy of 89.42% and exhibited consistent precision, recall, and F1-score values of 0.89, indicating its reliability in text classification tasks. The other two models had accuracy of 87.87% and 88.57% respectively. Our model outperformed and gained better score in all the metrics compared to these models (Fig. 3).

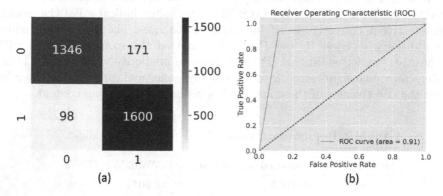

**Fig. 3.** (a) Confusion matrix (b) ROC curve of the proposed model. AUC score of the proposed model was 0.91.

### 4.4    Performance Comparison with Different Architectures of the Proposed Model

We have tweaked our proposed model with different combinations of the underlying architectures. We have used a sequence of LSTM, GRU, Convolutional layers in our proposed model. We have made some alterations to observe the change in the results by adding some layers and removing them as well as mentioned in Table 4. We have removed the two convolutional layers and the max pooling in a combination and this alteration produced a result of 91.56% accuracy while gaining sensitivity of 0.9187 and specificity of 0.9121. Again, including an additional LSTM layer, reducing a dense layer and adding an additional dense layer we analyzed the results. In this comparison, our proposed model's architecture outperformed the other ones.

**Table 4.** Performance Comparison with Different Architectures

| Architecture | Accuracy | Sensitivity | Specificity |
|---|---|---|---|
| Proposed Combination | **91.63%** | **0.9423** | **0.8873** |
| Reducing Convolution layers | 91.56% | 0.9187 | 0.9121 |
| Adding a LSTM Layer | 90.05% | 0.9079 | 0.9021 |
| Adding a Dense Layer | 91.58% | 0.9164 | 0.9161 |
| Removing a Dense Layer | 91.50% | 0.9321 | 0.8954 |

By adjusting various parameters, we developed our proposed model. The LSTM, GRU, CNN, and Dense layers were tried out in many ways before we figured out this model. In different units, filters, neuron ranges, we experimented with different combinations of 0 to 3 layers of BiLSTM, LSTM, BiGRU, GRU, Convolution, and dense layers. The results are displayed in Table 5.

**Table 5.** Model Tuning Range

| Layer | No of Layers tried | Tuning Range |
|---|---|---|
| BiLSTM | 0–3 | 100–300 units |
| LSTM | 0–3 | 100–300 units |
| Convolutional Layer | 0–2 | 50–500 filters |
| Dense layers (ReLU) | 1–3 | 100–1000 neurons |
| Dense layers (Softmax Activation) | 1 | 2 neurons |

### 4.5   Text Analysis in the Feature Level

The reasons behind classifying a text as cyberbullying or non-cyberbullying depends on the analysis of the text. The percentage of each word carries an indication to whether the text will fall into cyberbullying category or not. It has been seen that the significant words that make the sentence a cyberbullying text, is found in greater percentage in the training set of all cyberbullying texts as in Fig. 4.

**Fig. 4.** The explanation of the predictions of a cyberbullying and a non-cyberbullying texts.

## 5    Conclusion

The rising rate of cyberbullying necessitates immediate action to safeguard people from its adverse effects. To minimise the consequences of cyberbullying, real-time cyberbullying detection is essential. By harnessing the strength of advanced neural networks and large quantities of social media data, deep learning algorithms provide a promising solution for automated cyberbullying identification. In order to provide a safer online environment for everyone, this research study intended to explore and assess current developments in deep learning models for cyberbullying detection and proposing a promising model with a satisfactory accuracy of 91.63%. To the best of our knowledge this work is unique and not published anywhere. There is still a lot of rooms for future studies and improvement such as building a much larger dataset and using better and more furnished tools like stemmers, embedding techniques, lemmatises etc. Therefore, in spite of providing a satisfactory result, this study opens up new opportunities for this research field by highlighting the potential in precisely recognising and classifying cases of cyberbullying.

## References

1. Social Media in Bangladesh - 2023 Stats Platform Trends - OOSGA – oosga.com
2. Ahammed, S., Rahman, M., Niloy, M.H., Chowdhury, S.M.H.: Implementation of machine learning to detect hate speech in Bangla language. In: 2019 8th International Conference System Modeling and Advancement in Research Trends (SMART), pp. 317–320. IEEE (2019)
3. Ahmed, M.F., Mahmud, Z., Biash, Z.T., Ryen, A.A.N., Hossain, A., Ashraf, F.B.: Cyberbullying detection using deep neural network from social media comments in Bangla language. arXiv preprint arXiv:2106.04506 (2021)
4. Arreerard, R., Senivongse, T.: Thai defamatory text classification on social media. In: 2018 IEEE International Conference on Big Data, Cloud Computing, Data Science & Engineering (BCD), pp. 73–78. IEEE (2018)
5. Aurpa, T.T., Sadik, R., Ahmed, M.S.: Abusive Bangla comments detection on Facebook using transformer-based deep learning models. Soc. Netw. Anal. Min. **12**(1), 24 (2022)
6. Banik, N., Rahman, M.H.H.: Toxicity detection on Bengali social media comments using supervised models. In: 2019 2nd International Conference on Innovation in Engineering and Technology (ICIET), pp. 1–5. IEEE (2019)
7. Belal, T.A., Shahariar, G., Kabir, M.H.: Interpretable multi labeled Bengali toxic comments classification using deep learning. In: 2023 International Conference on Electrical, Computer and Communication Engineering (ECCE), pp. 1–6. IEEE (2023)
8. Cecillon, N., Labatut, V., Dufour, R., Linarès, G.: Abusive language detection in online conversations by combining content-and graph-based features. Front. Big Data **2**, 8 (2019)
9. Chakraborty, P., Seddiqui, M.H.: Threat and abusive language detection on social media in Bengali language. In: 2019 1st International Conference on Advances in Science, Engineering and Robotics Technology (ICASERT), pp. 1–6. IEEE (2019)

10. Das, A.K., Asif, A.A., Paul, A., Hossain, M.N.: Bangla hate speech detection on social media using attention-based recurrent neural network. J. Intell. Syst. **30**(1), 578–591 (2021). https://doi.org/10.1515/jisys-2020-0060

11. Gers, F.: Learning to forget: continual prediction with LSTM. In: 9th International Conference on Artificial Neural Networks: ICANN 1999 (1999). https://doi.org/10.1049/cp:19991218

12. Ghosh, T., Chowdhury, A.A.K., Banna, M.H.A., Nahian, M.J.A., Kaiser, M.S., Mahmud, M.: A hybrid deep learning approach to detect Bangla social media hate speech. In: Hossain, S., Hossain, M.S., Kaiser, M.S., Majumder, S.P., Ray, K. (eds.) Proceedings of International Conference on Fourth Industrial Revolution and Beyond 2021, pp. 711–722. Springer, Cham (2022). https://doi.org/10.1007/978-981-19-2445-3_50

13. Gordeev, D.: Automatic detection of verbal aggression for Russian and American imageboards. Procedia. Soc. Behav. Sci. **236**, 71–75 (2016)

14. Haidar, B., Chamoun, M., Serrhrouchni, A.: Multilingual cyberbullying detection system: detecting cyberbullying in Arabic content. In: 2017 1st Cyber Security in Networking Conference (CSNet), pp. 1–8. IEEE (2017)

15. Haidar, B., Chamoun, M., Serrhrouchni, A.: A multilingual system for cyberbullying detection: Arabic content detection using machine learning. Adv. Sci. Technol. Eng. Syst. J. **2**(6), 275–284 (2017)

16. Social Media User Statistics: How Many People Use Social Media? searchlogistics.com. https://www.facebook.com/mattwoodwarduk. https://www.searchlogistics.com/learn/statistics/social-media-user-statistics/. Accessed 12 July 2023

17. Huan, J.L., Sekh, A.A., Quek, C., Prasad, D.K.: Emotionally charged text classification with deep learning and sentiment semantic. Neural Comput. Appl. **34**(3), 2341–2351 (2021). https://doi.org/10.1007/s00521-021-06542-1

18. Ibrohim, M.O., Budi, I.: Multi-label hate speech and abusive language detection in Indonesian twitter. In: Proceedings of the Third Workshop on Abusive Language Online, pp. 46–57 (2019)

19. Karim, M.R., Chakravarthi, B.R., McCrae, J.P., Cochez, M.: Classification benchmarks for under-resourced Bengali language based on multichannel convolutional-LSTM network. In: 2020 IEEE 7th International Conference on Data Science and Advanced Analytics (DSAA), pp. 390–399. IEEE (2020)

20. Kumar, A., Sachdeva, N.: Multi-input integrative learning using deep neural networks and transfer learning for cyberbullying detection in real-time code-mix data. Multimedia Syst. **28**(6), 2027–2041 (2022)

21. Lane, J.: The 10 most spoken languages in the world (2023). https://www.babbel.com/en/magazine/the-10-most-spoken-languages-in-the-world

22. Luan, Y., Lin, S.: Research on text classification based on CNN and LSTM. In: 2019 IEEE International Conference on Artificial Intelligence and Computer Applications (ICAICA), pp. 352–355. IEEE (2019)

23. Luo, X.: Efficient English text classification using selected machine learning techniques. Alex. Eng. J. **60**(3), 3401–3409 (2021)

24. Malik, P., Aggrawal, A., Vishwakarma, D.K.: Toxic speech detection using traditional machine learning models and BERT and fasttext embedding with deep neural networks. In: 2021 5th International Conference on Computing Methodologies and Communication (ICCMC), pp. 1254–1259. IEEE (2021)

25. Mohammed, A., Kora, R.: An effective ensemble deep learning framework for text classification. J. King Saud Univ.-Comput. Inf. Sci. **34**(10), 8825–8837 (2022)

26. Pawar, R., Raje, R.R.: Multilingual cyberbullying detection system. In: 2019 IEEE International Conference on Electro Information Technology (EIT), pp. 040–044. IEEE (2019)

27. Pratiwi, N.I., Budi, I., Jiwanggi, M.A.: Hate speech identification using the hate codes for Indonesian tweets. In: Proceedings of the 2019 2nd International Conference on Data Science and Information Technology, pp. 128–133 (2019)

28. Ptaszynski, M., Pieciukiewicz, A., Dybała, P.: Results of the poleval 2019 shared task 6: first dataset and open shared task for automatic cyberbullying detection in polish twitter (2019)

29. Ritu, S.S., Mondal, J., Mia, M.M., Al Marouf, A.: Bangla abusive language detection using machine learning on radio message gateway. In: 2021 6th International Conference on Communication and Electronics Systems (ICCES), pp. 1725–1729. IEEE (2021)

30. Sazzed, S.: Abusive content detection in transliterated Bengali-English social media corpus. In: Proceedings of the Fifth Workshop on Computational Approaches to Linguistic Code-Switching, pp. 125–130 (2021)

31. Team, B.: All the latest cyberbullying statistics for 2023 (2023). https://www.broadbandsearch.net/blog/cyber-bullying-statistics

32. Yuvaraj, N., et al.: Automatic detection of cyberbullying using multi-feature based artificial intelligence with deep decision tree classification. Comput. Electr. Eng. **92**, 107186 (2021)

# Bangladeshi Native Vehicle Classification Employing YOLOv8

Siraj Us Salekin[1], Md. Hasib Ullah[1], Abdullah Al Ahad Khan[1],
Md. Shah Jalal[1], Huu-Hoa Nguyen[2], and Dewan Md. Farid[1(✉)]

[1] Department of Computer Science and Engineering, United International
University, United City, Madani Avenue, Badda, Dhaka 1212, Bangladesh
ssalekin213059@mscse.uiu.ac.bd, dewanfarid@cse.uiu.ac.bd
[2] College of Information and Communication Technology, Can Tho University,
3/2 Street, Ninh Kieu District, Can Tho City, Vietnam
nhhoa@ctu.edu.vn
https://cse.uiu.ac.bd/profiles/dewanfarid/

**Abstract.** Traffic congestion poses a significant challenge in Bangladesh
due to the growing number of vehicles. To tackle the obstacle needs an
effective intelligent system that can reduce the traffic congestion. With a
vision of building such an intelligent system, this paper presents a study
on vehicle classification using the YOLO (You Only Look Once) v8 transfer learning model, customized for Bangladeshi native vehicles. Besides,
we propose a transfer learning model-based system that helps to analyse
the video footage of vehicle movements from the elevated viewpoints of
foot over-bridges. Initially, the Bangladeshi Native Vehicle Image dataset
is gathered, processed, and used to train the model. Once the model is
trained and evaluated, the model is integrated into the vehicle detection
system. The system detects and tracks the vehicles, providing practical
traffic volume and movement insights. After the result analysis, We have
found a high mean average precision (mAP) of 91.3 % using intersection
over union (IoU). The model's performance enables proactive measures
to reduce congestion and optimise traffic flow. To build an efficient transportation network, this system can assist the Bangladesh Road Transport
Authority (BRTA) and Bangladesh Police Traffic Division to address the
challenges of increasing traffic and enhance traffic management.

**Keywords:** Deep Transfer Learning · YOLOv8 · Vehicle Classification

## 1 Introduction

Dhaka, the capital of Bangladesh, is grappling with severe traffic congestion resulting from haphazard growth and insufficient planning. Over the past
10 years, average driving speeds have dramatically declined from 21 kilometers
per hour to a mere 6 kilometers per hour. If the current trend persists, it could
further plummet to just 4 kilometers per hour by 2035, slower than the average walking pace. This congestion wastes 3.2 million working hours per day

N. Thai-Nghe et al. (Eds.): ISDS 2023, CCIS 1949, pp. 185–199, 2024.
https://doi.org/10.1007/978-981-99-7649-2_14

and imposes billions of dollars in economic costs annually [3]. The poor traffic management is causing delays and leading to a staggering 40% fuel wastage, amounting to daily losses of Bangladeshi BDT 41.5 million ($483,872). This fuel consumption and traffic congestion issue is adversely affecting the economy. Dhaka commuters face long travel times, with just 3.8 kilometers taking up to one and a half hours. The country is estimated to lose 1.38 billion BDT ($160 million) daily due to traffic jams. Insufficient road capacity, with only 7% of the required road capacity available, and an overload of vehicles contribute to traffic congestion. Road accidents, mental discomfort, and loss of valuable time are additional consequences of heavy traffic [9]. Experts recommend government intervention to establish discipline in the public road transport system. So we propose a method that can facilitate reducing traffic congestion in Bangladesh. To develop the system, we need Object Detection which is related to Computer Vision and is also a part of Deep Learning.

Object detection is an advancing field in computer vision, focusing on identifying and locating objects in images or videos. It's complex due to variations in appearance, lighting, occlusion, and cluttered backgrounds. Two main approaches exist: two-stage detectors proposing object locations and then classifying them, and one-stage detectors predicting object categories and bounding them in a single pass. Deep learning models like Convolutional Neural Networks (CNNs) are commonly used to extract features and classify objects. Challenges include detecting small objects, handling clutter, and enhancing robustness to changes in object appearance. Deep learning, a subset of machine learning, employs artificial neural networks to analyse large datasets and make predictions. It has gained popularity for solving complex problems in computer vision, natural language processing, and speech recognition. Deep learning algorithms recognise patterns by adjusting neuron weights through back-propagation. However, training and deploying deep learning models require substantial data and computational resources, and their inner workings may be hard to comprehend. Ongoing research addresses these challenges and enhances deep learning capabilities [5]. Also, to make it more accessible, researchers developed the Transfer Learning technique.

Transfer learning, a technique in deep learning, utilities knowledge gained from one task to improve performance in another. It reduces data and computation needs for training deep learning models. Pre-trained models like VGG, ResNet, YOLOv8, and BERT serve as common benchmarks. Transfer learning offers advantages such as faster training, higher accuracy, and learning from limited data. Challenges include selecting appropriate pre-trained models and managing domain discrepancies. Research focuses on refining pre-training methods, enhancing adaptation and fine-tuning, and deepening the understanding of transfer learning principles [1]. YOLOv8, developed by Ultralytics, is a cutting-edge computer vision pre-trained deep-learning model succeeding YOLOv5. This model offers built-in capabilities for object detection, classification, and segmentation tasks. It provides an accessible Python package and command line interface for easy usage. YOLOv8 enables the way of training its model on the custom data set to maximise the result for greater context [13].

Due to the efficiency of transfer learning, we have used the YOLOv8 deep-learning model for vehicle classification on the Bangladeshi Vehicle Annotated image data set named "Poribohon-BD" [12]. It requires data preparation, model training, evaluation, and deployment to achieve accurate and efficient vehicle detection results using our system. The annotated images are fed into the model and adjusted its weights and biases to optimise the detection performance. Once the model has been trained and evaluated, it is used to detect vehicles in new images or videos. And we have used our custom deep-learning model to develop a system that can detect traffic volume for both sides of the road from a given recorded video of the road from elevated viewpoints. We are hopeful that our system can help the authority of Bangladesh to give a proper insight into traffic congestion to take necessary steps on a particular road. Also, they can use our model in surveillance cameras for real-time vehicle detection in Bangladesh.

## 2   Literature Review

Transfer learning is used in Image Processing a lot these days. Deep Neural Networks are effective in recognizing intricate image features. The dense layers are responsible for image detection, and adjusting the higher layers does not impact the fundamental logic. Some related articles about machine learning, deep learning, and transfer learning in computer vision are discussed in this section.

Tabassum et al. developed a transfer learning method using YOLOv5 for detecting local vehicles on Bangladeshi roads using 9000 annotated images. The method achieved a 73% IoU score and a 55 frame per second speed after 56600 iterations. This approach holds promise for traffic management applications in Bangladesh [11].

Yiren et al. used deep neural networks to tackle vehicle detection and classification challenges in road images. Their YOLO detection model achieved 93.3% precision and 83.3% recall, comparable to the state-of-the-art DPM(Deformable Part Models) method had 94.4%. They explored fine-tuning and feature-extraction ways, and proposed techniques for addressing poor lighting conditions. This approach holds the potential for limited dataset training and can be extended for traffic development and planning purposes [18].

Shaoyong Yu et al. developed a deep-learning approach for classifying vehicles in complex transportation scenes. Their model uses a Faster R-CNN method for vehicle detection and a joint Bayesian network for classification. The classification model achieved 89% accuracy, but misclassifying non-vehicle regions may affect overall accuracy. Future work aims to improve detection accuracy, and speed, and incorporate feature classifiers for similar-looking vehicles [15].

Chen et al. developed a real-time vehicle classification model using AdaBoost and deep convolutional neural networks, achieving 99.50% accuracy in five vehicle groups. The model efficiently identifies vehicle images in 28ms and has low storage requirements, with training taking only 8 min. This model holds potential for intelligent transport systems and real-time traffic supervision [2].

Mahibul Hasan et al. developed a Bangladeshi model using transfer learning into ResNet50 and data augmentation to classify native vehicle types, achieving 98% accuracy in 13 standard classes. The Deshi-BD dataset used 10,440 images for training. This approach has strong generalization capabilities and the potential to address road traffic accidents in Bangladesh [4].

Maungmai et al. developed a vehicle classification system using Convolutional Neural Networks to classify vehicle characteristics, including type and color, from cropped images. The method outperformed Saripan et al.'s method and a deep neural network, with over 80% accuracy in type classification and 1.8% in color classification. Future research should explore different input image sizes, deeper CNN structures, and hyper-parameters [6].

Zhou et al. used deep neural networks to detect and classify vehicles in a public dataset. They achieved a 95.6% vehicle detection precision rate using state-of-the-art methods, including Alexnet-based methods. The method successfully detected and classified 714 passenger vehicles and 226 other vehicles, proving useful in multi-lane highway scenarios [17].

Yiren et al. also developed a visual attention-based CNN model for image classification, using a processing module to highlight specific parts of an image and weaken others. They improved image classification by computing information entropy and guiding reinforcement learning agents to select critical parts. The model was tested on a surveillance-nature dataset and showed better performance than large-scale CNNs in vehicle classification tasks. The VGG58 model (ImageNet-Based) showed a larger performance boost on Vehicle-58 (4%) than Vehicle-5 (3%) custom models. The accuracy rate was about 3% higher than large-scale CNNs, demonstrating the effectiveness of visual attention in improving image classification performance [16].

Vijayaraghavan et al. developed a convolution neural network for detecting and classifying vehicles using an entire image as input and a bounding box with feature class probabilities. The model outperformed state-of-the-art Fast R-CNN and cars, with an accuracy of 87% and 76%, respectively [14].

Neupane et al. developed a large training dataset, domain-shift problem, and real-time multi-vehicle tracking algorithm using deep learning. They created a 30,000-sample dataset and fine-tuned YOLO networks. The YOLOv5 model achieved 95% accuracy and performed well under various conditions. Future work should validate vehicle speed and compare internet network speed effects on deep learning models for road safety [7].

Overall, all the authors discussed and proposed the various uses of deep learning, and transfer learning for vehicle detection, and classification. Therefore an updated deep learning and transfer learning model will be beneficiary in the context of vehicle detection in the current world. Table 1 showcases some of the relevant research on vehicle detection and classification. After going through above mentioned research, we have found that there is a lack of implementing the latest YOLOv8 model for vehicle detection and tracking in the context of

Bangladesh. YOLOv8 is the updated deep neural network model for object detection which is showing promising results in custom image datasets for Transfer Learning. So, what we have done in our research is listed below:

Native vehicle detection in Bangladesh.
Native vehicle tracking in Bangladesh.
No. of Traffic volume & movement insights.
A highly accurate Transfer Learning Model which can be used in real-time native vehicle tracking in Bangladesh.
Providing a tool that can be used by the Road Transport Authority of Bangladesh (BRTA) to reduce traffic congestion.

## 3  Methodology

In Fig. 1, we have illustrated how the proposed system is built and the process of detecting Bangladeshi native vehicles on given input videos. Now, a detailed process is discussed in this section.

### 3.1  Dataset Description

Poribohon-BD [12] is a vehicle dataset consisting of 15 Bangladeshi vehicles, including a bicycle, boat, bus, car, CNG, easy-bike, horse-cart, launch, leguna, motorcycle, rickshaw, tractor, truck, van, and wheelbarrow. The dataset consists of 9058 JPG images depicting a wide variety of poses, angles, illumination, and weather conditions, as well as backgrounds. Each image is accompanied by an annotation file in XML format that indicates the exact positions and labels of each object. Data augmentation techniques were utilized to ensure that image counts for each vehicle classification were comparable. The faces of people were obscured in order to protect their privacy. The dataset is organized into separate folders for each vehicle classification, with an additional folder titled'Multi-class Vehicles' containing images and annotations of vehicles of multiple classes. Poribohon-BD is compatible with widely-used CNN architectures such as YOLO, VGG-16, R-CNN, and DPM. It functions as a valuable resource for Bangladeshi vehicle classification and detection research. But to train the model of YOLOv8 using this dataset is not possible because YOLOv8 expects PyTorch TXT annotated data. So, we have to re-annotate the dataset.

### 3.2  Dataset Processing

In the Yolov8 transfer learning process, PyTorch TXT annotation format is needed. XML annotation is not the right format. The available dataset is not in the right format to work with the latest Yolov8 transfer learning model.

So, we used Roboflow and that is an online tool that simplifies the conversion of XML annotation image files to PyTorch TXT format. By uploading your dataset and configuring settings, Roboflow exports the dataset with transformed annotations. The converted dataset, including images and TXT annotation files, is ready for PyTorch-based training [8]. This user-friendly process facilitates the integration of annotated data into machine learning workflows. This way we have re-annotated our data so that it can be used in the YOLOv8 object detection model. That is why we say our data is customized. Also in Fig. 1, this dataset is mentioned as custom data. In the future, anyone can access the customised dataset for further research.

**Table 1.** Research findings based vehicle classification models and its accuracy.

| Author | Task | Method | Model | Accuracy |
|---|---|---|---|---|
| Tabassum et al. (2020) [11] | Classification | Transfer Learning | YOLOv5 | 73% |
| Yiren et al. (2016) [18] | Classification, Detection | Transfer Learning | Darknet + Coffee Framework and YOLO with DMP | 93.3%, 94.4% |
| Shaoyong Yu et al. (2017) [15] | Classification | Deep-Learning | R-CNN | 89% |
| Chen et al. (2018) [2] | Classification | Deep-Learning | AdaBoost + CNN | 99.50% |
| Mahibul Hasan et al. (2021) [4] | Classification | Transfer Learning | ResNet50-Based | 98% |
| Maungmai et al. (2019) [6] | Classification | Deep Learning | CNN | 80% |
| Zhou et al. (2016) [17] | Classification, Detection | Transfer Learning | AlexNet-Based | 95.6% |
| Yiren et al. (2016) [16] | Classification | Guided Reinforcement Learning | Imagenet-Based | > 90% |
| Vijayaraghavan et al. (2019) [14] | Classification, Detection | Deep-Learning | Fast-RNN | 87% |
| Neupane et al. (2022) [7] | Detection, Real-Time Tracking | Transfer-Learning | YOLOv5 | 95% |

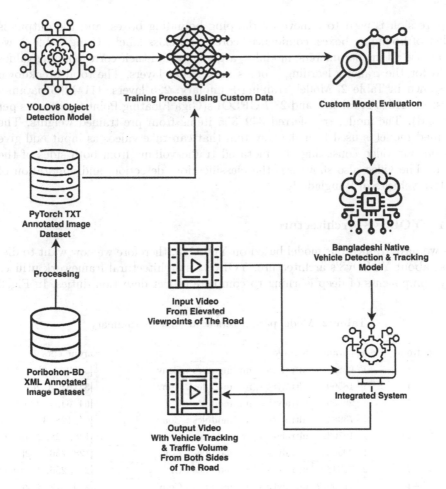

**Fig. 1.** Overview of the methodology.

## 3.3   Model Training

Once Dataset is processed, it is used by YOLOv8 object detection model. Basically, a training process is done on the custom dataset. Afterward, the trained model is evaluated. If the evaluation is not satisfactory then the training epochs are increased. In our case, 150 epochs are satisfactory to get our desired output. YOLOv8 algorithm divides an image into a grid. In YOLOv8, we utilized images that were 640 pixels wide by 640 pixels height. YOLOv8 employs a unique approach in which each grid cell spans an area of 32 pixels wide and 32 pixels height, and the number of grid cells can be calculated by dividing 640 by 32. That brings us to 20. It's the same as dividing the image into 20 × 20 grids (each grid is 32 pixels wide and 32 pixels height). These options may be changed. However, for our study, we left the default settings alone. Each grid cell predicts bounding boxes and confidence scores for objects that may be present in the cell. Non-maximum suppression is used to remove overlapping bounding boxes, and the output is a list of bounding boxes, confidence scores, and class labels. Using Table 2, we showcase all the parameters, modules, arguments and their corresponding values used for the transfer learning process in different layers. The total breakdown is shown in Table 2. Model Training Summary: 225 layers, 11141405 parameters, 11141389 gradients, and 28.7 GFLOPs(Giga Floating Point Operations per Second). The model transferred 349/355 items from pre-trained weights. The trained model is used to make a system that can take videos as input and give an output video consisting of annotated traffic volume from both sides of the road. The video also showcases the classification, detection, and annotation of native vehicles of Bangladesh.

### 3.4  YOLOv8 Architecture

As we have trained our model based on YOLOv8, therefore we now want to discuss about YOLOv8's architecture. YOLOv8's architectural framework utilizes key components of deep learning to complete object detection duties. In Fig. 2

**Table 2.** Model parameters & arguments summary

| index | from | n | params | module | arguments |
|---|---|---|---|---|---|
| 0 | −1 | 1 | 928 | ultralytics. nn. modules . Conv | [3, 32, 3, 2] |
| 1 | −1 | 1 | 18560 | ultralytics. nn. modules . Conv | [32, 64, 3, 2] |
| 2 | −1 | 1 | 29056 | ultralytics. nn. modules . C2f | [64, 64, 1, True] |
| 3 | −1 | 1 | 73984 | ultralytics. nn. modules . Conv | [64, 128, 3, 21] |
| 4 | −1 | 2 | 197632 | ultralytics. nn. modules . C2f | [128, 128, 2, True] |
| 5 | −1 | 1 | 295424 | ultralytics. nn. modules . Conv | [128, 256, 3, 2] |
| 6 | −1 | 2 | 788480 | ultralytics. nn. modules . C2f | [256, 256, 2, True] |
| 7 | −1 | 1 | 1180672 | ultralytics. nn. modules . Conv | [256, 512, 3, 2] |
| 8 | −1 | 1 | 1838080 | ultralytics. nn. modules . C2f | [512, 512, 1, True] |
| 9 | −1 | 1 | 656896 | ultralytics. nn. modules. SPPF | [512, 512, 5] |
| 10 | −1 | 1 | 0 | torch. nn. modules. upsampling. Upsample | [None, 2, Nearest] |
| 11 | [−1, 6] | 1 | 0 | ultralytics. nn. modules . Concat | [1] |
| 12 | −1 | 1 | 591360 | ultralytics. nn. modules . C2f | [768, 256, 11] |
| 13 | −1 | 1 | 0 | torch. nn. modules. upsampling. Upsample | [None, 2, Nearest] |
| 14 | [−1, 4] | 1 | 0 | ultralytics. nn. modules . Concat | [1] |
| 15 | −1 | 1 | 148224 | ultralytics. nn. modules . C2f | [384, 128, 11] |
| 16 | −1 | 1 | 147712 | ultralytics. nn. modules . Conv | [128, 128, 3, 21] |
| 17 | [−1,12] | 1 | 0 | ultralytics. nn. modules . Concat | [11] |
| 18 | −1 | 1 | 493056 | ultralytics. nn. modules . C2f | [384, 256, 11] |
| 19 | −1 | 1 | 590336 | ultralytics. nn. modules . Conv | [256, 256, 3, 2] |
| 20 | [−1,9] | 1 | 0 | ultralytics. nn. modules . Concat | [1] |
| 21 | −1 | 1 | 1969152 | ultralytics. nn. modules . C2f | [768, 512, 1] |
| 22 | [15,18,21] | 1 | 2121853 | ultralytics. nn. modules . Detect | [15, [128, 256, 512] |

the architecture of YOLOv8 is illustrated in detail [10]. The core, also known as the Backbone, is comprised of convolutional layers that extract relevant image features. The SPPF(Feature (feature selection) level, followed by a sequence of convolutional levels, manages characteristics at multiple dimensions. Simultaneously, the Upsample level boosts the resolution of these characteristic maps. To improve detection precision, the C2f module combines superior-level characteristics with context-based data. The last portion Detection module employs a combination of convolution and linear strata to transform high-dimensional characteristics into the resultant object categories and bounding boxes. The YOLOv8 architectural framework is designed to be quick and resourceful while maintaining superior precision in detection. Regarding the blueprint legend, the rectangles represent the strata, with marks denoting the stratum type (Conv, Upsample, etc.) and any significant parameters (channel count, kernel size, etc.). The directional arrows represent the data transition between strata, with the arrow orientation representing the data movement from one stratum to the following stratum.

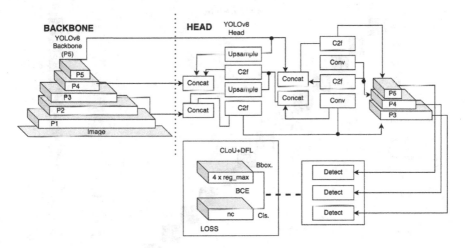

**Fig. 2.** YOLOv8 Architecture

## 3.5   Integrated System

After developing the customized YOLOv8 model, we have used the model to create the system that takes any video from the elevated viewpoints of foot over-bridges in Bangladesh and gives native vehicle detection result including the number of vehicles that passes the road. The number of volume indicates the number of vehicles from the both side of the road. Finally, a video consisting of number of vehicles from both side of the road will be the output. The default behavior of our model allows simultaneous detection of up to 300 vehicles. To modify this behavior and enhance detection capabilities, it is imperative to customize the settings within YOLOv8's detection module before initiating the transfer learning process. It's important to note that achieving improved detection performance through these adjustments is contingent upon employing robust hardware infrastructure. Otherwise, the model will continue to exhibit its default behavior of detecting a maximum of 300 objects. The model can be used to analyze video frames, identifying detected vehicles with timestamps. Tally the vehicle counts for each timestamp to create a dataset. To find the peak vehicle time, analyze the data, potentially using charts or graphs. This reveals the timestamp with the highest vehicle count. In our research, we didn't employ this method, but it's an option for future use.

## 4    Result Analysis

In this section, the proposed model is evaluated based on multiple factors like confusion matrix, precision-recall curve, and validation set results.

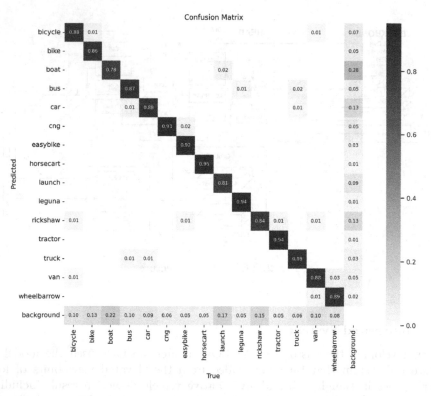

**Fig. 3.** Confusion matrix of the proposed model.

In the context of object detection algorithms such as YOLO, "mAP @5" refers to the mean average precision at an intersection over union (IoU) threshold of

**Table 3.** Validation set result.

| Class | Instances | Box(P) | R | mAP50 | mAP(50–95) |
|---|---|---|---|---|---|
| all | 5238 | 0.903 | 0.854 | 0.913 | 0.663 |
| bi-cycle | 307 | 0.885 | 0.827 | 0.914 | 0.62 |
| bike | 379 | 0.913 | 0.831 | 0.898 | 0.634 |
| boat | 732 | 0.79 | 0.701 | 0.784 | 0.446 |
| bus | 334 | 0.902 | 0.83 | 0.902 | 0.667 |
| car | 624 | 0.882 | 0.848 | 0.903 | 0.646 |
| cng | 433 | 0.931 | 0.91 | 0.953 | 0.732 |
| easy-bike | 274 | 0.943 | 0.912 | 0.96 | 0.749 |
| horse-cart | 62 | 0.872 | 0.935 | 0.957 | 0.743 |
| launch | 370 | 0.887 | 0.759 | 0.848 | 0.581 |
| leguna | 370 | 0.931 | 0.902 | 0.961 | 0.807 |
| rickshaw | 609 | 0.872 | 0.783 | 0.872 | 0.584 |
| tractor | 102 | 0.947 | 0.931 | 0.961 | 0.743 |
| truck | 341 | 0.947 | 0.886 | 0.93 | 0.701 |
| van | 390 | 0.919 | 0.856 | 0.917 | 0.692 |
| wheelbarrow | 131 | 0.925 | 0.893 | 0.931 | 0.603 |

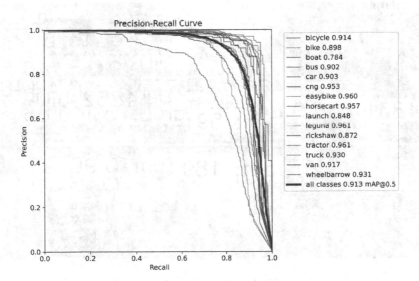

**Fig. 4.** Precision-recall curve of the proposed model.

**Fig. 5.** Vehicle Detection by the model

**Fig. 6.** Vehicle Tracking by the model

0.5. Mean Average Precision (mAP) is a common evaluation metric used to assess the performance of object detection models. It measures both the precision and recall of the model across different classes and IoU thresholds. Table 3 showcases these evaluations of our model on the validation set. In our case, an mAP of 0.913

@5 means that, on average, the object detection model achieved a precision of 0.913 when considering an IoU threshold of 0.5. This indicates that the model's predictions are highly accurate and had a good overlap with the ground truth bounding boxes. It's important to note that the mAP value is typically reported per class, so "all classes" indicates that this performance metric is an average across all the classes in the dataset being evaluated. Figure 5 illustrates the precision-recall curve that showcases the effectiveness of our model in detecting native Bangladeshi vehicles. Also, Fig. 4 showcases the confusion matrix that indicates how much accurate our model is to classify native Bangladeshi vehicles. A classification example using our model is shown in Fig. 6. mAP, IoU, Precision-Recall Curve and Confusion matrix are used to evaluate the model. One thing to consider is that the light change in the morning or afternoon will affect the detection of the vehicles as most of the images used in the training are captured in broad daylight or using flashlight. However, sufficient low light images are also used and the accuracy of the model will decrease slightly which is negligible in real life. Finally, we can say that our model is robust enough to be used in a system that can detect Bangladeshi native vehicles. Eventually, the system takes videos as input and successfully gives an output video consisting of annotated traffic volume from both sides of the road. Finally, a video consisting of number of vehicles from both side of the road is the output of the integrated system. An image is included to showcase the classification, detection, and annotation of native vehicles of Bangladesh in a video. An example of that is showcased in Fig. 6.

## 5   Conclusion

Our research offers a promising strategy for addressing the expanding problem of traffic congestion in Bangladesh. We have effectively developed a vehicle detection system based on the YOLOv8 model of transfer learning that has been customised for Bangladesh. This system, which is incorporated with elevated viewpoints and video footage analysis, enables the accurate detection and tracking of vehicles, thereby providing valuable information about traffic volume and movement. Through the collection and processing of the Bangladeshi Native Vehicle Image dataset, we trained the model to attain a remarkable mean average precision (mAP) of 91.3%. The YOLOv8 model utilized in this research is a potent instrument for detecting objects in images or videos. It is intended to derive significant image features using convolutional layers. Then, these features are analyzed at various levels in order to capture specific characteristics and improve their resolution. In addition, the model includes a module that integrates various features to enhance detection precision. Then the model is used in the vehicle detection system. The capability of our system to detect local vehicles in video footage and to provide traffic volume data, including vehicle counting, is a significant step toward enhancing traffic management in Bangladesh. This instrument can aid the Bangladeshi Road Transport Authority in regulating traffic and addressing the challenges posed by growing traffic congestion. Our research proposes a practicable and efficient method for constructing an effective transportation network in Bangladesh. By utilising advanced technology and deep learning algorithms, we can make informed decisions to reduce congestion, optimise traffic flow, and enhance the nation's transportation infrastructure. In the future, we are hopeful to utilise the mentioned model in this paper in real-time vehicle detection system in Bangladesh. We also believe, our study will help researcher to build a proper vehicle classification and detection system to tackle the issue of traffic congestion in their respective countries.

**Acknowledgements.** We appreciate the support for this research received from the a2i Innovation Fund of Innov-A-Thon 2018 (Ideabank ID No.: 12502) from a2i-Access to Information Program - II, Information & Communication TechnologyDivision, Government of the People's Republic of Bangladesh and Institute for Advanced Research (IAR), United International University (Project Code: UIU/IAR/01/2021/SE/23).

# References

1. Baheti, P.: What is transfer learning? [examples & newbie-friendly guide] (2021). https://www.v7labs.com/blog/transfer-learning-guide
2. Chen, W., Sun, Q., Wang, J., Dong, J.J., Xu, C.: A novel model based on AdaBoost and deep CNN for vehicle classification. IEEE Access **6**, 60445–60455 (2018)
3. Fan, Q.: Toward great dhaka. World Bank (2017). https://www.worldbank.org/en/news/speech/2017/07/19/toward-great-dhaka
4. Hasan, M.M., Wang, Z., Hussain, M.A.I., Fatima, K.: Bangladeshi native vehicle classification based on transfer learning with deep convolutional neural network. Sensors **21**(22), 7545 (2021)
5. IBM: What is deep learning? (nd). https://www.ibm.com/topics/deep-learning
6. Maungmai, W., Nuthong, C.: Vehicle classification with deep learning. In: 2019 IEEE 4th International Conference on Computer and Communication Systems (ICCCS), pp. 294–298. IEEE (2019)
7. Neupane, B., Horanont, T., Aryal, J.: Real-time vehicle classification and tracking using a transfer learning-improved deep learning network. Sensors **22**(10), 3813 (2022)
8. Roboflow: Give your software the power to see objects in images and video (2023). https://roboflow.com/
9. Sakib, S.N.: Bangladesh loses (2021). https://www.aa.com.tr/en/asia-pacific/bangladesh-loses-40-of-fuel-due-to-poor-traffic-management/2449934
10. Solawetz, J.: What is YOLOv8? the ultimate guide (2023). https://blog.roboflow.com/whats-new-in-yolov8/
11. Tabassum, S., Ullah, M., Al-Nur, N., Shatabda, S.: Native vehicles classification on bangladeshi roads using CNN with transfer learning (2020). https://doi.org/10.1109/TENSYMP50017.2020.9230991
12. Tabassum, S., Ullah, M., Al-Nur, N., Shatabda, S.: Poribohon-BD: bangladeshi local vehicle image dataset with annotation for classification. Data in Brief **33**, 106465 (2020). https://doi.org/10.1016/j.dib.2020.106465
13. Ultralytics: revolutionizing the world of vision AI (2023). https://ultralytics.com/
14. Vijayaraghavan, V., Laavanya, M.: Vehicle classification and detection using deep learning. Int. J. Eng. Adv. Technol. **9**(1S5), 24–28 (2019)

15. Yu, S., Wu, Y., Li, W., Song, Z., Zeng, W.: A model for fine-grained vehicle classification based on deep learning. Neurocomputing **257**, 97–103 (2017)
16. Zhao, D., Chen, Y., Lv, L.: Deep reinforcement learning with visual attention for vehicle classification. IEEE Trans. Cogn. Dev. Syst. **9**(4), 356–367 (2016)
17. Zhou, Y., Cheung, N.: Vehicle classification using transferable deep neural network features. CoRR abs/1601.01145 (2016). http://arxiv.org/abs/1601.01145
18. Zhou, Y., Nejati, H., Do, T.T., Cheung, N.M., Cheah, L.: Image-based vehicle analysis using deep neural network: a systematic study. In: 2016 IEEE International Conference on Digital Signal Processing (DSP), pp. 276–280. IEEE (2016)

# Monitoring Attendance and Checking School Uniforms Using YOLOv8

Khang Nhut Lam[1]([✉]), Trong Thanh La[1], Khang Duy Nguyen[1],
Man Minh Le[1], Vy Trieu Truong[1], and Andrew Ware[2]

[1] Can Tho University, Can Tho, Vietnam
lnkhang@ctu.edu.vn,
{trongb2014957,khangb2005843,manb2003791,vyb2014962}@student.ctu.edu.vn
[2] University of South Wales, Pontypridd, UK
andrew.ware@southwales.ac.uk

**Abstract.** The performance of students is affected by various factors, including attendance tracking and school uniform checking. This study develops a system for automatically tracking student attendance and checking school uniforms using the pre-trained YOLOv8-based models. The system consists of six models: YOLOv8Students for detecting humans, YOLOv8Face and ArcFace for identifying students, YOLOv8Shirts and YOLOv8Pants for detecting and predicting types of shirts and pants, respectively, and YOLOv8Card for detecting student ID cards. Our research addresses practical concerns in educational institutions. The experimental results show that the models perform fairly well in optimal lighting conditions.

**Keywords:** attendance · face recognition · pants detection · school uniform · shirt detection · ID card detection · student attendance · YOLOv8

## 1 Introduction

Tracking attendance and checking the school uniforms of students are important factors in the university environment. The term "school uniforms" in the paper refers to the dresses specified in the academic regulations of a university, specifically Can Tho University (CTU) of Vietnam. According to Decision 1813/QD-DHCT dated June 18, 2021, on academic regulations of CTU, students are required to wear student ID cards as well as neat, polite, and discreet clothes when entering the university environments. Particularly, students at CTU must wear student ID cards, sleeved shirts, long pants, and sandals. Based on our observation, some students do not follow the school's dress regulations, which may have a negative impact on the campus's aesthetics and the classroom's atmosphere. In addition, students will not be permitted to take the final exam if they are absent for more than 20% of the lecture hours. The task of checking attendance and school uniforms is time-consuming and monotonous. To the best

N. Thai-Nghe et al. (Eds.): ISDS 2023, CCIS 1949, pp. 200–207, 2024.
https://doi.org/10.1007/978-981-99-7649-2_15

of our knowledge, we have not found any application that supports both student attendance monitoring and school uniform checking. This study proposes an approach based on machine learning for building such a system.

The remaining parts of this paper are organized as follows. Section 2 presents related work. The system architecture is proposed in Sect. 3. Sections 4 and 5 discuss our data collection and experimental results. Section 6 concludes the paper.

## 2  Related Work

Several attendance checking applications utilize face recognition or barcode scanning on student ID cards. Saparkhojayev and Guvercin [15] used an RFID reader to read the student ID card and a web-camera to simultaneously take a picture of the student, which was then sent to a computer and stored in a database. Alghamdi [2] built a mobile application to recognize students using a Turck device to detect the RFID tags found on their university ID cards within a distance of 3 m. Islam et al. [8] reported that tracking attendance using RFID technology is slower, less effective, and more expensive than using a smartphone. Therefore, they built a system comprising a mobile application supporting an interface for teacher-student communication and attendance tracking. Attendance data was stored in the databases of the mobile phone and a web sever.

Some studies use biometrics for attendance checking. Kadry and Smaili [16] utilized wireless iris recognition, micro-controller and RF wireless techniques for developing an attendance tracking system. However, this attendance checking system seems unsuitable for a university environment. Rufai et al. [14] designed a biometric access system for screening exams and checking attendance. Biometric information is collected from each student during enrollment, and all information of each student is detected and compared with the stored data during the examination.

Neural networks have been intensively developed and are utilized to automatically monitor student attendance. Filippidou and Papakostas [7] built a real-time checking attendance model by fine-tuning pre-trained convolutional neural networks (CNNs) for recognizing the faces of students. In particular, the CNN-based attendance checking model captures images of the classroom using a webcam, detects faces using the MTCNN algorithm, and then recognizes students' faces using a CNN-based model. Dang [5] introduces a smart attendance system by enhancing the FaceNet facial recognition model based on a MobileNetV2 backbone with an SSD subsection. The author performs experiments on mobile devices and achieves an accuracy of 95% on a small dataset. The YOLO model [13] is utilized for checking attendance. Alon et al. [3] built a student attendance system using YOLOv3, which obtained an accuracy of 94%. Mardiana et al. [10] developed a YOLOv5-based system for checking library attendance. Other approaches consist of Local Binary Pattern Histogram face recognizer to identify students [12], and QR code [1,9,11].

# 3   System Architecture

This study uses the pre-trained YOLOv8[1] models to detect and predict objects in a frame. Figure 1 presents the system architecture. At the beginning of each class, the monitor will start the application, select the class information, and activate the webcam to begin the process of verifying attendance and uniforms. The application will automatically create a file for storing attendance and uniform verification data. After completing the attendance checking process, the monitor can easily export the .csv file containing attendance information.

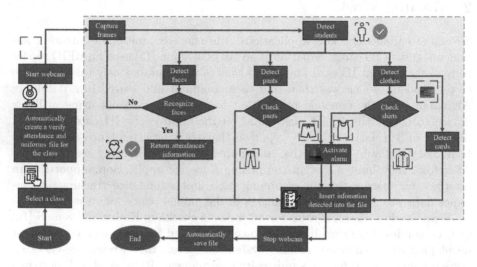

**Fig. 1.** The system architecture.

Figure 2 shows the YOLOv8-based models for monitoring attendance and checking uniforms in detail. First, when an object enters the frame, the YOLOv8-based student detection model (named YOLOv8Student) trained on the MS COCO dataset detects and localizes a human. Then, the student's face is detected using the YOLOv8-based facial recognition model (referred to as YOLOv8Face), which is trained on the WIDER FACE dataset [17], and identified using ArcFace [6] by returning the most comparable face found in the database. The YOLOv8-based models for checking shirts and pants, called YOLOv8Shirt and YOLOv8Pants, respectively, are used to detect and predict the types of shirts and pants. After checking the shirt, another YOLO-v8-based model is utilized to detect and verify the student ID card located in front of the student's chest. The system will send out a notification signal if it detects that the student does not follow the university's requirements.

# 4   Data Collection

Our datasets comprise images taken by ourselves, extracted from the Internet, and provided by Roboflow Universe as follows:

---

[1] https://docs.ultralytics.com/.

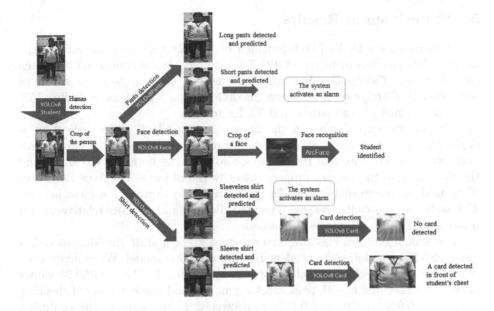

**Fig. 2.** YOLOv8-based models for monitoring attendance and checking uniforms.

- Attendance dataset: has 80 images of 11 volunteers, who are students at the College of Information and Communication Technology (CICT) of CTU.
- Shirt and pant datasets: obtain from Roboflow[2]. The shirt dataset consists of 517 images including 96 images of jackets and vests, 69 images of sleeved shirts, 97 images of sleeveless shirts, 168 images extracted from the Internet, and 87 images taken by ourselves. The pant dataset has 297 images, consisting of 48 images of short pants, 92 images of long pants, 27 images of pantalones, 121 images from the Internet, and 9 images gathered manually.
- Student card dataset: comprises 155 images collected manually from the ID cards of students at CICT. We observe that the student cards are rectangular, and can be worn horizontally or vertically. This dataset includes 42 horizontal card images, 106 vertical card images, and 7 unspecified-dimension card images.

The Make Sense toolkit[3] is utilized to label images, and the Roboflow Project is used to manage and augment images. We employ several levels of augmentation, including horizontal flipping, 12-degree rotation, adjusting brightness, and applying Gausian blur. After the augmentation process, the total number of images used for experiments is 1,343 images of shirts (labeled with "sleeved shirt" and "sleeveless shirt"), 796 images of pants (labeled with "long pants" and "short pants"), and 395 images of student cards.

---

[2] https://universe.roboflow.com/sookmyung-women-university/.

[3] https://www.makesense.ai/.

# 5   Experimental Results

Our experiments with YOLOv8 provided by Ultralytics[4] are performed in the Google Colab environment with GPU T4. The models are trained with a batch size of 16 and 150 epochs. For other parameters, we use the same values as suggested by Ultralytics. Each dataset is divided into 3 portions, including 75% for training, 20% for validation, and 5% for testing.

We perform experiments in the classrooms at CICT. The application is deployed on a computer located inside a classroom, opposite the classroom's main entrance. We note that there must be no obstacles between the camera and the door to help the camera properly identify the objects. The Asus ROG Eye S[5] is used as an external camera for monitoring and capturing student images. This webcam is cost-effective, has a compact design, and provides relatively good image quality with 60 frames per second.

The school uniform checking system consists of a shirt checking model, a pants checking model, and a student ID card checking model. We evaluate each school uniforms checking model, as presented in Table 1. The mAP50-95 values of the shirt checking model, pant checking model, and student ID card checking model are 0.902, 0.859, and 0.687, respectively. Figure 3 shows the confusion matrices for predicting students wearing shirts, pants, and ID cards.

**Table 1.** Experimental results of the school uniform checking model.

| Models | Precision | Recall | mAP50-95 |
|---|---|---|---|
| Shirt checking model | 0.998 | 0.993 | 0.902 |
| Pant checking model | 0.977 | 0.974 | 0.859 |
| Student ID card checking model | 0.996 | 1.00 | 0.687 |

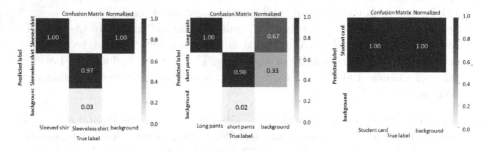

**Fig. 3.** Confusion matrix of predicting wearing shirts (left side); wearing pants (middle); and wearing student cards (right side).

When the distance between the camera and the objects is between 1.3 m and 2.9 m, the models are able to detect and identify the objects perfectly. Specifically,

---

[4] https://docs.ultralytics.com/.
[5] https://rog.asus.com/streaming-kits/rog-eye-s-model/.

the models perform the best when this distance is between 1.8 m and 1.9 m. The models become less accurate when this distance is less than 1.3 m or greater than 3.0 m.

We conduct two scenarios for evaluating the school uniform checking model in real-life. Experiments are performed in both natural (sunlight) and LED (classroom) lights. Occasionally, students enter the classroom without turning on the lights; hence, we also conduct experiments using only natural light. Our application performs more accurately in well-lit conditions. In particular, the facial recognition model achieves an accuracy of 85% in optimal lighting conditions. Without LED lighting and under cloudy conditions, this model's accuracy significantly drops to 20%. Table 2 presents the accuracies of models in different lighting conditions. Figure 4 illustrates the lights' impact on our application's accuracy.

**Table 2.** The accuracies of models in different light conditions

| Model | Optimal lighting conditions | Dark lighting conditions |
| --- | --- | --- |
| Face recognition model | 85% | 20% |
| Shirt checking model | 99% | 99% |
| Pant checking model | 97% | 94% |
| Student ID card checking | 87% | 27% |

**Fig. 4.** Example of the light's effect on the application: from left to right, the brightness decreases, causing the accuracies of the models to go down.

Currently, the model cannot distinguish between real and fake student ID cards. In other words, if students are wearing cards that resembles the shape of student ID cards, the system predicts that they are wearing student ID cards. We will enhance the model to recognize student ID cards and extract the information on them in real time. In addition, students are not permitted to wear short pants, but may wear skirts. Our model currently does not differentiate between short pants and skirts. For future work, we may utilize a skirt dataset [4] to help the model study an additional type of dress allowed to be worn at CTU. Moreover,

**Fig. 5.** Some situations that the models do not perform well: a parking ticket can be recognized as a student ID card (left side); attendance can be checked with a wrong name due to the student's lowered head (middle); the student's uniform in the middle cannot be recognized because it is covered (right side).

when a student wears a mask, raises or lowers his head, the facial recognition model cannot identify him correctly. When a group of students enters the class, the model performs very well if no one is obscured; otherwise, the obscured objects will not be detected correctly. Figure 5 illustrates some examples in which models do not perform well. After completing the attendance checking progress, the monitor can easily export the .csv file containing attendance information as shown in Fig. 6.

| | A | B | C | D | E | F | G |
|---|---|---|---|---|---|---|---|
| 1 | StudentID | StudentName | Pants | Shirt | Card | DateTime | |
| 2 | B2003791 | Le Minh Man | Long pants | Sleeved | Verified | 14/08/2023 09:55:51 | |
| 3 | B2014957 | La Thanh Trong | Long pants | Sleeved | | 14/08/2023 09:55:48 | |
| 4 | B2005843 | Nguyen Duy Khang | Long pants | Sleeved | Verified | 14/08/2023 09:55:51 | |
| 5 | B2014962 | Truong Trieu Vy | Long pants | Sleeveless | Verified | 14/08/2023 09:39:21 | |
| 6 | B2014748 | Lam Hoang Khang | Long pants | Sleeved | Verified | 14/08/2023 09:55:49 | |
| 7 | B2017065 | Duong Thi Yen Nhi | Long pants | Sleeved | Verified | 14/08/2023 09:32:42 | |

**Fig. 6.** A spreadsheet for checking the attendance and school uniforms of students.

# 6    Conclusion

This paper details a system for automatically checking the attendance and school uniforms of students using YOLOv8. The main goal of this system is to eliminate the disadvantages of checking attendance manuals, which are tedious and time-consuming. The system performs fairly well in optimal lighting conditions. Our model is simple, accurate, and deployable at other institutions. More work is required to enable the system to work in less favorable lighting conditions.

# References

1. Abdellatif, H., Sirasanagandlae, S.R., Al-Mushaiqri, M., Sakr, H.F.: Location-linked QR code as a safe tool for recording classroom attendance during COVID-19 pandemic: perspectives of medical students. Med. Sci. Educ. **32**(5), 971–974 (2022)
2. Alghamdi, S.: Monitoring student attendance using a smart system at Taif University. Int. J. Comput. Sci. Inf. Technol. (IJCSIT) **11** (2019)
3. Alon, A., Casuat, C., Malbog, M., Marasigan, R., Gulmatico, J.: A YOLOv3 inference approach for student attendance face recognition system. Int. J. Emerg. Trends Eng. Res. **8**(2), 384–390 (2020)
4. Computacional, V.: Skirt dataset (2022). https://universe.roboflow.com/vision-computacional-jddzr/skirt-kwpqv
5. Dang, T.V.: Smart attendance system based on improved facial recognition. J. Robot. Control (JRC) **4**(1), 46–53 (2023)
6. Deng, J., Guo, J., Xue, N., Zafeiriou, S.: ArcFace: additive angular margin loss for deep face recognition. In: Proceedings of the IEEE/CVF Conference on Computer Vision and Pattern Recognition, pp. 4690–4699 (2019)
7. Filippidou, F.P., Papakostas, G.A.: Single sample face recognition using convolutional neural networks for automated attendance systems. In: 2020 Fourth International Conference On Intelligent Computing in Data Sciences (ICDS), pp. 1–6. IEEE (2020)
8. Islam, M.M., Hasan, M.K., Billah, M.M., Uddin, M.M.: Development of smartphone-based student attendance system. In: 2017 IEEE Region 10 Humanitarian Technology Conference (R10-HTC), pp. 230–233. IEEE (2017)
9. Liew, K.J., Tan, T.H.: QR code-based student attendance system. In: 2021 2nd Asia Conference on Computers and Communications (ACCC), pp. 10–14. IEEE (2021)
10. Muhammad, M.A., Mulyani, Y., et al.: Library attendance system using YOLOv5 faces recognition. In: 2021 International Conference on Converging Technology in Electrical and Information Engineering (ICCTEIE), pp. 68–72. IEEE (2021)
11. Nuhi, A., Memeti, A., Imeri, F., Cico, B.: Smart attendance system using QR code. In: 2020 9th Mediterranean Conference on Embedded Computing (MECO), pp. 1–4. IEEE (2020)
12. Raj, A.A., Shoheb, M., Arvind, K., Chethan, K.: Face recognition based smart attendance system. In: 2020 International Conference on Intelligent Engineering and Management (ICIEM), pp. 354–357. IEEE (2020)
13. Redmon, J., Divvala, S., Girshick, R., Farhadi, A.: You only look once: unified, real-time object detection. In: Proceedings of the IEEE Conference on Computer Vision and Pattern Recognition, pp. 779–788 (2016)
14. Rufai, M., Adigun, J., Yekini, N.: A biometric model for examination screening and attendance monitoring in Yaba college of technology. World Comput. Sci. Inf. Technol. J. **2**(4), 120–124 (2012)
15. Saparkhojayev, N., Guvercin, S.: Attendance control system based on RFID-technology. Int. J. Comput. Sci. Issues (IJCSI) **9**(3), 227 (2012)
16. Seifedine, K., Mohamad, S.: Wireless attendance management system based on iris recognition. Sci. Res. Essays **5**(12), 1428–1435 (2010)
17. Yang, S., Luo, P., Loy, C.C., Tang, X.: WIDER FACE: a face detection benchmark. In: IEEE Conference on Computer Vision and Pattern Recognition (CVPR) (2016)

# Topic Classification Based on Scientific Article Structure: A Case Study at Can Tho University Journal of Science

Hai Thanh Nguyen, Tuyet Ngoc Huynh, Anh Duy Le,
and Tran Thanh Dien(✉)

Can Tho University, Can Tho, Vietnam
{nthai.cit,ldanh,thanhdien}@ctu.edu.vn

**Abstract.** With a massive amount of stored articles, text-based topic classification plays a vital role in enhancing the document management efficiency of scientific journals. The articles can be found faster by filtering out the appropriate topic and speeding up to determine appropriate reviewers for the review phase. In addition, it can be beneficial to recommend related articles for the considered manuscript. However, fetching entire documents for the process can consume much time. Especially, Can Tho University Journal of Science (CTUJS) is a multidisciplinary journal with many topics. Therefore, it is necessary to evaluate various common structures in an article. Extracted sections can be short but efficient in determining the article's topic. In this study, we explore and analyze the paper structure of articles obtained from CTUJS for topic classification using Support Vector Machine (SVM), Random Forest (RF), and Naive Bayes (NB). The results show that Random Forest outperforms Naïve Bayes and SVM regarding performance and training time. As shown, the topic classification performance based on the section of "Method" can reach 0.53 compared to the whole content of the paper with 0.61 in accuracy.

**Keywords:** Paper Structure · Scientific journal · Topic classification

## 1 Introduction

Text classification has been the most widespread problem of NLP, and is often used for spam detection [1], topic classification, etc. In scientific journals, managing many articles requires a scientific and time-saving management process. Automatic classification of topics is quite essential to help manage articles easier. A faster search based on selected topics to filter and label topics before detecting similar text saves processing costs than scanning the entire data warehouse. Review is an indispensable process for a scientific article to be approved for publication. The classification of documents by topic helps the article be evaluated by the right reviewers in the main areas of the article, improving high-quality articles.

N. Thai-Nghe et al. (Eds.): ISDS 2023, CCIS 1949, pp. 208–215, 2024.
https://doi.org/10.1007/978-981-99-7649-2_16

The case study at CTUJS is a multidisciplinary journal that provides information on Can Tho University's scientific research and introduces domestic and foreign scientific research. Therefore, there are both English and Vietnamese scientific articles. In the previous work, there have been studies on automatic topic classification. However, the classification results in English documents are rather positive, and for Vietnamese, it is still limited, and no feasible results have been found. Therefore, this study proposes a method to classify topics based on the structure of scientific articles in Vietnamese and English.

## 2 Related Work

Text classification in many studies aimed to efficiently find the necessary documents and save time than searching for irrelevant data. The authors in [2] also proposed topic classification approaches through Support Vector Machines, Naïve Bayes, and k-Nearest Neighbors and preprocessed input data, extracting information, and vectorizing. According to [3], authors described some main steps in text classification, including document preprocessing, feature extraction/selection, model selection, training, and classifier testing. The text preprocessing stage converts the original textual data to a raw data structure, where the most significant text features that distinguish between text categories are identified [4]. The study in [5] presented a method to indicate similar Vietnamese documents from English Documents.

The work in [6] investigated the process of text classification, the process of different methods of weighing and measuring terms, and compares other classification techniques, including Naïve Bayes classifiers (NBC), SVM, and k-Nearest Neighbors (kNN), KNC [7] - a combination of KNN and other three classifiers (C4.5 algorithm, NBC and SVM), etc. According to the authors in [8], Naïve Bayes is quite effective in data mining tasks, but it gives pretty bad results when used for automatic text classification tasks.

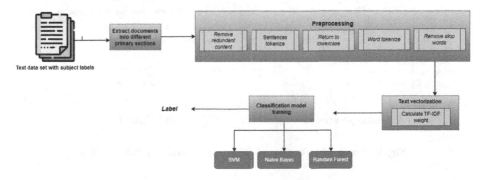

**Fig. 1.** The main steps in topic classification tasks based on article structure.

## 3   Methods

Figure 1 illustrates several of the main steps for topic classification based on article structure with the main steps as follows. Step 1 collects articles, each with a
corresponding topic. Input data are Vietnamese documents presented based on
the labeled scientific paper structure. In Step 2, we extract the documents with
various primary sections of typical article structure such as Abstract, Introduction, Method, Results (Experiments), and Conclusion. Preprocessing data by
sentence splitting, word splitting, removing redundant characters in sentences
(such as punctuation marks, mathematical formulas, etc.), removing stop words,
etc. are performed in Step 3. In Step 4, After preprocessing, we vectorize the
data set corresponding to an article that will be a feature vector, including the
weights of words. In step 5, we leverage some classification models. After preprocessing and vectorizing each document, we perform the classification tasks.
In the next section, we will detail the steps above.

### 3.1   Data Collection

This is one of the essential process steps in the text classification model. It determines the complete elements of the text classification system. Experimental data
were collected from the system of CTUJS[1]. The collected data of Vietnamese
scientific articles include 1371 articles, divided into 17 different topics described
in Fig. 2a.

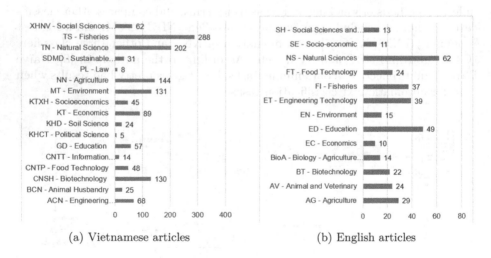

(a) Vietnamese articles       (b) English articles

**Fig. 2.** Topic distribution of Vietnamese and English articles

The English documents for classification include 348 articles, divided into 13
different topics described in Fig. 2b. In this study, the data collected was not

---

[1] https://ctujs.ctu.edu.vn/.

evenly distributed among the topics. The number of Law or Political Science papers in the Vietnamese data section is less than ten. In contrast, the number of papers on others, such as Fisheries or Agriculture, is much higher than others, including 144 articles on Agriculture and 288 articles on Fisheries. As for the English data section, topics such as Information Technology or Political Science have very few samples, with only 1–2 articles. The difference in the number of articles can lead to the situation that topics with too few articles will have low accuracy. In contrast, topics with a rich number of articles will give high accuracy. Therefore, to avoid reducing the classification efficiency, this study omitted topics with less than three articles.

## 3.2   Extracting Article Structure

As the guidelines from Springer Nature publisher [9], they described a standard structure of the body of research papers, including four primary sections: Introduction, Materials and Methods, Results, Discussion, and Conclusions. Many articles have followed these guidelines. Such structures are prevalent in most scientific journals. We briefly present "Materials and Methods" as "Method" and "Discussion and Conclusions" as "Conclusion" in the Experiment section. Besides, we also consider "Abstract" in classification tasks. For articles published in the CTUJS, although most of the articles have the standard structure as above, some papers may have sections with slight differences in names. In this study, we analyzed the published articles' structure in CTUJS and observed the positions of each section. The Abstract section is a summary of the article and is named "Abstract", placed at the top of the article. The introduction is usually located at the beginning of the paper, right after the Abstract, and usually gives an overview of the main problem that the research paper needs to solve. Methods are usually techniques and approaches to problem-solving, usually after the "Introduction". Results section is often called "Results" or "Experimental Results", presenting the results obtained after applying the proposed method, usually right after the Methods section. The Conclusion section, which is usually called "Conclusion" or "Discussion and Conclusion", is the conclusion of the article (before References). The content of this section is to discuss the results of the problem and summarize the main ideas of the study and future directions. Based on the above characteristics, we proceed to extract the sections for the classification task.

## 3.3   Data Preprocessing

Data preprocessing in natural language processing is an important step that affects the performance and accuracy of the classification model. In this problem, data (which are structured articles of scientific articles collected from CTUJS) will be preprocessed through the following steps:

- **Extract text:** We remove distracting content such as author information, title, etc. Only keep the main content, including the summary and main content of the article (from introduction to conclusion).

- **Split sentences:** After removing redundant content, proceed to split the document into a list of sentences, keeping only sentences with a length of more than four words (because often these sentences are meaningless).
- **Remove special characters and math formulas:** Keep only Alphabet characters, and remove special characters, math characters, and punctuation marks.
- **Convert content to lowercase:** Due to the nature of the computer processor, it will understand uppercase and lowercase letters differently, so it is necessary to convert content to lowercase to reduce the number of features, increase the accuracy in the classifying topics.
- **Split words:** use the Underthesea library[2] to separate words in Vietnamese text - this is a relatively popular Vietnamese processing support library.
- **Eliminate stop words (stop words):** Stop words appear frequently but are meaningless in a sentence. Removing the stop word increases the performance and reduces the number of features of the text topic classification model. The classification model will remove the 100 most frequently occurring words in the dataset to ensure classification accuracy.

### 3.4   Classification Algorithms

We conduct a topic classification test using three classification algorithms: SVM, Naive Bayes, and Random Forest. Naive Bayes classifier [10] is an algorithm belonging to the class of statistical algorithms. SVM [11] is a supervised learning algorithm that can be used for either classification or prediction tasks. In addition, we also perform the classification with Random Forest. Random forest [12] is a popular machine learning algorithm belonging to supervised learning techniques, widely used in the ML field's classification and regression problems. We use the default hyperparameters for the Naïve Bayes model, for SVM we train the model with C=1.0 and kernel=sigmoid, while for RF we run with n_estimators = 100. All three models are chosen max_df = 0.8

## 4   Experimental Results

The scenarios include a test script for word separation and topic classification of Vietnamese and English scientific articles using three techniques SVM, Naïve Bayes, and Random Forest.

### 4.1   Environmental Settings and Dataset

In both Vietnamese and English data sets, they are divided into two parts: 80% for the training set and 20% for the test set, using Stratified sampling. This study used the data of scientific articles in English and Vietnamese collected from CTUJS for classification. The collected Vietnamese scientific article data includes 1371 articles, divided into 17 different topics, and the English data for classification includes 348 articles, divided into 13 different topics. Detailed data distribution has been mentioned in Sect. 3.1 of this study.

---

[2] https://github.com/undertheseanlp/underthesea.

## 4.2 Experimental Results

We conduct problem classification on three proposed models on data from Vietnamese and English articles. Table 1 summarizes the three algorithms' classification performance results of Vietnamese/English articles. Although Naïve Bayes has the fastest training time, It has the lowest accuracy on both the training set and test set. The accuracy on the SVM and Random Forest test sets is quite different, 60% and 62%, respectively, but SVM has the disadvantage that it takes longer to train. In the case of English articles, Random Forest has the highest accuracy on the test set of the reviewed models (55%) and takes 0.92 s to train. SVM and Naïve Bayes have relatively low accuracy, and Naïve Bayes has the fastest training speed. In general, the accuracy of English data is relatively low, partly due to the limited number of samples.

**Table 1.** Topic classification performance with algorithms on Vietnamese articles

|  | Algorithm | Accuracy | Training time (seconds) |
|---|---|---|---|
| Vietnamese articles | Naïve Bayes | 0.40 | 1.56 |
|  | SVM | 0.60 | 14.12 |
|  | Random Forest | 0.62 | 3.58 |
| English articles | Naïve Bayes | 0.41 | 0.45 |
|  | SVM | 0.41 | 1.19 |
|  | Random Forest | 0.55 | 0.92 |

In the second scenario, the data to experiment is to separate English scientific articles into five parts: Abstract, Introduction, Method, Result, and Conclusion to assess the classification ability based on each component of a text-structured article. The classification model used is Random Forest as reported in Table Table 1. Table 2 presents the performance of each topic based on different sections using the Random Forest algorithm. Fisheries and Education topics that have high classification efficiency in all sections, from 72–93%. The classification based on the whole paper gives the highest accuracy. However, it takes longer to train than other sections. If we ignore entire-paper-based classification because of the long time for processing, the classification based on the "Method" section can also be a much better alternative than the other sections.

## 4.3 Discussion

Through the experimental results above, it can be concluded that Random Forest outperforms the models considered in this study. Accuracy, F1 score in each topic, and training time proved the model's effectiveness. Moreover, under the same training conditions, with a limited number of samples such as datasets of English scientific papers, Naïve Bayes and SVM also give much worse classification performance than Random Forest. Random Forest is known as an ensemble

**Table 2.** Topic classification Performance Comparison in F1-scores on English articles using Random Forest. Some rows include Average (AVG) Accuracy (ACC) on all topics, macro AVG, and weighted AVG, respectively.

| Topic | Abstract | Introduction | Method | Result | Conclusion | All sections |
|---|---|---|---|---|---|---|
| AG | 0.00 | 0.29 | 0.00 | 0.00 | 0.67 | 0.60 |
| AV | 0.00 | 0.75 | 0.57 | 0.33 | 0.75 | 0.75 |
| BT | 0.40 | 0.40 | 0.75 | 0.33 | 0.40 | 0.75 |
| BioA | 0.00 | 0.00 | 0.00 | 0.00 | 0.00 | 0.50 |
| EC | 0.00 | 0.67 | 0.00 | 0.00 | 0.00 | 0.00 |
| ED | 0.82 | 0.83 | 0.76 | 0.78 | 0.72 | 0.80 |
| EN | 0.00 | 0.00 | 0.00 | 0.00 | 0.00 | 0.00 |
| ET | 0.18 | 0.17 | 0.50 | 0.25 | 0.20 | 0.46 |
| FI | 0.78 | 0.78 | 0.70 | 0.53 | 0.86 | 0.93 |
| FT | 0.29 | 0.33 | 0.67 | 0.33 | 0.33 | 0.33 |
| NS | 0.47 | 0.39 | 0.53 | 0.44 | 0.41 | 0.62 |
| SE | 0.00 | 0.00 | 0.00 | 0.00 | 0.00 | 0.00 |
| SH | 0.00 | 0.00 | 0.00 | 0.00 | 0.00 | 0.00 |
| **AVG ACC** | 0.44 | 0.49 | 0.53 | 0.41 | 0.49 | 0.61 |
| **Macro AVG** | 0.23 | 0.35 | 0.34 | 0.23 | 0.33 | 0.44 |
| **Weighted AVG** | 0.34 | 0.43 | 0.46 | 0.34 | 0.44 | 0.56 |
| Training time (s) | 0.52 | 0.47 | 0.58 | 0.60 | 0.38 | 0.80 |

approach [12] that can integrate and consider multiple decision trees to select the best one, while SVM and Naïve Bayes are single models. Therefore, in many cases, SVM and Naïve Bayes may not resolve complex relationships in the content of the articles as effectively as the ensemble-based techniques such as Random Forest. Therefore, Random Forest can be applied in severe data shortage conditions and still maintain accuracy. Moreover, "Method" is a crucial section in a typical research article. It may contain special terms which can be key to discriminating the research topic, so topic classification based on "Method" can obtain the best result among all considered sections. "Conclusion" can give more information on future directions compared to "Abstract", hence, it achieves better performance than "Abstract". In addition, "Introduction" which usually reveals the specific context of the research also exhibits an informative section to indicate the topic of the article.

## 5   Conclusion

In this study, we collected 1371 scientific papers in Vietnamese and 348 in English from the CTUJS to serve the work of topic classification. The results show that the Random Forest model has a better classification ability than the SVM or Naïve Bayes models, even for missing data. Extracting sections in articles gives very positive results, improving time efficiency in topic classification. Giving users more options is to choose one of the elements in an article for quick classification. Besides, training time is also an advantage of Random Forest. In addition, we analyzed the potential benefits of each separated section in a typical research article structure to perform the topic classification. As shown from the experiments, the performances vary depending on the topic. The section "Method" can discriminate the topic of articles, while "Introduction" obtained approximate performance compared to "Conclusion".

# References

1. Ghanem, R., Erbay, H.: Spam detection on social networks using deep contextualized word representation. Multimedia Tools Appl. **82**(3), 3697–3712 (2022). https://doi.org/10.1007/s11042-022-13397-8
2. Dien, T.T., Loc, B.H., Thai-Nghe, N.: Article classification using natural language processing and machine learning. In: 2019 International Conference on Advanced Computing and Applications (ACOMP). IEEE (2019). https://doi.org/10.1109/ACOMP.2019.00019
3. Yang, Y., Liu, X.: A re-examination of text categorization methods. In: Proceedings of the 22nd Annual International ACM SIGIR Conference on Research and Development in Information Retrieval, pp. 42–49 (1999)
4. Kadhim, A.I.: An evaluation of preprocessing techniques for text classification. Int. J. Comput. Sci. Inf. Secur. (IJCSIS) **16**(6), 22–32 (2018)
5. Nguyen, H.T., Le, A.D., Thai-Nghe, N., Dien, T.T.: An approach for similarity vietnamese documents detection from English documents. In: Future Data and Security Engineering. Big Data, Security and Privacy, Smart City and Industry 4.0 Applications, pp. 574–587. Springer Nature Singapore (2022). https://doi.org/10.1007/978-981-19-8069-5_39
6. Kadhim, A.I.: Survey on supervised machine learning techniques for automatic text classification. Artif. Intell. Rev. **52**(1), 273–292 (2019). https://doi.org/10.1007/s10462-018-09677-1
7. Hao, P., Ying, D., Longyuan, T.: Application for web text categorization based on support vector machine. In: 2009 International Forum on Computer Science-Technology and Applications. IEEE (2009). https://doi.org/10.1109/ifcsta.2009.132
8. Kim, S.B., Han, K.S., Rim, H.C., Myaeng, S.H.: Some effective techniques for naive bayes text classification. IEEE Trans. Knowl. Data Eng. **18**(11), 1457–1466 (2006). https://doi.org/10.1109/tkde.2006.180
9. Structuring your manuscript | Springer - International Publisher, https://www.springer.com/gp/authors-editors/authorandreviewertutorials/writing-a-journal-manuscript/author-academy/10534936
10. Webb, G.I., Keogh, E., Miikkulainen, R., Miikkulainen, R., Sebag, M.: Naïve bayes. In: Encyclopedia of Machine Learning, pp. 713–714. Springer, US (2011). https://doi.org/10.1007/978-0-387-30164-8_576
11. Cortes, C., Vapnik, V.: Support-vector networks. Mach. Learn. **20**(3), 273–297 (1995)
12. Ho, T.K.: Random decision forests. In: Proceedings of 3rd International Conference on Document Analysis and Recognition, vol. 1, pp. 278–282. IEEE (1995)

# Fake News Detection Using Knowledge Graph and Graph Convolutional Network

Vy Duong Kim Nguyen[(✉)] and Phuc Do

University of Information Technology (UIT),
Ho Chi Minh National University, Ho Chi Minh City, Vietnam
vyndk.15@grad.uit.edu.vn, phucdo@uit.edu.vn

**Abstract.** Detecting fake news problems has been a topic of research noticed in recent years. However, at present, fake news detection research has only been conducted with English data sets, no research has been conducted with Vietnamese datasets. Therefore, it is essentially important and urgent to do research on fake news detection with the Vietnamese dataset. In this research, we proposed a detecting fake news method using a knowledge graph (KG) combined with a semi-supervised learning Graph Convolutional Networks (GCN) to predict whether the news is real or fake. Our research is implemented by collecting data including real and fake news from the datasets of online newspapers in Vietnamese including vnexpress, tuoitre, etc., and VFND-vietnamese-fake-news. In addition, we will not label news that has not been verified for authenticity. We build a knowledge graph for the dataset by embedding words into the dataset using the Glove library and building a knowledge graph using the Word Mover's Distance algorithm (WMD) combined with the K-nearest-neighbor algorithm (KNN). Detecting real or fake news models with input knowledge graph built and GCN algorithm and finally perform the tests with datasets. With our research, we have built a knowledge graph for the Vietnamese data set of real and fake news collected from 2018 to 2023 combined with the VFND-vietnamese-fake-news dataset. Our proposed method can resolve to detect fake news problems by the GCN algorithm. Research's results have a high precision of up to 85%, in the case of labeled data accounting for 50% of the input dataset.

**Keywords:** Semi-supervised classification · Graph Convolutional Network · K-Nearest Neighbor · Word Mover's Distance

## 1 Introduction

Accompanied by the strong development of the Internet today, the quantity of news is increasing, by various communication media, along with the quantity of fake, and false news also increasing. This leads to many serious consequences, affecting faulty social life in many different fields. Consequently, fake news detection problems are a topic of concern and research implemented in the recent past. Researchers in the field of fake news detection mention that this is an extremely important problem today. However, there are many current practical challenges such as the quantity of news exploding. The news needs to be cross-checked for the origin, and content of news. Besides, we need to demonstrate all relevant facts to determine whether the news is real or fake. Nonetheless, identifying trust in any news is very difficult in most scenarios. The real or fake news must match the specific topics of news in the actual world such as sports, culture, society, economy, law, medicine, health, etc. The fake news detection framework based on KG approach (see Fig. 1) is divided into the following stages:

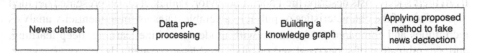

**Fig. 1.** Stages in fake news detection framework

In addition, Graph Neural Networks (GNN) is a group of new techniques based on focusing on deep learning techniques on graph-based structure. Before being applied GNN to fake news methodology, GNNs also had been completely implemented in different machine learning algorithms as well as NLP methodology including object detection, sentiment analysis, and machine translation with many successes. Moreover, the fast development of many GNNs has been gained by enhancing many different variants of GCNs in many real-world applications such as recommendation systems, healthcare, e-commerce, marketing, traffic management, selling retail, etc.

In favor of resolving the circumstantial that labeled datasets for fake news detection in the actual world are commonly extremely limited and sparse. In this research, we propose a method using KG and semi-supervised classification to detect the unlabeled news is fake or real. The main contributions of our research can be summarized as follows:

- Developing Vietnamese dataset fake news, using a collection of news collected from Internet newspapers such as vnexpress, tuoitre, baomoi, etc. We combine VFND-Vietnamese-fake-news dataset. Finally, we have a dataset with a wide variety of different topics, using input data for the proposed method.
- Using the word embeddings method to get implicit representations of the dataset about all news into the lower-dimensional Euclidean space. After that, we proceed to catch a recapitulative resemblance between news by a graph visualization schema.
- Implementing the fake news detection subject using a semi-supervised classification method to detect whether a new article is real or fake, using Graph Neural Network techniques that are able to practice proficiently with extremely limited labeled data.
- Accomplishing a general evaluation of the proposed method in our research with the real fake news dataset, representing that the proposed method has more outstanding than previous content-based approaches, and also requires fewer labels in the dataset.

## 2  Related Works

The research topic of detecting fake news has grown into an emerging subject in modern communication media. Current veracity analysis methods can be broadly classified into a couple of sections [6, 7]: utilizing news data for handling social information/context, extracting linguistic, syntactic, lexical features or visual characteristics from news articles to obtain particular features as well as interesting patterns that usually happen in fraudulent documents.

Visual features are used to recognize fake photos that are intentionally created to create realism in fake news. Context-based authenticity analysis methods include the characteristics of user data on social sites and social systems [8]. The profile of the user's social site is used to estimate the user's features and reliability. Traits derived from their social posts describe their social behavior. Pathak et al. [9] presented detailed discussions of the characteristics of publicly available datasets for authenticity analysis techniques and compared performance with modern machine learning, deep learning, and a combination of supervised and unsupervised learning methods.

The modern survey of semi-supervised and unsupervised methods in the reliability analysis of web content is detailed by Saini et al. [10]. Another research using heterogeneous Graph Neural Networks for the AAAI2021-COVID-19 English dataset to detect fake news on social networks is proposed by Andrea Stevens Karnyoto et al. [11]. Nikhil Mehta et al. [12] proposed a method named Continually Improving Social Context Representations using the Graph Neural Networks method.

Research [11] and [12] are state-of-the-art research in the recent past using the Graph Neural Networks method to detect fake news problems and up to 85% accuracy in the author's research. However, they are completed with datasets of a specific topic about COVID-19 news which does not cover the other topics in the real world.

## 3 Methodology

### 3.1 Document Embeddings

Document embeddings in Euclidean space is a method that helps transform documents into multi-dimensional vectors by using the word embeddings method. In, word embedding is a constructing process of equivalence operations, impressions in the mathematical structure of each word from the right corpus.

A news article embedding is typically trained using unsupervised algorithms and constructed through transfer learning from large-scale unlabeled datasets. This process involves converting the words in a document into vectors of a specified dimension within a specific vector space. The conversion from words to vectors employs a neural network approach to capture the essential values of these vectors. The number of dimensions for a vector can range from a small integer to several hundred. By visualizing the vectors, words with similar or related meanings exhibit proximity, indicating that the overall semantic meaning of words has been captured through an appropriate learning method. The proposed approach in this research involves utilizing comprehensive word vectors to embed all the words within each news article in the datasets, visualizing them in Euclidean space. To word vectors into an n-dimensional space, the procedure ensures that similar words are situated closer to one other, while words with different connotations are positioned farther apart. One of the notable advantages of the GloVe method, in comparison to other embedding methods like Word2Vec, is its emphasis on global statistics rather than local statistics.

## 3.2 Building a Knowledge Graph (KG)

This classification step is mandatory with graph data input and is accomplished by building up a resemblance graph visualizing the proximity among various nodes, i.e., for our research, it is every news article. In most scenarios in the actual world, we can visualize any graph in the form of a k-nearest-neighbor graph. Especially, toward every news article, we prospect K-Nearest-Neighbors in a news article embedding space. The K Neighbors can be confirmatory by manipulating WMD amidst every news article with the others. The KNN is defined as a graph in which two nodes "p" and "q" exist on one edge amidst them if node "p" is a node of the k nearest vertices in all other different neighbors and reverse. Any of the nearest KNNs represent one locality in n-dimensional space and is specified to use one "close relationship" anywhere proximity normally supposed in distance to measure words as a WMD [13]. Therefore, for a given problem in vector space, we build the KNN graph of the points by manipulating the distance amongst each couple of news articles and relating each news article to the nearest k points.

The WMD measure is a function of the distance between the articles. In our research, a news article is represented by a article, with the lowest text classification error rate. An accurate representation of the distance between two documents is useful in document compensation, text clustering, multilingual document matching, and many others. With two news documents are "A" and "B", in which |A| and |B| as the count of exclusive words that appear within both of them completely [13].

Every word "W" of news article "A" is transformed into any word of news document "B". $T_{ij} >= 0$ represents the separation that the word "W" in news document "A" has to move to transform to word "W" in news document "B". T is the streaming matrix stray. Toward turn to news document "A" totally in news document "B", we determine the accomplished outflow from $i$ to j must match B represented mathematically in Eq. (1) as follows:

$$\sum_j T_{ij} = A_i \text{ and } \sum_j T_{ij} = B_j \tag{1}$$

The WMD separation among two news documents is meant as the smallest weighted accumulative cost mandatory in favor of transforming every word from news document "A" to news document "B" expressed in Eq. (2):

$$WMD(A, B) = \sum_{ij} T_{ij} c(w_i, w_j) \tag{2}$$

The official WMD amidst two news documents is meant the optimized explanation value to the sequence transport clarification, this is a distinctive circumstance of the Earth's distance in Eqs. (3)–(5):

$$min_{T \geq 0} \sum_{i=1}^{|A|} \sum_{j=1}^{|B|} c(w_i, w_j) T_{ij} \tag{3}$$

$$\sum_{j=1}^{|B|} c(w_i, w_j) T_{ij} = A_i \quad \forall i \in (1, 2, 3, .., |A|) \tag{4}$$

$$\sum_{j=1}^{|A|} c(w_i, w_j) T_{ij} = B_j \quad \forall i \in (1, 2, 3, .., |B|) \tag{5}$$

In this research, we will build a KG model for real and fake news datasets; each node is performed by all entities of a news article, and each edge is the distance between that paper and other news articles. Likely a graph is an impression as the KNN graph. Specifically, with each news article, we intend to the nearest neighbor in a specific embedding space. The k-nearest neighbor can be defined by computing the distance of Word Mover among a news article and others. A KG usually details all entities and the relationships amidst them in the dataset. A KG can be determined as KG = {E, R, S}, in which E manifests the set of entities, R is the set of relationships, and S is the triple set [13]. A news article "A" is a set of entities in this news article, including the title, the full content, and the label that is determined true or false, which may be unlabeled. A KG is also effective with content fact-checking problems and detecting fake or real news articles. As the final result of building a KG, we have a KG with vertices representing entities in news articles, edges are each relationship between these entities. The main mission of content fact-checking is to check if a target triple (h, r, t) is correct and strongly an established KG. The object of detecting fake news problems is to check if a target news article is correct strongly firm its title and main content and other homologous KG. The final KG is used for input data for the semi-supervised classification method of detecting fake news (see Fig. 2).

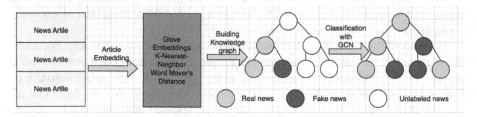

**Fig. 2.** Knowledge Graph with Glove Embeddings for fake news detection

### 3.3 Graph Convolutional Network (GCN) for Graph Classification

Neural Network algorithms have achieved huge success in recent years. Nonetheless, proximate variants of Neural Networks only could be practiced with ordinary or Euclidean data, while most data in the actual world have graph-based structure essentials, meaning they are non-Euclidean data. As a result, the data structure revolution has guided modern improvements in Graph Neural Networks.

The GCN method is one of the noticeable divergent Graph Neural Networks techniques being applied in many actual applications using non-Euclidean actual world data based on graph-based structures. GCN is an improved model of a convolutional neural network that operates straightway with graph-based structures. The preference for a convolutional methodology is encouraged through a localized primary approximation of spectral graph convolutions. The model scale is linear to the count of edges of the histogram using a powered layer-spread rule based on a primary estimation of the spectral textures on the graph. The GCN model can learn hidden layer representations encoding the pair node features and graph-based structure to a degree helpful with semi-supervised algorithms. The major purpose of GCN is to find out the function of features on a graph by taking an input character matrix X of size N x D in which N is the count of nodes, and D is the count of features. Matrix "A" is an adjacency matrix to represent every input graph structure.

## 4  Results and Discussion

We used accuracy, precision, recall, and F1-score metrics to evaluate the performance of the proposed model with the dataset. Tests were performed with two other values of K (including K = 3, K = 5), and different ratios of the labeled training data from 20% to 50%. GCN involves layers = 5 to feature extraction with intermediate indiscreet ReLU effect and dropout layers. GCN model with 5 layers is trained for 1000 epochs, dropout = 0.5, learning rate = 0.005, hidden = 16, and $weight\_decay = 5^{e-4}$ (see Table 1).

**Table 1.** Performance evaluation results of the model with K = 3 and K = 5

| %<br>Labeled<br>Data | K = 3 | | | | K = 5 | | | |
|---|---|---|---|---|---|---|---|---|
| | Accuracy | Precision | Recall | F1-score | Accuracy | Precision | Recall | F1-score |
| 20% | 60.5% | 63% | 62.7% | 62.9% | 63.5% | 66.8% | 64.6% | 65.6% |
| 30% | 70% | 74% | 73.8% | 72.8% | 73% | 77.9% | 75.6% | 76.9% |
| 40% | 75.1% | 78.8% | 77.5% | 77.8% | 76% | 81.9% | 80.8% | 81.3% |
| 50% | 80% | 82% | 81.7% | 81.6% | 84.7% | 85.1% | 83.7% | 83.9% |

Based on the results of Table 1, we hold that with K = 5, the measures to evaluate the performance of the model will be better than that of K = 3. Besides, as the amount of labeled data increases, the precision of the detecting fake news method also increases. The proposed method in our research achieves the highest precision 85.1% in the case of labeled data accounting for 50% of the total input data in our research. With the results of our research, when comparing with the detecting fake news results from other research in the recent past, we have the following (see Table 2):

When looking into the comparison results, we find that the results of the research model are superior to other fake news methods when testing the same fake news dataset in the research. This result is achieved by applying GCN and WMD to calculate the distance between news, helping the KG to have high accuracy. We believe the proposed method in the research has a high effect in detecting fake news, especially with the Vietnamese dataset. This is a step that creates a very important premise in fake news detection with the Vietnamese dataset, something that no previous research has done. Moreover, when compared with the state-of-the-art method, the construction of a proficient dataset, with a wide variety of many different topics, our proposed method has better results than method 5, which is the most state-of-the-art method of detecting fake news in the recent past but it only detects real or fake news with Covid-19 topic. Moreover, integration

**Table 2.** Comparison of research results with other fake news prediction methods

| Method | Accuracy (%) | Precision (%) | Recall (%) | F1-score (%) |
|---|---|---|---|---|
| Method 1as [16] | 82.5 | 83 | 83.4 | 83.6 |
| Method 2 as [17] | 81.2 | 82.3 | 80.9 | 81.5 |
| Method 3 as [18] | 78.3 | 79.4 | 80.1 | 80.5 |
| Method 4 as [19] | 80.1 | 81.2 | 82.2 | 81.8 |
| Method 5 as [11] | 82.3 | 81.5 | 81.9 | 82.5 |
| Proposed method in our research | **84.7** | **85.1** | **83.7** | **83.9** |

between actual, and contextual depictions, gives knowledge of graph-based depictions with the best performance.

## 5 Conclusion and Future Works

Fake news is one of the burning problems and causes many severe consequences in today's society. Nonetheless, one of the most difficult in detecting fake news is collecting data and labeling each news article in such a huge data block today for the input of supervised learning algorithms. Therefore, we proposed a method for our research. The main benefaction of the research included developing detection datasets in Vietnamese, building KG models for Vietnamese datasets, and detecting fake news for Vietnamese data using KG. The KG for fake news detection with Vietnamese news articles is an important key in our research, with the advantage is to fully representing the meaning of the entities in each news article and the relationship between the entities, making it more effective to verify the accuracy of the news article than the article data many times. Besides, this is also the first complete KG dataset for detecting fake news in the Vietnamese language. Based on the implementation results, clearly shows that our proposed method achieves a precision is 85.1% with the dataset developed in our research, and the labeled ratio is 50%.

The drawback of our research is input dataset is collected from online newspapers but does not include news articles from social media. To improve the performance of the model, we collect data from different newspapers and social networks such as Zalo, Facebook, etc. with many different topics to diversify input data. Besides, we will experiment with other Vietnamese data pre-training models like PhoBert, to enhance the precision of the model in scenarios where the labeled data is too small compared with the total input dataset in the model. We also aim to improve the semi-supervised learning algorithm that classifies real or fake news by training with various parameters so that we can improve the precision of our proposed method.

## References

1. Villela, H.F., Corrêa, F., Ribeiro, A., Rabelo, A., Carvalho, D.: Fake news detection: a systematic literature review of machine learning algorithms and datasets. J. Interact. Syst. **14**, 47–58 (2023)

2. Koloski, B., Perdih, T.S., Robnik-Šikonja, M., Pollak, S., Škrlj, B.: Knowledge graph informed fake news classification via heterogeneous representation ensembles. Neurocomputing **496**, 208–226 (2022). https://doi.org/10.1016/j.neucom.2022.01.096

3. Kim, G., Ko, Y.: Graph-based fake news detection using a summarization technique. In: Proceedings of the 16th Conference of the European Chapter of the Association for Computational Linguistics (2021)

4. Bondielli, A., Marcelloni, F.: A survey on fake news and rumor detection techniques. Inf. Sci. **497**, 38–55 (2020)

5. Phan, H.T., Nguyen, N.T., Hwang, D.: Fake news detection: a survey of graph neural network methods. Appl. Soft Comput. **139**, 110235 (2023). https://doi.org/10.1016/j.asoc.2023.110235

6. Zhang, Q., Guo, Z., Zhu, Y., Vijayakumar, P., Castiglione, A., Gupta, B.B.: Content-based fake news detection with machine and deep learning: a systematic review. Elsevier **168**, 31–38 (2023)

7. Sadeghi, F., Bidgoly, A.J., Amirkhani, H.: Fake news detection on social media using a natural language inference approach. Multimedia Tools Appl. **81**(23), 33801–33821 (2022). https://doi.org/10.1007/s11042-022-12428-8

8. Meel, P., Vishwakarma, D.K.: Fake news, rumor, information pollution in social media and web: a contemporary survey of state-of-the-arts, challenges and opportunities. Expert Syst. Appl. **153**, 112986 (2020)

9. Rana, R.T., Meel, P.: Rumor propagation: a state-of-the-art survey of current challenges and opportunities. In: 2nd International Conference on Intelligent Communication and Computational Techniques (ICCT), Jaipur, India (2021)

10. Saini, N., Singhal, M., Tanwar, M., Meel, P.: Multimodal, semi-supervised and unsupervised web content credibility analysis frameworks. In: IEEE 4th International Conference on Intelligent Computing and Control Systems (ICICCS), Madurai , India (2020)

11. Karnyoto, A.S., Sun, C., Liu, B., Wang, X.: Augmentation and heterogeneous graph neural network for AAAI2021-COVID-19 fake news detection. Int. J. Mach. Learn. Cybern. **13**, 2033–2043 (2022). https://doi.org/10.1007/s13042-021-01503-5

12. Mehta, N., Pacheco, M.L., Goldwasser, D.: Tackling fake news detection by continually improving social context representations using graph neural networks. In: Proceedings of the 60th Annual Meeting of the Association for Computational Linguistics, vol. 1, pp. 1363–1380 (2022)

13. Ullah, A., Abbasi, R.A., Khattak, A.S., Said, A.: Identifying misinformation spreaders: a graph-based semi-supervised learning approach. In: MediaEval 2022: Multimedia Evaluation Workshop (2023)

14. Li, X., Lu, P., Hu, L., Wang, X., Lu, L.: A novel self-learning semi-supervised deep learning network to detect fake news on social media. Multimedia Tools Appl. **81**, 19341–19349 (2022)

15. Raj, C., Meel, P.: ConvNet frameworks for multi-modal fake news detection. Appl. Intell. 1–17 (2021)

16. Bali, A.P.S., Fernandez, M., Choubey, S., Goel, M., Roy, P.K.: Comparative performance of machine learning algorithms for fake news detection. In: International Conference on Advances in Computing and Data Sciences, Springer, Singapore (2021)

17. Agarwalla, K., Nandan, S., Nair, V.A., Hema, D.D.: Fake news detection using machine learning and natural language processing. Int. J. Recent Technol. Eng. (IJRTE) **7**(6), 844–847 (2022)

18. Karimi, H., Tang, J.: Learning hierarchical discourse-level structure for fake news detection. In: Proceedings of the 2022 Conference of the North American Chapter of the Association for Computational Linguistics: Human Language Technologies, vol. 1, pp. 3432–3442. Association for Computational Linguistics, Minneapolis, Minnesota (2022)
19. Vishwakarma, D.K., Varshney, D., Yadav, A.: Detection and veracity analysis of fake news via scrapping and authenticating the web search. Cognit. Syst. Res. **58**, 217–229 (2019). https://doi.org/10.1016/j.cogsys.2019.07.004

# Intelligent Systems

# PETSAI: Physical Education Teaching Support with Artificial Intelligence

Thanh Ma(✉), Thanh-Nhan Huynh, Viet-Chau Tran, Bich-Chung Phan,
Nguyen-Khang Pham, and Thanh-Nghi Do

CICT, Can Tho University, Can Tho City, Vietnam
{mtthanh,tvchau,pbchung,pnkhang,dtnghi}@ctu.edu.vn

**Abstract.** Physical Education Subject (PES) is a challenging problem
for conveying IT into education. From the perspective of Artificial Intelligence (AI) and realizing PES's abstractness in knowledge representation,
we pay attention to the idea of an intelligent system with a combination
breath of computer vision and speech recognition. Inspired by that vision,
we propose a support system called PETSAI for teaching the movements
of the PES. Our primary idea is to introduce a "audio-visual" framework
to compute the scores of the movements. To this end, we leverage human
pose estimation (HPE) to detect a student's skeleton for each movement
and utilize convolutional neural networks (CNN) for motion and gesture
classification. In addition, we also use the power of speech recognition
to control the entire framework. Two operations, the speech and image
processes, will be performed in parallel. Moreover, we also provide an
algorithm to calculate scores (from gestures to movements). Finally, we
implement the system on both platforms, including the website and desktop application.

**Keywords:** Physical Education Subject · Human Estimation Pose ·
Machine Learning · Speech Recognition · Convolutional neural network

## 1 Introduction

Education associated with technology [5,9] is a remarkable issue that most governments are concerned about. Applying information technology (IT for short)
to lectures will be helpful to aid lessons in becoming animated, intuitive, and
straightforward to transfer knowledge. In addition, during the recent COVID'19
pandemic, all subjects were taught and assessed through software applications
and information systems [10]. In general, most subjects are straightforward to
leverage IT in pedagogy. However, "physical education" is one of the "quite
abstract" subjects, which will be challenging to bring techniques in the lessons
[1,14], i.e., assignments, passive presentation, evaluation methods, and others.
Typically, the teacher utilizes the slideshow (i.e., PowerPoint) for each les-
son/class, which is complicated to leverage novel and advanced technologies for
the physical education subject (PES). Significantly, the assessment method for
the PES is typically eye-to-eye [20] and face-to-face [18].

N. Thai-Nghe et al. (Eds.): ISDS 2023, CCIS 1949, pp. 227–242, 2024.
https://doi.org/10.1007/978-981-99-7649-2_18

On the other hand, physical education (PE, for short) is an integral part of the school's comprehensive education, the most active measure to protect and enhance students' health, promote the complete development, rhythm, and balance of the body, enhance the quality and improve the mobility of students. But currently, most children are "lazy to exercise". Although they are "practicing", their "feeling" is uninterested and forced exercise, leading to ineffective training and not being maintained regularly. Children's health will be reduced not only physically but also mentally. This issue will significantly affect their study, work, and other activities in life.

From realizing the importance of PE and the potential of applying artificial intelligence (AI) to the educational environment [6,25], we pay attention to this aspect. Moreover, to the best of our knowledge, no research currently utilizes AI to support teaching physical education subjects *(in Vietnamese)*. Hence, we propose a framework to reinforce teaching for the PES. To this end, we take inspiration and motivation from studies on supporting Yoga [3,7,11] and material arts (i.e., kung-fu) [13,23,26] using AI to help practitioners. Our main idea is a computing score system to evaluate each exercise movement. Through the system, students can practice at home or assist teachers in grading. Namely, we use the Human Estimation Pose [19,28] to find skeletons on the body. Then, based on the details of the bone structures, we propose an algorithm to calculate the score for each exercise movement. Furthermore, we apply the convolutional neural network (CNN) [17,30] to identify/classify movements to aid in automatic scoring. In addition, to control the entire system, we aim to use speech recognition to operate the change of movements.

Regarding the PES, there is a comprehensive multiplicity of exercises at different levels of education. i.e., elementary, middle, and high school. For this investigation, we concentrate predominantly on the "continuous exercises" of the PES program for elementary school students in Vietnam. We work on the real dataset captured from a 2D camera. Another remarkable point is that we calculate scores over a series of actions for each movement (on multi-frames). It will also be a challenge for computing scores on a continuous sequence of frames[1]. In this paper, our primary contributions are: (1) Collecting a set of movement pictures/videos for the gestures classification; (2) Each movement consists of several gestures. We provide eight CNN classification models for each movement; (3) Introduce an audio-visual framework to operate the entire system; (4) A scoring algorithm based on the angle of skeletons has been proposed.

This research focuses on operating the 2D camera instead of utilizing the Microsoft Kinect Camera (MKC) [27]. To deploy our system in practice in Vietnam, we select ordinary 2D cameras that will be more suitable for students because the cost of MKC is high. Knowing that 2D space lacks information will be an exciting discovery for research. We will also indicate the constraints for the system to work efficiently. Moreover, each movement includes many gestures. We do not classify all gestures we conduct on building the classification models

---

[1] We will discuss in detail in Sect. 4.

for motions of each movement. The reason is that several gestures are the same and quite similar. We mention this problem in detail in the following sections.

The remainder of this paper is structured as follows: We briefly present a fundamental background in Sect. 2. Next, Sect. 3 describes the proposed framework. Then, Sect. 4 introduces how to compute movement scores. Section 5 presents the experiment and the results of the summary models. Finally, Sect. 6 shows the conclusions and future works.

## 2   Background

In this section, we introduce the movements considered for this study. Moreover, we also provide the fundamental knowledge of detecting the skeleton and machine learning models. Then, we discuss Speech Recognition applying our framework.

### 2.1   Physical Education in Vietnam

We concentrate on the exercises of "*Physical Education Book $3^2$,*" compiled to develop children's qualities and abilities. The scopes and forms of exercise are represented according to the Vietnamese Ministry of Education's standards. They are suitable for students in grade 3. The teaching content keeps the motto of taking learners as the center, promoting maximum capacity and qualities of learners.

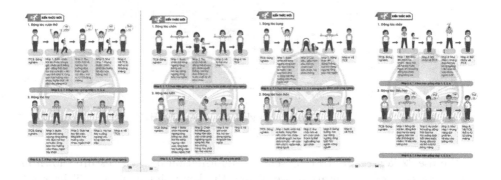

**Fig. 1.** Eight movements of grade 3's physical exercise

This paper explores and represents our idea on eight sequence movements presented in Fig. 1: *(1) Stretching, (2) Hands, (3) Legs, (4)Waist, (5)Abs, (6)Full Body, (7) Jumps, and (8)Body Conditioning.* We denote $\Phi$ as a movement and $\Omega$ as a set of movements ($\Omega = \{\Phi_1, \ldots, \Phi_8\}$). In the teaching program, each movement has many gestures. We denote a gesture as $\alpha$. Then, we have

---

[2] https://chantroisangtao.vn/he-tai-nguyen/giao-duc-the-chat-3-sach-hoc-sinh-bo-sach-giao-khoa-chan-troi-sang-tao/.

$\Phi_1 = \{\alpha_1, \ldots, \alpha_i\}$ where $i$ is the number of gesture in each $\Phi_1$. The pre-order of $\alpha$ is significant. We write $\alpha_2 < \alpha_3$ to say that $\alpha_2$ must be done before $\alpha_3$. To apply AI, we propose machine learning models to classify each movement's gestures. Note that we do not classify all gestures because multiple motions may overlap if putting them together. Hence it will not make sense when the gesture is duplicated. Especially, they lose the order of gestures in a series of motions, i.e., $\alpha_1 < \alpha_2 < \alpha_3$.

To identify the skeleton, we introduce a foundation of detecting the poses in the next subsection.

## 2.2  Human Pose Estimation

Human pose estimation (HPE) [2] is one of the heart of a computer vision task to represent a person's posture in a graphical configuration. A pose estimation model is responsible for identifying and distinguishing the joints of the human body. The links between key-points are called pairs. We have a human skeleton when connecting key-points and pairs. Placing many skeletons in one frame is also known as multi-person pose estimation. In this paper, we only play with one person in each frame. Based on reliability and comparison between algorithms [8], we decide to choose the Mediapipe library[3] to use in our framework.

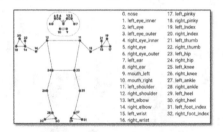

**Fig. 2.** Thirty-two joints of the human skeleton from the HPE of the Mediapipe

We use 32 key-points to represent the information of each gesture shown in Fig. 2. For ease of presentation, we denote the key-point as $\psi$. Let $\Psi$ be a set of key-points ($\Psi = \{\psi_i\}$). In general, we just use key-points at hands and legs, i.e., right-hand includes $\psi_{12}, \psi_{14}, \psi_{16}$. We highlight that these key-points will be utilized to compute the scores. They do not use to determine the movements. We identify the gestures of each movement based on the deep learning models presented in the following sub-section.

---

[3] https://developers.google.com/mediapipe.

## 2.3   Classification with Convolutional Neural Network

Convolutional neural network (CNN) [15] was developed from an original idea inspired by how the human eye works, breaking down images into convolutions for neurons to process. In 2015, CNNs were introduced by Bengio, Le Cun, Bottou, and Haffner. Their first model was named LeNet-5 [16]. In this paper, we investigate the CNN models with various different architectures, i.e., Lenet [16], Mobilenet [22], Densenet [29], and others. A result comparison table of models (i.e., accuracy, Recall, Precision, F1, and consumption time) will be provided to evaluate the models' effectiveness. Since each movement has a different sufficient classifier, hence, we build eight classification models that are efficient and plausible. Now, we denote $\Delta$ as a classifier. We define the prediction result $\mathfrak{R}$ from a frame $f$ (i.e., gesture) as follows: $\mathfrak{R} = \Delta(f_i)$. Here, $\mathfrak{R}$ will be one of $\alpha_i$ in each movement.

To operate the entire framework, we leverage the voice speech technique shown in the following section.

## 2.4   Speech Recognition

Speech recognition (SR) [12,21], or speech-to-text (STT), is the power of a machine or program to determine words spoken audibly and convert them into readable text. More refined applications can deal with natural speech, diverse accents, and various languages. The SR utilizes multiple computer science, linguistics, and computer engineering studies. Many SR applications have been deployed in real life, i.e., Google Assistant, Siri, Alexa, and others [4]. In this paper, we select "google assistant" (GA for short) to implement for our framework. We choose GA because it supports the Vietnamese language and is familiar to Vietnamese individuals.

GA is a virtual assistant software application invented by Google that is preeminent and available on mobile and home automation devices. Based on AI, the GA can engage in two-way conversations. Most of the speech recognition are supported in English. In addition to Apple's Siri and Amazon's Alexa, the GA allows working with Vietnamese. The benefits of the GA [24] are (1) Searching for information by voice saves time. (2) Speech-to-text is software that works by listening to sound and providing a transcript.

Note that we apply the GA to command/control the system. We define a list of "Movements" (in Vietnamese) to identify the keywords for each movement. We explicitly present this in the next section.

## 3   PETSAI Framework

We introduce the speech-visual framework to compute the scores for each movement. Our framework is a combination of various AI techniques. It includes seven steps as presented in Fig. 3. The idea of the framework consists of two main parts, including (1) Using speech to request movements, i.e., student A says "Hand movement" *(in Vietnamese)*. Then, our system determines the movement name $\Phi$ from the fixed keyword list ($\Phi \in \Omega$) and loads the classification model $\Delta$ for the gestures $\alpha_i$ of the hand movement to transfer into the image part $\alpha_i = \Delta(f_i)$ where $f_i$ is a sequence of image frames; (2) Regarding the "vision" part, we use the image classifier to identify gestures. Then, we compute the score for each gesture.

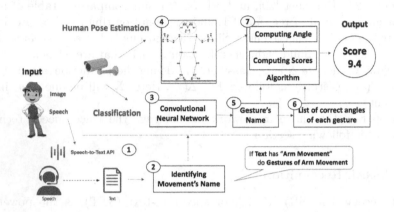

**Fig. 3.** PETSAI Framework for supporting physical education teaching

An important point is that we are not only calculating the *local score* on a frame *(corresponding to a gesture $\alpha$)* denoted as $\mathcal{LS}$, but also we carry out the median of scores of the $n$ frames to determine the *glocal score* of a gesture denoted $\mathcal{GS}$. Namely, we have $\mathcal{GS} = median(\{\mathcal{LS}_1, \ldots, \mathcal{LS}_n\})$. Furthermore, for each movement (a.k.a. a serials of gestures $\alpha$), we calculate the score of each movement $\Phi = \{\alpha_i\}$, denoted $\mathcal{SM}$, as follows: $\mathcal{SM} = average(\{\mathcal{GS}_i \mid i \leq \mid \Phi \mid\})$. Here, an explanation why we use the *median* and *average* function will be presented in Sect. 4. Now, we explicitly present seven steps of our PETSAI framework (see them in Fig. 3 as follows:

1. We use "Google Assistant" to conduct *Speech-to-Text*. Namely, the student says a sentence, then the system uses the GA API to convert the sentence into text, i.e., *Text* = "Use the leg movement";
2. From the sentence collected in Step 1, if the movement name belongs to the speech sentence, the new movement will be updated. Here we build a specified list of the keywords of the movement's name in Vietnamese (a.k.a. $\Omega = \{\Phi_1, \ldots, \Phi_8\}$). For example, we assume that $\Phi_4$ = "leg movement", if $\Phi_4 \in Text$ then we will conduct $\Phi_4$ in the next step;
3. After identifying the movement's name, i.e., $\Phi_4$, we use the CNN model $\Delta$ selected to classify gestures ($\Phi_4 = \{\alpha_1, \alpha_2, \alpha_3, \alpha_4\}$). For example, at $25^{th}$ frame $f_{25}$, we obtain $\alpha_3 = \Delta(f_{25})$;
4. In Step 4, we detect the human skeleton by the HPE. Here, we use Mediapipe library (mentioned above) to determine the joints;

5. From the step 3, we identify the gesture's name in Step 5;
6. After we have a specific gesture, we collect that gesture's "correct" angles. Here, one question arises: "how are the correct angles?". To answer this question, we remind that camera 2D's weakness is lacking depth dimension information. Hence, we fix the camera and get the expert's opinion to have the correct angles;
7. Finally, in Step 7, after obtaining the gesture name $\alpha$ (i.e., $\alpha_3$ in Step 3), determining the "correct" angle of that gesture, and identifying the joints, we propose an algorithm to compute the score.

In this paper, we set up several constraints to have a set of correct angles from experts. Namely, the distance between the student and the camera is about 2.0 to 2.5 m. The camera is positioned on the right side at an angle of 250 to assist the system in recognizing the movements. We have to move the camera angle because if the camera faces the students, some movements will not be visible because of overlap. For example of the correct angle $X$ of *Hand-Gesture-1*: $X = Angle(\psi_{11}) = 95$, $X = Angle(\psi_{13}) = 170$, $X = Angle(\psi_{12}) = 95$, $X = Angle(\psi_{23}) = 167$, and others.

To understand clearly how to compute the score of movements, we present in detail in the next section.

## 4    Movement Score Computation

In this section, we provide how to compute the score of each gesture (called "local" score). Before doing this, we need to compute the point of each joint. Moreover, we also present a ScoCa algorithm proposed for "global" score.

### 4.1    Angle Calculation and Local Score

We compute the score based on the angle. Hence, we cover the formula for calculating the angle between joints in this subsection. Namely, when we have two points A, B, we can identify two vectors $\vec{u}$ and $\vec{v}$. To compute the (directed) angle from vector 1 $u(x, y)$ to vector 2 $v(x, y)$ illustrated in Fig. 4, we conduct the following formula[4]: $\varphi = atan2(v_y, v_x) - atan2(u_y, u_x)$

We use this formula to compute the angles of joints. Namely, based on the coordinates of the image plane, we determine the position of the joints. Now, we calculate the angle between three adjacent points from the joint position coordinates. For example, the left hand has three joints, including joint 11 ($\psi_{11}$), joint 13 ($\psi_{13}$), and joint 15 ($\psi_{15}$). We have two vectors $u$ (of joints 13 and 15) and $v$ (between joints 11 and 13).

Next, we compute the score of each joint based on the angles. We denote $\omega^{Joint}$ as the score of each joint which is computed as follows: $\omega^{Joint}(\psi, \alpha) = \omega^{max} - \frac{|X_\psi^\alpha - \varphi_\psi|}{\beta} \times C$, where $\omega^{max}$ is the maximum score. Its default value will be

---

[4] https://wumbo.net/formulas/angle-between-two-vectors-2d/.

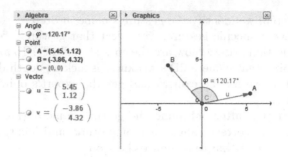

**Fig. 4.** Computing the angle $\varphi$ between two vectors $u$ and $v$

10; $X_{\psi}^{\alpha}$ is a correct angle of the joint $\psi$ in the gesture $\alpha$; $\varphi_{\psi}$ is the current angle of the $\psi$(the above formula); $\beta$ is the allowable error margin, e.g., the correct angle is 45. We assume that $\beta = 5$. Then, $\omega^{Joint} = 10$ if $\varphi \in [40, 50]$; $C$ is a minus coefficient, i.e., if $\mid X - \varphi \mid > \beta$ then we will deduct points by the coefficient $C$.

*Example 1.* Now, we consider hand-gesture3 ($\alpha_{H3}$). Then, we assume that a correct angle of the joint $\psi_{13}$ is $X_{\psi_{13}}^{\alpha_{H3}} = 90$ and a current angle is $\varphi_{\psi_{13}} = 82$. We compute on a scale of 10, $\omega^{max} = 10$ and the allowable error margin $\beta = 5$. So, we have $\omega^{Joint}(\psi_{13}, \alpha_{H3}) = 10 - \frac{|90-82|}{5} \times 1.0 = 10 - 1.6 = 8.4$.

Now, we compute the score of gesture for the entire body denoted $\omega^{G}$. The simple idea is to compute the average of all joints. We compute as follows: $\omega^{G}(\alpha) = Average(\{\omega^{Joint}(\psi_i, \alpha) \mid \psi_i \in \Psi\})$ After we compute the score of each gesture, now we will compute the score of movement in the following sub-section.

## 4.2    ScoCa Algorithm from Local to Global

Our ScoCa algorithm consists of two parts calculating in parallel, including (1) audio computation and (2) Visual Computation. Regarding the (1), we use the GA API to transfer "Speech" to "Text". We determine the movement key-word in "Text". Then, we load the gesture classification model to support the visual computation side. For the (2), we calculate from the local score (on all joints) to the global score (on all gestures of movement). Then, when we fully obtained the global score on all gestures, we give a computation of the movement score. Now, an following explanation in detail of two algorithms will be declared and presented.

Algorithm 1, named AudioComputation, focuses crucially on determining the classification model for the movement requested from the speech. The input of Algorithm 1 is a speech recorded from microphone. Algorithm 1's output is a classification model of the movement ($\Phi$) found ($\Delta_\Phi$). For example, if $\Phi$ searched in $Text$ is "Leg Movement", the gesture classifier of the $\Phi$ will be called by the $getModel()$ function in line 7. At the same time, Algorithm 2 is running. However, it is still waiting for $\Delta_\Phi$ found a model ($\Delta \neq \emptyset$). Remind that both algorithms are running in parallel with two distinct threads.

---

**Algorithm 1:** AudioComputation(micro)

1  #run threading 1
2  $\Delta_\Phi \longleftarrow \emptyset$
3  **begin**
4  $\quad$ *Text* $\longleftarrow$ *GoogleAssistant(micro)* // to transfer Speech-to-Text
5  $\quad$ **for** $\Phi \in \Omega$ **do**
6  $\quad\quad$ **if** $\Phi \in Text$ **then**
7  $\quad\quad\quad$ $\Delta_\Phi \longleftarrow getModel(\Phi)$
8  $\quad\quad\quad$ break

---

**Algorithm 2:** VisualComputation($f$, numF=20, MinF=30, MaxF=100)

1  #run threading 2
2  **begin**
3  $\quad$ **if** $\Delta_\Phi \neq \emptyset$ **then**
4  $\quad\quad$ *img* $\longleftarrow$ *Preprocessing(f)*;// f is the current frame
5  $\quad\quad$ $\alpha \longleftarrow \Delta_\Phi(img)$ // Image Classification for each gesture
6  $\quad\quad$ $L \longleftarrow L \cup \alpha$ // collect gesture names to identify the change between gestures
7  $\quad$ **if** $|L| > numF$ **then**
8  $\quad\quad$ $i \longleftarrow i+1$ // counting the frame number
9  $\quad\quad$ $k \longleftarrow Majority(L)$
10 $\quad\quad$ **if** $MinFr \leq i \leq MaxFr$ **then**
11 $\quad\quad\quad$ $LS \longleftarrow \omega^G(\alpha)$ // compute the local score on each gesture
12 $\quad\quad\quad$ $S \longleftarrow S \cup LS$ // collect all local scores in range (30-100, default)
13 $\quad\quad$ **if** $k$ *is changed* **then**
14 $\quad\quad\quad$ $GS \longleftarrow Median(S)$
15 $\quad\quad\quad$ $L, S \longleftarrow \emptyset, i \longleftarrow 0$
16 $\quad$ **if** $\forall \alpha \in \Phi$ *obtained* $GS$ **then**
17 $\quad\quad$ $SM = average(\{GS_j \mid j \leq |\Phi|\})$

---

Algorithm 2 focuses on image frame processing to calculate the score. When $\Delta_\Phi$ is obtained, the system first performs image (frame $f$) pre-processing in line 4. Then, we use the model $\Delta_\Phi$ loaded in Algorithm 1 to predict *gesture names* of the $\Phi$ in line 5. Here, we expect sufficient information; hence, we collect the result of 20 frames (*numF* = 20) into $L$ (line 6 and 7). Then, we decide the gesture name based on the $L$'s majority technique in line 9. Moreover, since our framework allows playing with multi-frames, we only grade in a constraint/limited range to expect the best result as in line 10 (i.e., $i \in [30, 100]$).

Regarding the score computation, we first consider the local-global score on each gesture from line 11 to 14. This $LS$ is computed from the formula 4.1. After ultimately collecting a list of scores of all frames in the predicted gesture (line 12), we compute the median of the list to get the best score (in line 14). The reason for choosing the median method is to avoid cases where misidentifying of the gestures from classification model will strongly affect the final score. For example, we have the following list of scores: $S = \{8.0, 8.0, 8.0, 9.0, 1.0, 8.0, 8.0\}$. If taking the median, the score will be 8.0. In the case of taking the average, the score of 1.0 will have a "big" impact on the final score result. Namely, the average score will be 7.1. A remarkable point that the global score of gesture will be finalised when changing the next gesture (see in line 13). Finally, we compute the score of movement. It will be an average of all gestures in that movement (see in line 16 and 17).

To sum up, we have three stages of the score computation, including (1) score computation of one frame for the gesture *(called "local score", $\mathcal{LS}$ with joint score $\omega^{joint}$ and gesture score $\omega^{G}$)*; (2) score calculation of multi-frames for the gesture *("global score", $\mathcal{GS}$)*; (3) score computation of movement *($\mathcal{SM}$)(with many gestures)*. Both algorithms run continuously on two threads and have a $P^{NP}$ complexity.

## 5    Dataset and Experimental Result

This section details the collected dataset and provides an empirical comparison of the classification models. In addition, we also show the user interface that we have implemented and deployed in Python language. Finally, the advantages and disadvantages of our system also will be discussed. Our dataset and implementation is published in Github[5].

### 5.1    Dataset and Implementation Environment

We collect 34,571 of images for all the movements to implement our framework. In particular, each movement has about 2900 – 5000 images depending on the number of gestures. The author group gathers, accumulates, and collects the dataset through video recording and image capturing. Experts and teachers who directly teach the PES have verified and validated our dataset. We remind that camera placement angles are $250^{o}$. The practitioner stands about 2 m–2.5 m from the camera. The lighting conditions are good without being tied to direct sunlight.

**Table 1.** Collected Dataset for PESTAI.

| ID | Movement Name | Number of Gesture | Nbr of Images |
|---|---|---|---|
| 1 | Stretching | 3 | 2,953 |
| 2 | Hands | 5 | 4,953 |
| 3 | Legs | 4 | 3,953 |
| 4 | Waists | 5 | 4,953 |
| 5 | Abs (Belly) | 5 | 4,953 |
| 6 | Full Body | 5 | 4,953 |
| 7 | Jump | 4 | 3,900 |
| 8 | Body Conditioning | 4 | 3,953 |
| | | | 34,571 |

---

[5] https://github.com/thanhnhan10/PETSAI.

To evaluate our approach and the training model, we implement the AI libraries *(i.e., mediapipe (version 0.9.0.1), Speech-recognization (version 3.10.0))* and run the experiments on the computer with the following configuration: AMD Ryzen 7 6800HS, Ceator Edition 3.20 GHZ, 16 GB RAM, Windows 10 OS.

## 5.2   Result of Classification Models

In this study, we build eight gesture classification models corresponding to eight movements of the PES. To evaluate the classification model, we run the model with many different parameters and CNN network architecture. Each movement has a corresponding classification model. Here, we work 06 machine learning models, including CNN *(MobileNet, Basic CNN, Densenet121, LeNet-5)*, SVM, and KNN.

For the training process, we split the dataset into two parts 80% for the training set and 20% for the testing set. Because the number of gestures is little on each movement, the models' accuracy achieves high and expected results. Note that we have experimented with numerous different sets of parameters. However, we only present the results of the parameter sets that obtain the highest accuracy in this paper. The gesture classification's experimental results for each movement are shown in Table 2 and 3 (due to limited space, we only show a few results. The rest of the results will be presented at the end of the paper). Note that, for evaluating the classification results objectively, we take the average of 10 experimental runs for each model with 04 evaluation indicators *(F1, Accuracy, Precision, Recall)*. On the other hand, we also compared the models in terms of testing and training times. After experimenting and writing the results, we realize that the running time of each model in all movements is not much different. Therefore, we aggregate and average the running times of all eight movements to present in this paper (see Fig. 5).

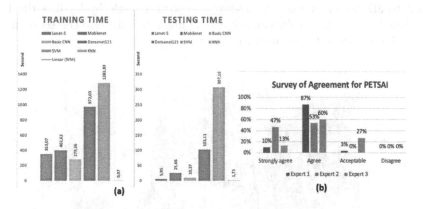

**Fig. 5.** (a) Comparison of Training and Testing Time of the classification models. (b) An evaluation survey for PETSAI

**Table 2.** Results of the gesture classification (two first movements)

| ID | Models | Parameter | F1 | Precision | Recall | Accuracy |
|----|--------|-----------|-----|-----------|--------|----------|
| Stretching movement | | | | | | |
| 1 | CNN (Mobilenet) | Epoches: 30 | 98.97 | 99.01 | 98.98 | 99.00 |
| 2 | CNN (Basic) | Epoches: 20 | 98.81 | 98.84 | 98.84 | 98.82 |
| 3 | CNN (Densenet121) | Epoches: 40 | 99.15 | 99.17 | 98.15 | 99.18 |
| 4 | CNN (Lenet) | Epoches: 30 | 99.80 | 99.83 | 98.81 | 99.84 |
| 5 | SVM | $C = 10000, \gamma = 0.001$ | 94.60 | 95.46 | 94.29 | 94.29 |
| 6 | KNN | $K = 10$ | 63.32 | 66.72 | 61.76 | 63.75 |
| Hands movement | | | | | | |
| 1 | CNN (Mobilenet) | Epoches: 30 | 99.56 | 99.54 | 99.53 | 99.55 |
| 2 | CNN (Basic) | Epoches: 40 | 98.51 | 98.54 | 98.53 | 98.50 |
| 3 | CNN (Densenet121) | Epoches: 40 | 99.47 | 99.46 | 98.45 | 99.47 |
| 4 | CNN (Lenet) | Epoches: 30 | 99.92 | 99.93 | 98.90 | 99.93 |
| 5 | SVM | $C = 10000, \gamma = 0.001$ | 93.45 | 93.46 | 93.29 | 93.37 |
| 6 | KNN | $K = 10$ | 57.15 | 64.04 | 57.12 | 55.10 |
| Legs movement | | | | | | |
| 1 | CNN (Mobilenet) | Epoches: 40 | 98.59 | 98.60 | 98.53 | 98.62 |
| 2 | CNN (Basic) | Epoches: 40 | 98.89 | 97.78 | 98.90 | 98.87 |
| 3 | CNN (Densenet121) | Epoches: 40 | 95.53 | 95.68 | 95.26 | 95.46 |
| 4 | CNN (Lenet) | Epoches: 10 | 99.90 | 99.78 | 98.86 | 99.89 |
| 5 | SVM | $C = 10000, \gamma = 0.001$ | 96.03 | 95.36 | 96.15 | 96.22 |
| 6 | KNN | $K = 10$ | 64.05 | 56.01 | 56.34 | 58.22 |

In general, although the testing time of KNN is the lowest, the accuracy results are not high. Therefore, to balance between accuracy and execution time, we choose the CNN model with Lenet-5 architecture to implement our approach.

## 5.3    Application and Discussion

We implement the framework and allow the user to edit/modify the parameters from the user interface. Moreover, to overcome/handle the weakness of 2D compared to 3D, we allow users to update the correct angle by loading the Excel file. Based on this file, users can adjust the angle to calculate the score in such a way that it can suit the camera position.

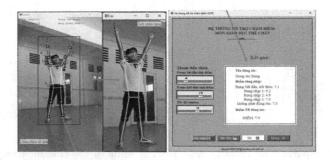

**Fig. 6.** PETSAI Application for Physical Education Teaching Support

**Table 3.** Results of the gesture classification

| ID | Models | Parameter | F1 | Precision | Recall | Accuracy |
|----|--------|-----------|-----|-----------|--------|----------|
| Waists movement | | | | | | |
| 1 | CNN (Mobilenet) | Epochs: 40 | 98.24 | 98.54 | 98.23 | 98.25 |
| 2 | CNN (Basic) | Epochs: 40 | 94.42 | 94.68 | 98.44 | 94.50 |
| 3 | CNN (Densenet121) | Epochs: 40 | 96.73 | 96.83 | 96.71 | 96.73 |
| 4 | CNN (Lenet) | Epochs: 30 | 99.87 | 99.83 | 98.85 | 99.87 |
| 5 | SVM | $C = 10000, \gamma = 0.001$ | 96.63 | 97.36 | 97.09 | 97.12 |
| 6 | KNN | $K = 10$ | 60.25 | 64.99 | 58.53 | 59.51 |
| Abs movement | | | | | | |
| 1 | CNN (Mobilenet) | Epochs: 40 | 99.09 | 99.13 | 99.24 | 99.17 |
| 2 | CNN (Basic) | Epochs: 40 | 99.13 | 99.15 | 99.03 | 99.20 |
| 3 | CNN (Densenet121) | Epochs: 40 | 98.18 | 98.20 | 98.23 | 98.21 |
| 4 | CNN (Lenet) | Epochs: 30 | 100.0 | 100.0 | 100.0 | 100.0 |
| 5 | SVM | $C = 10000, \gamma = 0.0001$ | 97.55 | 98.47 | 98.29 | 98.47 |
| 6 | KNN | $K = 10$ | 56.08 | 63.01 | 53.58 | 54.98 |
| Full Body movement | | | | | | |
| 1 | CNN (Mobilenet) | Epochs: 40 | 99.39 | 99.42 | 99.44 | 99.44 |
| 2 | CNN (Basic) | Epochs: 30 | 99.10 | 99.09 | 99.12 | 99.08 |
| 3 | CNN (Densenet121) | Epochs: 40 | 98.63 | 98.59 | 98.40 | 98.51 |
| 4 | CNN (Lenet) | Epochs: 30 | 99.93 | 99.91 | 99.89 | 99.90 |
| 5 | SVM | $C = 10000, \gamma = 0.001$ | 95.03 | 96.46 | 95.29 | 95.29 |
| 6 | KNN | $K = 10$ | 56.98 | 63.25 | 55.14 | 54.69 |
| Jump movement | | | | | | |
| 1 | CNN (Mobilenet) | Epochs: 30 | 98.59 | 98.61 | 99.54 | 99.59 |
| 2 | CNN (Basic) | Epochs: 30 | 98.72 | 98.77 | 98.81 | 98.68 |
| 3 | CNN (Densenet121) | Epochs: 40 | 95.40 | 95.67 | 95.38 | 95.41 |
| 4 | CNN (Lenet) | Epochs: 40 | 99.97 | 99.95 | 99.93 | 99.94 |
| 5 | SVM | $C = 10000, \gamma = 0.01$ | 97.73 | 98.36 | 97.98 | 97.92 |
| 6 | KNN | $K = 10$ | 65.15 | 65.05 | 58.27 | 58.21 |
| Body Conditioning Movement | | | | | | |
| 1 | CNN (Mobilenet) | Epochs: 30 | 96.23 | 96.31 | 96.21 | 96.32 |
| 2 | CNN (Basic) | Epochs: 30 | 98.75 | 99.00 | 98.77 | 98.75 |
| 3 | CNN (Densenet121) | Epochs: 40 | 95.36 | 95.45 | 95.32 | 95.34 |
| 4 | CNN (Lenet) | Epochs: 40 | 99.75 | 99.72 | 99.56 | 99.75 |
| 5 | SVM | $C = 100000, \gamma = 0.001$ | 95.37 | 95.45 | 95.30 | 95.42 |
| 6 | KNN | $K = 10$ | 55.01 | 59.29 | 53.35 | 53.28 |

For the display, we provide 03 discrete interfaces, as shown in Fig. 6, including (1) Interface to show the local-global score; (2) The interface of the practitioner's vision/image; (3) Interface scores in detail and allow modifying the parameters. The weakness of our approach is to determine the camera angle and determine the correct angle to aid in scoring. Moreover, because of the continuity/sequence of the gestures in each movement, if the practitioner does not perform in the correct order, the system will give "0 point" to the user and stop the grading session. To evaluate the PETSAI, we surveyed the consensus of experts. Our assessment survey has four levels, including *strongly agree, agree, acceptable, and disagree*. Namely, we investigated 30 students *(corresponding to 30 sessions)* with examination/review by three experts. The survey results are presented in Fig. 5 (b). From the survey's outcome, most experts agree with the grading results of PETSAI.

Furthermore, we have also measured the application run time and algorithm time in 10 times. The result is that the average application time for a gesture is 3.82s, and the algorithm's average time is 2.17s. In fact, the time depends on the speed of the exerciser.

## 6   Conclusion

In this paper, we introduced an audio-visual framework to support physical education teaching *(in Vietnamese)*. Therein, we provided eight models of gesture classification for the movements in the grade 3 physical education program. A dataset of the movement images and the implementation source of our approach have also been published on GitHub. Moreover, we proposed an algorithm to compute scores for each movement. Furthermore, we also built applications on both web and desktop platforms. For future work, we will work with 3D models of the HPE to improve the score computation. A regression model for scores will be the next direction for this work. We also plan to implement the system for primary schools in Vietnam.

**Acknowledgements.** This research has received support from the European Union's Horizon research and innovation program under the MSCA-SE (Marie Skłodowska-Curie Actions Staff Exchange) grant agreement 101086252; Call: HORIZON-MSCA-2021-SE-01; Project title: STARWARS (STormwAteR and WastewAteR networkS heterogeneous data AI-driven management).

## References

1. Almusawi, H.A., Durugbo, C.M., Bugawa, A.M.: Innovation in physical education: teachers' perspectives on readiness for wearable technology integration. Comput. Educ. **167**, 104185 (2021)
2. Andriluka, M., Pishchulin, L., Gehler, P., Schiele, B.: 2D human pose estimation: new benchmark and state of the art analysis. In: CVPR'14, pp. 3686–3693 (2014)
3. Ashraf, F.B., Islam, M.U., Kabir, M.R., Uddin, J.: YoNet: a neural network for yoga pose classification. SN Comput. Sci. **4**(2), 198 (2023)
4. Berdasco, A., López, G., Diaz, I., Quesada, L., Guerrero, L.A.: User experience comparison of intelligent personal assistants: Alexa, google assistant, Siri and Cortana. UCAml **2019**, 51 (2019)
5. Burbules, N.C., Fan, G., Repp, P.: Five trends of education and technology in a sustainable future. Geogr. Sustain. **1**(2), 93–97 (2020)
6. Chen, L., Chen, P., Lin, Z.: Artificial intelligence in education: a review. IEEE Access **8**, 75264–75278 (2020)
7. Chiddarwar, G.G., Ranjane, A., Chindhe, M., Deodhar, R., Gangamwar, P.: AI-based yoga pose estimation for android application. Int. J. Inn. Sci. Res. Tech. **5**, 1070–1073 (2020)
8. Chung, J.L., Ong, L.Y., Leow, M.C.: Comparative analysis of skeleton-based human pose estimation. Future Internet **14**(12), 380 (2022)

9. Cloete, A.L.: Technology and education: challenges and opportunities. HTS: Theological Stud. **73**(3), 1–7 (2017)
10. Daniel, S.J.: Education and the COVID-19 pandemic. Prospects **49**(1), 91–96 (2020)
11. Garg, S., Saxena, A., Gupta, R.: Yoga pose classification: a CNN and MediaPipe inspired deep learning approach for real-world application. J. Ambient Intell. Human. Comput., 1–12 (2022)
12. Jelinek, F.: Statistical Methods for Speech Recognition. MIT Press, Cambridge (1998)
13. Kamel, A., Liu, B., Li, P., Sheng, B.: An investigation of 3D human pose estimation for learning Tai Chi: a human factor perspective. Int. J. Hum.-Comput. Interact. **35**(4–5), 427–439 (2019)
14. Krause, J.M.: Physical education student teachers' technology integration self-efficacy. Phys. Educ. **74**(3), 476 (2017)
15. Lawrence, S., Giles, C.L., Tsoi, A.C., Back, A.D.: Face recognition: a convolutional neural-network approach. IEEE Trans. Neural Netw. **8**(1), 98–113 (1997)
16. LeCun, Y., et al.: LeNet-5, convolutional neural networks. AI **1**(5), 14 (2015)
17. Li, Z., Liu, F., Yang, W., Peng, S., Zhou, J.: A survey of convolutional neural networks: analysis, applications, and prospects. IEEE Trans. Neural Netw. Learn. Syst. (2021)
18. Mata, C., Onofre, M., Costa, J., Ramos, M., Marques, A., Martins, J.: Motivation and perceived motivational climate by adolescents in face-to-face physical education during the COVID-19 pandemic. Sustainability **13**(23), 13051 (2021)
19. Munea, T.L., Jembre, Y.Z., Weldegebriel, H.T., Chen, L., Huang, C., Yang, C.: The progress of human pose estimation: a survey and taxonomy of models applied in 2D human pose estimation. IEEE Access **8**, 133330–133348 (2020)
20. Quay, J.: The importance of context to learning: physical education and outdoor education seeing eye to eye. In: The Proceedings of ACHPER Interactive Health and Physical Education Conference, pp. 1–15 (2002)
21. Shadiev, R., Hwang, W.Y., Chen, N.S., Huang, Y.M.: Review of speech-to-text recognition technology for enhancing learning. J. Educ. Technol. Soc. **17**(4), 65–84 (2014)
22. Sinha, D., El-Sharkawy, M.: Thin MobileNet: an enhanced MobileNet architecture. In: UEMCON'19, pp. 0280–0285. IEEE (2019)
23. Thành, N.T., Công, P.T., et al.: An evaluation of pose estimation in video of traditional martial arts presentation. J. IT&C **2019**(2), 114–126 (2019)
24. Tulshan, A.S., Dhage, S.N.: Survey on virtual assistant: Google assistant, Siri, Cortana, Alexa. In: Thampi, S.M., Marques, O., Krishnan, S., Li, K.-C., Ciuonzo, D., Kolekar, M.H. (eds.) SIRS 2018. CCIS, vol. 968, pp. 190–201. Springer, Singapore (2019). https://doi.org/10.1007/978-981-13-5758-9_17
25. Yang, W.: Artificial intelligence education for young children: why, what, and how in curriculum design and implementation. Comput. Educ.: Artif. Intell. **3**, 100061 (2022)
26. Zhang, W., Liu, Z., Zhou, L., Leung, H., Chan, A.B.: Martial arts, dancing and sports dataset: a challenging stereo and multi-view dataset for 3D human pose estimation. Image Vis. Comput. **61**, 22–39 (2017)
27. Zhang, Z.: Microsoft Kinect sensor and its effect. IEEE Multimedia **19**(2), 4–10 (2012)

28. Zheng, C., et al.: Deep learning-based human pose estimation: a survey. ACM Comput. Surv. (2020)

29. Zhu, Y., Newsam, S.: DenseNet for dense flow. In: ICIP'17, pp. 790–794. IEEE (2017)

30. Zhu, Z., Lin, K., Jain, A.K., Zhou, J.: Transfer learning in deep reinforcement learning: a survey. IEEE Trans. PAMI (2023)

# A Graph-Based Approach for Representing Wastewater Networks from GIS Data: Ensuring Connectivity and Consistency

Omar Et-Targuy[1,2,3,5]([✉]), Ahlame Begdouri[2], Salem Benferhat[1],
Carole Delenne[3,5], Thanh-Nghi Do[4], and Truong-Thanh Ma[4]

[1] CRIL, CNRS UMR 8188, University of Artois, Lens, France
ettarguy@cril.fr
[2] LSIA, Université Sidi Mohamed Ben Abdellah (USMBA), Fez, Morocco
[3] HSM, University of Montpellier, CNRS, IRD, Montpellier, France
[4] CICT, Can Tho University, Can Tho, Vietnam
[5] Inria, Team Lemon, Montpellier, France

**Abstract.** Wastewater network management is a critical aspect of ensuring public health and environmental sustainability. Geographic Information Systems (GIS) are widely employed for managing and analyzing wastewater network data. However, storing data in separate databases in GIS poses challenges in accurately representing the connectivity of the wastewater network components, which are connected in reality. To address this issue, this paper proposes a novel graph-based approach for representing wastewater networks. In this approach, each component of the network, such as manholes, structures, pumps, etc., is represented as a node in the novel graph, while the pipes represent the connections between them. By adopting this graph-based representation, the true interconnected nature of the wastewater network can be effectively captured and visualized. This paper begins by discussing the limitations of the traditional separate databases approach in GIS for representing wastewater networks. Subsequently, the graph-based algorithm for transforming the GIS representation into a graph-based representation is presented. To validate the proposed approach, real-world wastewater network data is employed, demonstrating its ability to capture the network's connectivity and facilitate network analysis.

**Keywords:** Wastewater networks · Graph-based representation · Connectivity

## 1 Introduction

Efficient wastewater management plays a vital role in ensuring public health and environmental sustainability. Geographic Information Systems (GIS) have become a popular tool for managing and analyzing geographic data associated with wastewater networks. However, the traditional approach of storing each

N. Thai-Nghe et al. (Eds.): ISDS 2023, CCIS 1949, pp. 243–257, 2024.
https://doi.org/10.1007/978-981-99-7649-2_19

component in separate databases can result in inconsistencies and inaccuracies, especially when it comes to representing the connections between these components.

In reality, wastewater networks are interconnected systems where various components work together to transport wastewater from its source to treatment plants. For example, manholes serve as access points to the pipes, facilitating maintenance and inspection activities. These connections between different components are crucial for the proper functioning of the network.

In GIS, each component of the wastewater network is typically stored in its own database, e.g. manholes in one database and pipes in another. While this approach allows for efficient management of individual components, it lacks the ability to capture and represent the connections between them. As a result, over time, as updates and modifications are made to the network, the databases become unsynchronized, leading to a loss of the true connectivity and relationships between the components.

To address the challenges associated with representing wastewater networks, a comprehensive and integrated approach is needed. A graph-based representation proves to be highly effective in capturing the interconnected nature of the network. By treating each component, such as manholes, pumps and structures as nodes, and representing the connections between them as edges which represents pipes in wastewater networks,

The Graph-Based Representation is a versatile method adept at illustrating data interrelationships in diverse domains. In [3], a pioneering graph-based spatial approach is introduced, employing Minimum Planar Graphs from Delaunay triangulations to assess habitat connectivity effectively. This innovation facilitates wildlife habitat network analysis and visualization, aiding in conservation decisions. In [4], the authors conduct a systematic survey of graph-based text representation, highlighting its superiority over the bag-of-words model in capturing structural and semantic information. The use of graph-based representation leads to enhanced performance in information retrieval, term weighting and ranking across various text applications. Additionally, in [5], a novel graph-based approach for multiview image representation is proposed. This method utilizes graph connections to describe pixel proximity in 3D space, achieving a compact and controllable representation that is well-suited for coding and reconstructing multiple views. In [1], a knowledge graph-based data representation is proposed for Industrial Internet of Things enabled manufacturing. [2] presents a global graph for modeling social interactions among pedestrians.

The paper is organized as follows. In Sect. 2, an overview of wastewater networks data is provided, and the limitations of representing them in a SIG using the traditional separate databases approach are discussed. Section 3 presents the graph-based algorithm, which transforms the GIS representation into a interconnected graph-based representation. In Sect. 4, experiments and results from applying the graph-based algorithm to real-world wastewater networks data are presented, with a focus on highlighting the algorithm's advantages and limitations.

Finally, Sect. 5 concludes the paper by summarizing the benefits of the graph-based representation and its implications for wastewater network management.

## 2    Wastewater Networks in GIS

Wastewater networks comprise a variety of interconnected components (see Fig. 1) that collectively ensure the efficient and safe transportation of wastewater from its sources to treatment plants. Some of the essential components of wastewater networks include:

- **Pipes:** sewer pipes are the primary components of a wastewater network. They are used to transport wastewater from its source, such as residential, commercial or industrial buildings, to the treatment plants.
- **Manholes:** manholes are access points in the wastewater network that allow for maintenance and inspection activities. They are usually constructed at regular intervals along the sewer pipes. Manholes provide entry points for workers to access the sewer pipes and perform tasks like cleaning, repairs and inspections.
- **Pumps:** pumps are used to increase the pressure and flow of wastewater, especially in areas where gravity flow is not possible.
- **Fittings:** fittings are used to connect and join different sections of sewer pipes in wastewater networks.

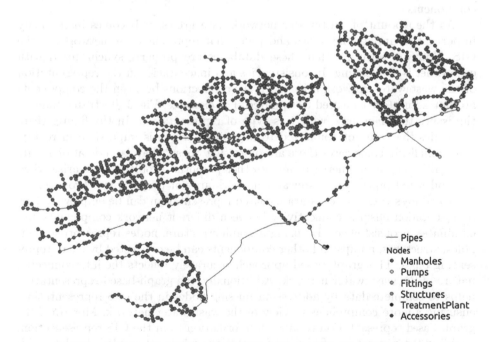

**Fig. 1.** Example of wastewater networks in GIS representation.

One characteristic of wastewater network data is the geographical information associated with each component, representing its position on the Earth. To store, visualize and manage this spatial data, a Geographic Information System (GIS) is commonly employed. A GIS allows for the integration, analysis and interpretation of spatial data within the context of wastewater network management.

In the context of wastewater networks, two primary geometries are used to represent the network components in GIS (see Fig. 1). The first geometry is the point geometry, which represents specific locations such as manholes, structures, pumps and other components. The second geometry is the line geometry, which represents the pipes that form the network infrastructure.

In addition to the geographical information, each component in the wastewater network is also characterized by attribute information. These attributes provide additional details and characteristics of the components, such as pipe diameter, material, flow capacity and other relevant data.

In a GIS system, the data related to each component of the wastewater network is typically stored in separate databases, as illustrated in Fig. 1. Each component, such as pipes, manholes, pumps, fittings and structures, has its own database, allowing for easy management and analysis of the data.

However, in reality, these components are interconnected and have spatial relationships with each other; e.g., manholes serve as access points to pipes, and pumps are connected to specific pipes within the network. Therefore, it is crucial to establish and maintain the appropriate connections between these components.

As the amount of wastewater network data grows, it becomes increasingly important to update the nodes and pipes that represent these networks in the GIS. However, ensuring that these databases are properly synchronized with each other is challenging. Inconsistencies and inaccuracies in the representation of the wastewater network arise when the connections between the components are not accurately captured or updated. For example, Fig. 2 illustrates some of the issues that can arise with this type of representation. In the figure, there are nodes that are not connected to any pipes, which is impossible in reality. Additionally, in some cases there are pipes that doesn't have a node at one of its extremities or both. There are many other potential problems that arise when the nodes and pipes of a wastewater network are not properly synchronized.

To address these issues, a graph-based representation can be employed. This representation ensures connectivity between different network components and eliminates inconsistencies. By using a graph structure, nodes representing manholes, structures, pumps and other components can be connected by edges representing pipes. This graph-based approach accurately reflects the interconnected nature of the wastewater network. Additionally, the graph-based representation helps to complete data by addressing missing nodes in the GIS representation, ensuring a more comprehensive view of the wastewater network. Moreover, the graph-based representation retains all information from the GIS representation, enabling the generation of GIS representations whenever needed. It also facili-

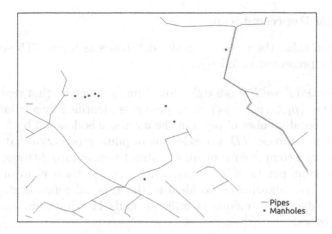

**Fig. 2.** Inconsistencies between databases

tates the integration of new information into the graph, making it a flexible and scalable solution for wastewater network management.

In the following section, the details of the graph-based representation and its benefits in maintaining the connectivity and integrity of wastewater network data will be presented.

## 3   Our Graph-Based Representation

Figure 3 depicts the general process of constructing a graph-based representation for wastewater network data. The inputs comprise a database of pipes and databases of nodes, encompassing various components such as manholes, structures, pumps, fittings and more - all represented by node geometry.

**Fig. 3.** General process

The algorithm begins by extracting the intersections between each pipe in the pipes database and the nodes in the nodes database. This ensures that the connections between pipes and nodes are accurately identified. After detecting the intersections, the resulting nodes are connected or inserted into the graph. During this step, verification is performed to determine if the extremities of each pipe have intersections. If no intersections are found, dummy nodes are added to ensure connectivity within the graph, with edges representing the pipes.

Overall, the output of this process is a graph-based representation that accurately depicts the relationships between the pipes and nodes in the wastewater network.

## 3.1   Formal Representation

The algorithm takes the pipes and nodes databases as inputs. These inputs can be formally represented as follows:

– The Pipes Database $PD$ is a collection of line geometries that represent sewer pipes $PD = \{p_1, ..., p_i, ..., p_k\}$. Each pipe $p_i$ is identified by a unique identifier $i$, and the total number of pipes in the database is denoted by $k$.
– The Nodes Database $ND$ is a collection of point geometries that includes all the point type components: manholes, structures, pumps, fittings, accessories and treatment plants $ND = \{n_1, ..., n_j, ..., n_m\}$. Each element $n_j$ has an attribute *type*, allowing us to identify the type of component. Each node $n_j$ is identified by a unique identifier $i$, and the total number of nodes is denoted by $m$.

The algorithm generates a graph denoted as $G = (V, E)$ as its final output, where:

– $V$ is a finite set of nodes that contains elements from $ND$ and dummy nodes that will be added to ensure connectivity and complete missing nodes.
– $E \subseteq V \times V$ is a finite set of edges that represents connections between two nodes, indicating that these connections correspond to the pipes from $PD$. Each edge in the graph represents the relationship between two nodes, reflecting the pipes that connect them in the wastewater network representation.

## 3.2   Proposal Algorithm

The graph construction algorithm (Algorithm 1) initializes an empty graph $G$ with sets of nodes $V$ and edges $E$. It iterates over each pipe $p_i$ in the pipes database $PD$. For each pipe $p_i$, the algorithm follows these steps:

---

**Algorithm 1.** Graph Construction

---

```
 1: procedure GRAPHCONSTRUCTION(PD, ND, Distance)
 2:     V, E ← {}, {}
 3:     G ← (V, E)
 4:     for i = 1 to k do
 5:         Buffer ← CreateBuffer(pᵢ, Distance)
 6:         Intersections ← ExtractIntersections(ND, Buffer)
 7:         Ext1 ← RandomExtrimity(pᵢ)
 8:         SortedInter ← DistanceAndSort(Intersections, Ext1)
 9:         SortedInter ← AddDummyNodes(SortedInter)
10:         G ← ConnectionToGraph(SortedInter, G)
11:     return G
```

---

**Step 1: Create a Buffer.** In GIS a "buffer zone" is defined by an area around a geographic feature that contains every points within a specified distance from that feature. In this step, a buffer is created around each pipe $p_i$ by applying a distance that accounts for the physical dimensions of the pipe and any potential shifting of the nodes that are near to the pipe.

For example, in Fig. 4, the blue line represents the pipe, and a buffer is created around it using a distance. The resulting yellow buffer represents the extended region around the pipe that encompasses its potential location within the given distance.

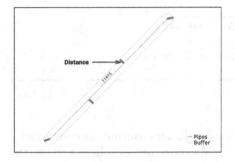

**Fig. 4.** Create a buffer (Color figure online)          **Fig. 5.** Intersections detection

**Step 2: Extract Intersections.** In a real-world of wastewater networks, pipes are connected to nodes. This step aims to detect intersections between the buffer created around the pipe $p_i$ and the nodes databases $ND$, to identify their connections.

The result is a set of nodes that must be connected to the corresponding pipe. For example, in Fig. 5, the intersection between the buffer of pipe 11415 and the $ND$ is detected, resulting in a set of nodes denoted as Intersections = {41132, 41142, 41140}. These nodes represent the set of nodes that must be connected to the pipe $p_i$, which is the pipe 11415 in this example.

**Step 3: Calculate Distances and Sort Intersections.** In this step of the algorithm, the Euclidean distance is calculated between one of the pipe's extremity $Extr1$ and each of the nodes intersecting the buffer (see Algorithm 2). The Euclidean distance is the straight-line distance between two points in a two-dimensional space; it is suitable for cases where the distances between points are relatively small on surface Earth, which is the case in our situation with nodes that are near to each other in a wastewater network.

Using Euclidean distance, the proximity of each intersection node to one of the extremities of the pipe, labeled as $Extr1$, can be determined. Sorting the intersection nodes based on their distances allows us to establish a logical order that reflects their positions relatively to the pipe's extremities. This step is essential for checking the intersections with the extremities of the pipe.

---

**Algorithm 2.** Calculate Euclidean Distance and Sort Intersections

---

1: **procedure** DISTANCEANDSORT($Intersections$, $Extr1$)
2:     $distances \leftarrow \{\}$
3:     **for** $node$ in $Intersections$ **do**
4:         $x, y \leftarrow$ coordinates of $node$
5:         $distance \leftarrow \sqrt{(x - Extr1.x)^2 + (y - Extr1.y)^2}$
6:         Append $distance$ to $distances$
7:     $SortedInter \leftarrow$ Sort intersections based on $distances$
8:     **return** $SortedInter$

---

**Step 4: Add Dummy Nodes.** This step is crucial for ensuring the connectivity of wastewater networks (see Algorithm 3). To construct graphs based on the intersections, it is essential to have a node near to each extremity of the pipe $p_i$. However, this condition is not always satisfied. It can occur that the set of intersections is empty ($|SortedInter| = 0$ in Algorithm 3), or that certain nodes are present but not located at the extremities of the pipe $p_i$.

In such scenarios, it becomes necessary to add dummy nodes to the extremities of the pipe. These dummy nodes serve as virtual representations of the missing nodes, facilitating the connection of the pipe to the overall wastewater network.

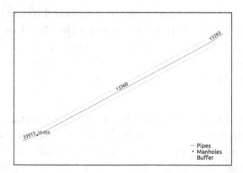

**Fig. 6.** Example of a pipe with only one node intersecting the buffer area in manholes database, thus one missing extremity.

In Fig. 6, it can be observed that there is only one node inside the buffer around pipe number 13360, which is near one extremity of the pipe. However,

**Algorithm 3.** Adding Dummy Nodes

---

1: **procedure** ADDDUMMYNODES(*SortedInter*, $p_i$, *Threshold*)
2:     $Ext1 \leftarrow$ the first extremity of $p_i$
3:     $Ext2 \leftarrow$ the second extremity of $p_i$
4:     **if** $|SortedInter| = 0$ **then**
5:         $SortedInter \leftarrow AddFirst(Type = Dummy, coord = (Ext1.x, Ext1.y))$
6:         $SortedInter \leftarrow AddLast(Type = Dummy, coord = (Ext2.x, Ext2.y))$
7:     **else if** $|SortedInter| = 1$ **then**
8:         $node \leftarrow$ node in $SortedInter$
9:         $distance1 \leftarrow EuclideanDistance(node, Ext1)$
10:        $distance2 \leftarrow EuclideanDistance(node, Ext2)$
11:        **if** $distance1 \geq Threshold$ and $distance2 \geq Threshold$ **then**
12:            $SortedInter \leftarrow AddFirst(Type = Dummy, coord = (Ext1.x, Ext1.y))$
13:            $SortedInter \leftarrow AddLast(Type = Dummy, coord = (Ext2.x, Ext2.y))$
14:        **else if** $distance1 \geq Threshold$ **then**
15:            $SortedInter \leftarrow AddLast(Type = Dummy, coord = (Ext1.x, Ext1.y))$
16:        **else if** $distance2 \geq Threshold$ **then**
17:            $SortedInter \leftarrow AddFirst(Type = Dummy, coord = (Ext2.x, Ext2.y))$
18:    **else if** $|SortedInter| >= 2$ **then**
19:        $firstNode \leftarrow$ first node in $SortedInter$
20:        $lastNode \leftarrow$ last node in $SortedInter$
21:        **if** $EuclideanDistance(firstNode, Ext1.x, Ext1.y) > Threshold$ **then**
22:            $SortedInter \leftarrow AddFirst(Type = Dummy, coord = (Ext1.x, Ext1.y))$
23:        **if** $EuclideanDistance(lastNode, Ext2.x, Ext2.y) > Threshold$ **then**
24:            $SortedInter \leftarrow AddLast(Type = Dummy, coord = (Ext2.x, Ext2.y))$
25:    **return** $SortedInter$

---

there is no node near the other extremity. To maintain connectivity with other pipes, a dummy node is necessary for the missing extremity. The introduction of this dummy node serves the purpose of ensuring the connection of the pipe with the remainder of the network while preserving the integrity of the wastewater flow representation.

**Step 5: Connecting to the Graph.** The final step involves connecting the sorted intersections into the graph. After completing the step 4, the sorted intersections set is guaranteed to have at least two nodes that must be connected to the extremities of the pipe $p_i$. These nodes can either be from the original node set $ND$ or dummy nodes that were added to ensure connectivity.

In this step the Algorithm 4 begins by selecting the first node in the set of sorted intersections as the source node and adds it to the set of vertices in the graph $G$. Next, the algorithm iterates through the remaining nodes in this set. For each node, it creates an edge between the current source node and the destination node. Each edge in the graph represents a pipe in wastewater networks.

In the illustrated example (Fig. 7), the buffer zone created around pipe number 11415 is taken into consideration. This buffer intersects three nodes, namely

**Algorithm 4.** Connecting to the graph

---
1: **procedure** CONNECTIONTOGRAPH($SortedInter$, $G$)
2:     $sourceNode \leftarrow$ first node in $SortedInter$
3:     $V \leftarrow V \cup \{sourceNode\}$
4:     **for** $node$ in $SortedInter$ excluding the first node **do**
5:         $destinationNode \leftarrow node$
6:         $E \leftarrow E \cup \{(sourceNode, destinationNode)\}$
7:         $sourceNode \leftarrow destinationNode$
8:     **return** $G$

---

Intersections = {41142, 41132, 41140}. These nodes represent the necessary connections to be established with pipe 11415. Before establishing these connections, a sorting process is performed and intersections with pipe extremities are verified. In this case, the intersections adequately cover the extremities, thus avoiding the need to introduce dummy nodes. Consequently, the graph can be constructed appropriately by connecting the sorted nodes to pipe 11415. It can be observed that a single pipe can result in the creation of two edges in the graph, as evidenced here with edges (41142, 41132) and (41132, 41140), owing to the connection of the middle node numbered 41132 to the pipe.

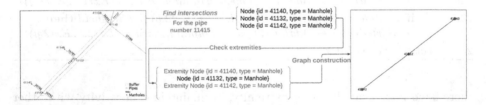

**Fig. 7.** Application of the algorithm: example with three nodes intersecting the buffer area including the pipe's extremities.

In the second example (Fig. 8), involving pipe number 13360, a different situation arises. Only one intersection is found between this pipe and the nodes database, which happens to be located at one extremity. Unfortunately, no intersection exists at the other extremity. To ensure the proper connectivity of pipe 13360 within the graph, a dummy node with the label 107965 is introduced. This dummy node acts as a placeholder, allowing the inclusion of the missing extremity in the graph representation. Therefore, the resulting graph reflects the connection between node 39488 and the dummy node 107965, representing pipe 13360.

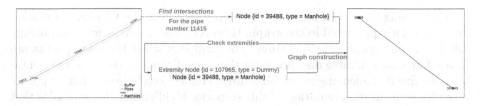

**Fig. 8.** Application of the algorithm: example with three nodes intersecting the buffer area including the pipe's extremities.

## 4   Experiments

The primary objective of our study is to develop a graph-based representation derived from GIS datasets of wastewater networks. As shown in Table 1, the results of experiments conducted on four distinct real datasets of wastewater networks are presented. These datasets all represent the same geographical zone but originate from different providers. Each dataset consists of multiple components, and for each of these datasets, both the traditional GIS representation, also known as the separate databases representation, and the innovative graph-based representation derived from the GIS representation are available. Additionally, the difference in the number of objects between the SIG and graph-based representations is presented.

**Table 1.** From GIS to graph: Comparison of components number

| Datasets | Representation | Pipes | Manholes | Pumps | Structures | Treatment plants | Accessories | Fittings | Dummy nodes |
|---|---|---|---|---|---|---|---|---|---|
| Dataset 1 | GIS | 43804 | 42007 | 217 | 760 | 14 | 0 | 451 | 0 |
|  | Graph | 43803 | 41976 | 217 | 760 | 14 | 0 | 451 | 405 |
|  | Differences | −1 | −31 | 0 | 0 | 0 | 0 | 0 | +405 |
| Dataset 2 | GIS | 47059 | 45411 | 245 | 801 | 14 | 427 | 0 | 0 |
|  | Graph | 47064 | 45410 | 245 | 800 | 14 | 427 | 0 | 188 |
|  | Differences | +5 | −1 | 0 | −1 | 0 | 0 | 0 | +188 |
| Dataset 3 | GIS | 35535 | 34492 | 0 | 236 | 0 | 119 | 0 | 0 |
|  | Graph | 35774 | 34434 | 0 | 141 | 0 | 27 | 0 | 1202 |
|  | Differences | +239 | −58 | 0 | −95 | 0 | −92 | 0 | +1202 |
| Dataset 4 | GIS | 37330 | 37683 | 0 | 285 | 0 | 216 | 0 | 0 |
|  | Graph | 37922 | 37518 | 0 | 182 | 0 | 129 | 0 | 38 |
|  | Differences | +592 | −165 | 0 | −103 | 0 | −97 | 0 | +38 |

Concerning dataset 1, most of the components from the GIS representation are successfully preserved in the graph. However, there are very few components missing in the graph representation. This happens when the components are located at a considerable distance from the nearest pipes in the GIS representation, making it challenging to establish direct connections between them. Figure 2 provides a visual representation of this scenario, highlighting the manholes that are not in close proximity to any pipes. The algorithm ensures that each pipe in the GIS representation is represented as an edge in the graph-based representation. However, in certain cases, the number of edges in the graph may be less than the number of pipes in the GIS representation. This can happen when there are duplicated pipes in the GIS representation, where two pipes have the same extremities. In such cases, the algorithm inserts only one of these duplicated pipes into the graph, resulting in a difference in the number of edges. In Fig. 9, it is evident that two pipes, namely 37974 and 16990, share the same extremities, specifically 32451 and 30017. As a result of removing these duplicated pipes during the graph-based representation, only one pipe with these extremities is inserted into the graph. This explains the negative difference in the number of edges compared to the pipes in the GIS representation for Dataset 1, where 21 duplicated pipes were encountered.

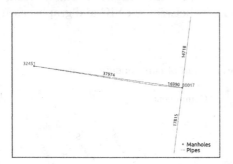

Fig. 9. Duplicated pipes in GIS

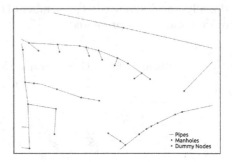

Fig. 10. Dummy nodes

Furthermore, a total of 405 dummy nodes were incorporated into the graph for dataset 1. Figure 10 illustrates an example of dummy nodes added to the graph. The nodes highlighted in rose color represent the dummy nodes. In Fig. 10, the GIS representation of the wastewater network is depicted, revealing the absence of nodes at the extremities of certain pipes. Without the presence of manholes or other components, it is challenging to establish a complete branch in the network. Hence, the addition of dummy nodes becomes necessary to ensure connectivity and complete the representation of the wastewater network data.

Indeed, the number of connected components in the resulting graphs is an important criterion for evaluating the accuracy of the graph-based representation. In real-world wastewater networks (see the Fig. 11), it is common to have

— Pipes

**Fig. 11.** Components in GIS representation

disconnected components due to various factors such as the physical layout of the network.

In some cases, the disconnected components in the GIS representation may be the result of inserting errors or shifting errors during data collection or digitization processes. These errors can lead to gaps in the network representation, making it challenging to accurately assess the connectivity of the wastewater network.

**Table 2.** Connected components

| Datasets | Representation | Connected components |
|---|---|---|
| Dataset 1 | Pipes (GIS) | 166 |
|  | graph | 151 |
| Dataset 2 | Pipes (GIS) | 475 |
|  | graph | 168 |
| Dataset 3 | Pipes (GIS) | 200 |
|  | graph | 158 |
| Dataset 4 | Pipes (GIS) | 177 |
|  | graph | 81 |

In the graph-based representation, one of the main goals is to identify and correct errors present in the GIS databases, which may lead to disconnected components in the resulting graph. The small threshold of 0.1 m used for the buffer distance is designed to capture and address these errors of small shifting or inserting errors.

By applying this threshold, the algorithm can identify nodes that are close to the pipes in the GIS representations but may not have a direct connection due to minor inaccuracies in their positions. The buffer distance helps in identifying these nearby nodes and ensures that they are connected by inserting appropriate edges in the graph. As a result of this error correction process, the number of components in the resulting graph is often less than the number of components in the GIS representation (see Table 2). This reduction in the number of components indicates that the graph-based representation is successfully identifying and addressing the errors in the GIS representation, resulting in a more accurate and coherent representation of the wastewater network.

In Fig. 12, an example of the graph-based representation of wastewater networks data is presented, where different colors of nodes symbolize specific types of components, with red indicating manholes. The edges represent the pipes, and their interconnections depict the real-world relationships found in wastewater networks. Furthermore, the graph-based representation retains all information about components from the GIS representation, encompassing both attribute and spatial details. This preservation enables us to generate GIS representations from the graph whenever required.

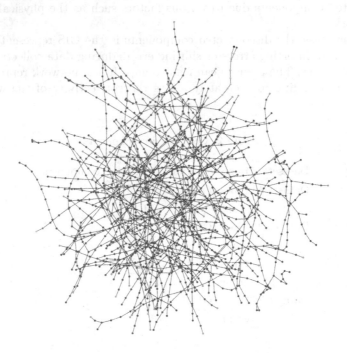

**Fig. 12.** Graph-based representation of wastewater networks

## 5    Conclusion

In this study, a graph-based approach for representing wastewater networks, derived from GIS datasets, has been proposed. Our graph-based representation accurately captures the interconnected nature of the network. By conducting experiments on real-world datasets, the efficiency and effectiveness of the method have been demonstrated. It showcases the ability to preserve most components from the GIS representation while resolving missing connections attributed to significant distances.

In future endeavors, our plans involve improving our graph-based representations by integrating water circulation within the pipes. This entails constructing oriented graph-based representations that reflect the flow direction and dynamics of wastewater through the network. Taking into account the directionality of pipes enables us to delve deeper into flow behavior analysis and optimize the network accordingly. Furthermore, since all the datasets we possess represent the same geographical zone, our objective is to explore the fusion of these oriented graphs. The fusion process will enable us to synthesize comprehensive representations that encompass the entire wastewater network for the given region.

**Acknowledgements.** This research has received support from the European Union's Horizon research and innovation programme under the MSCA-SE (Marie Skłodowska-Curie Actions Staff Exchange) grant agreement 101086252; Call: HORIZON-MSCA-2021-SE-01; Project title: STARWARS (STormwAteR and WastewAteR networkS heterogeneous data AI-driven management).

This research has also received support from the French national project ANR (Agence Nationale de la Recherche) CROQUIS (Collecte, représentation, complétion, fusion et interrogation de données hétérogènes et incertaines de réseaux d'eaux urbains).

# References

1. Liu, M., Li, X., Li, J., et al.: A knowledge graph-based data representation approach for IIoT-enabled cognitive manufacturing. Adv. Eng. Inform. **51**, 101515 (2022)
2. Zhou, H., Yang, X., Fan, M., et al.: Static-dynamic global graph representation for pedestrian trajectory prediction. Knowl.-Based Syst. **277**, 110775 (2023)
3. Fall, A., Fortin, M.-J., Manseau, M., O'Brien, D.: Spatial graphs: principles and applications for habitat connectivity. Ecosystems **10**(3), 448–461 (2007). https://doi.org/10.1007/s10021-007-9038-7
4. Sonawane, S.S., Kulkarni, P.A.: Graph based representation and analysis of text document: a survey of techniques. Int. J. Comput. Appl. **96**(19) (2014)
5. Maugey, T., Ortega, A., Frossard, P.: Graph-based representation for multiview image geometry. IEEE Trans. Image Process. **24**(5), 1573–1586 (2015)

# Image Recommendation Based on Pre-trained Deep Learning and Similarity Matching

Le Huynh Quoc Bao, Huynh Huu Bao Khoa, and Nguyen Thai-Nghe[(⊠)]

Can Tho University, Can Tho city, Vietnam
{lhqbao,ntnghe}@ctu.edu.vn, khoab1910658@student.ctu.edu.vn

**Abstract.** Recommender systems are widely used in many domains, especially in E-commerce. It can be used for attracting users by recommending appropriate products to them. There are many techniques in recommendation systems which can predict rating scores to recommend next products. In this work, we propose an approach for recommendation based on product images using pre-trained deep learning models and similarity matching. Specifically, in the proposed model, we have utilized the pre-trained deep learning models (e.g., the VGG16) to extract the image features. Then, based on the image features, we compute similarities between the products (e.g. using Cosine similarity). For recommending similar products to the users, we do the same tasks, e.g., extracting feature of the current image, computing its similarity with other images in the database, and generating a list of TOP-N (e.g. TOP-5) most similar products. Experimental results on two public data sets show that the approach can give good recommendations at more than 90% of accuracy.

**Keywords:** Image recommendations · Recommender systems · Feature extraction · Pre-trained deep learning

## 1 Introduction

With the rapid development of network technology, the number of digital images is increasing at an alarming rate, people's information needs are gradually shifting from text to images. However, it is very difficult for users to quickly find the images they are interested in from the large number of image libraries, as well as the products they need on e-commerce systems. With the idea of product search in online shopping system using image based approach, the user can provide, select or click an image and similar image-based products will be displayed to the user. In addition, the use of the Recommendation Systems (RS) help the users explore and exploit the information environment for better decisions [1]. Today's online shopping products are mainly images, most e-commerce stores use product images to advertise, arouse users' visual desires and encourage them to buy products. Thus, recommendation using images is necessary.

In this work, we propose an approach for recommendation based on product images using pre-trained deep learning models and similarity matching.

N. Thai-Nghe et al. (Eds.): ISDS 2023, CCIS 1949, pp. 258–270, 2024.
https://doi.org/10.1007/978-981-99-7649-2_20

## 2   Related Work

There are several related work in this domain, for examples, the authors in [2] have used convolutional neural network (CNN) models to extract features of product images. The system then uses the features to calculate the similarity between the images. They also analyzed four versions of the MovieLens dataset to demonstrate improvements in recommendation accuracy, including 100k, 1M, 10M, and 20M. The results of the experiment show a significant increase in accuracy compared to traditional methods.

The work in [3] has proposed a recommender system which is based on content-based image retrieval and consists of two main stages. In Phase 1, a product class learning approach is proposed. In Phase 2, the recommendation system recommends similar products. For Phase 1, the proposed approach creates a product model using Machine Learning (ML). The model is then used to find the test product type. From an ML point of view, using a Random Forest (RF) classifier and feature extraction, the paper uses JPEG coefficients. The dataset used for the demonstration consists of 20 product categories. For the suggestion on the image, the proposed RF model is evaluated for Phase 1 and Phase 2. In Phase 1, the evaluation of the proposed model has an accuracy of 75%. To improve performance, the RF model was integrated into the Deep Learning (DL) setup to achieve 84% prediction accuracy. Based on a customized assessment for Phase 2, the recommendation method is 98% accurate, demonstrating its effectiveness for product recommendations in real-world applications.

The paper in [4] shows a two-stage deep learning framework to recommend fashion images based on other similar input images. The CNNs are used to extract seemingly obvious and data-driven features. Stage 2 results are input for similarity-based recommendations using a ranking algorithm. The method is tested on a publicly available fashion data set. In the paper [5], the image classification algorithm is studied first. LReLU - Softplus activation function is formed by combining LReLU function and Softplus function, and CNN is improved. The paper proposes two models: (1) An image retrieval model based on the proposed "local sensitive hash algorithm". This model calculates the distance in hamming space for binary hashes generated by mapping. Euclidean distance is calculated inside the result set after similarity measurement to improve accuracy and image retrieval model is built; (2) An image suggestion model based on a hidden support vector machine (SVM) is proposed. This method combines image text information and image content information. In this paper, the overlap rate between the CNN-based recommendation model and the human recommendation algorithm was tested, and the matching level of the two suggested images reached 88%.

The article [6] follows the collaborative filtering method based on deep learning to discover the user's interest, understand the information between the user and the item, and make better recommendations. The paper used multilayer neural network to control the non-linearity between user and item communication from data. Tests to demonstrate that the HBSADE model performs better than existing methods over Amazon-b and Book-Crossing datasets. The article

[7] proposes a deep collaborative fitting algorithm. The article combines two methods to suggest: (1) using the advantages of convolutional neural network to effectively learn about non-linear interactions of users and items; (2) features of the collaborative filtering algorithm to model the linear interaction of users and items. The experimental results show that the proposed algorithm has significantly improved the accuracy on the public dataset (Yahoo! Movie).

The main idea of the paper [10] is to recognize feature classes from images by convolutional neural network to make suggestions. The proposed solution, users and images are placed in a semantic space represented by a graph. The work in [11] introduces a method that combines Collaborative Filtering with Content-Based Approach and CNN. It was tested on a Movies database, giving results that outperform modern methods in accuracy and precision, as well as effectively improving traditional Collaborative Filtering. In the paper [12], the main focus is on the combination of a recommendation system with visual features of the product. Done with the help of a deep architecture and a series of "convolution" operations that cause overlapping of edges and blobs in images. The paper shows that when the size problem is solved, the extracted features will serve as a good quality representation of the image. Experimental study compares with different linear and non-linear reduction techniques on the features of CNN to build a completely image-based recommender model.

The paper [13] develops a framework called Multimodal Graph Contrastive Learning (MGCL), which collects cooperation signals from interactions and uses visual and textual methods to in turn extract preference clues of the user in a particular manner. The key idea of MGCL involves two aspects: First, to alleviate noise contamination during graph learning, the authors construct three parallel graph convolution networks to independently generate three types of user and item rep- resentations, containing collaborative signals, visual preference clues, and textual preference clues. Second, to eliminate as much preference-independent noisy information as possible from the generated representations, the authors incorporate sufficient self- supervised signals into the model optimization with the help of contrastive learning, thus enhancing the expressiveness of the user and item representations. MGCL is not limited to graph learning schemes but can also be applied to most matrix factorization methods. The authors conducted extensive experiments on three public datasets to validate the effectiveness and scalability of MGCL1.

The paper [14] develops a hierarchical model for social contextual image recommendation. In addition to the basic user preference model in recommendation based on popularity matrix coefficients, the author identified three main aspects (i.e. upload history, social influence, and owner admiration) that affect each user's latent preferences, where each aspect summarizes a contextual element from the complex relationship between the user and the image. The authors then design a hierarchical network that naturally reflects the hierarchical relationships (components at each aspect level and facet level) of users' implicit preferences for aspects principal edge has been determined. Specifically, by using embeddings from state-of-the-art deep learning models tailored to each type of data, hierarchical attention networks can learn to attend to a variety of content. Finally, extensive experimental results on real-world datasets clearly demonstrate the superiority of the proposed model.

In this work, we have utilized the pre-trained deep learning models (e.g., the VGG16) to extract the image features. Then, based on the image features, we compute similarities between the products (e.g. using Cosine similarity) for recommendation. Although this approach is not personalized for each user, it could be a good choice for new user problem [15].

## 3    Proposed Approach

The proposed approach for image recommendation is presented in Fig. 1. In this model, we have utilized the pre-trained deep learning models (e.g., the VGG16 [8] or ResNet50 [9]) to extract the image features. The advantage of using pre-trained models is that it had been trained on larger dataset (e.g., the ImageNet with more than 14 million images of images), thus it already has very good parameters. However, this work does not use the pre-trained model for classification, but for image feature extraction, so we have to remove the last layers of the pre-trained models. After doing feature extraction, we compute similarities between the products (e.g. using Cosine or Pearson similarity). For testing new images, we do the same task as described above, then computing its similarity with other images in the database. Finally, a list of TOP-N (e.g. TOP-5) most similarity products is returned for recommendation.

## 4    Evaluation

### 4.1    Data Sets

Two published data sets in e-commerce/online systems are used for experiments as presented in Table 1. The first data set is **Style Color Images**[1]. This dataset has 2184 color images with 7 brands and 10 products in the .png format. The second dataset is the **Fashion Product Images Dataset**[2]. Both of these data sets are available for download. These data sets are divided to 80% for training and 20% for testing.

**Table 1.** Data set for experiments

| Dataset | Classes | Total image | Training image | Testing image |
|---|---|---|---|---|
| Style Color Images | 10 | 2184 | 1747 | 437 |
| Fashion Product Images | 42 | 18750 | 15000 | 3750 |

---

[1] https://www.kaggle.com/datasets/olgabelitskaya/style-color-images.
[2] https://www.kaggle.com/datasets/paramaggarwal/fashion-product-images-dataset.

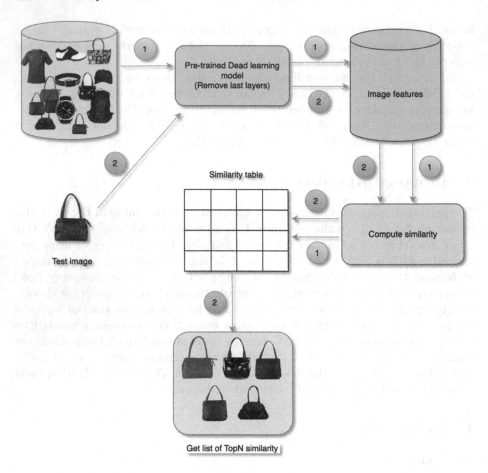

**Fig. 1.** Image recommendation with pre-trained deep learning and similarity matching (notation 1 is used for training phase and notation 2 is used for testing/recommendation phase)

## 4.2 Model Settings

In the proposed approach in Fig. 1, for image feature extraction phase using pre-trained deep learning model, this work has used the VGG16, VGG19 [8], and the RESNET50 [9] for comparison.

In the similarity computation phase, we have tested several similarity measures to investigate that which one can produce better results.

**Cosine Similarity:** Cosine similarity is a versatile technique for measuring the similarity between vectors, yielding values between zero and one for vectors with positive values. Geometrically, the cosine of an angle between two lines varies from 1 to 0 as the angle changes from $0°$ to $90°$. In the context of numerical vectors, cosine similarity does not take into account the order of values within the vectors. Hence, a result of 1 doesn't imply exact equivalence between two numerical vectors. Here's the mathematical formula used to compute cosine similarity between two given feature vectors:

$$\text{Similarity}(u,v) = \frac{r_u.r_v}{||r_u||.||r_v||}$$

Where $r_u$ represents the latent feature vector for item u, and $r_v$ corresponds to the latent feature vector for item v. The function Similarity(u,v) calculates the cosine similarity between image features of item u and v.

**Euclidean Distance:** Euclidean Distance, often simply referred to as the "distance", provides a measure of the straight-line distance between two points in Euclidean space. Named after the ancient Greek mathematician Euclid, it's one of the most commonly used distance metrics, especially in geometry and machine learning. Mathematically, for two points P(x1,y1) and Q(x2,y2) in a two-dimensional space, the Euclidean distance d between them is given by:

$$d(P,Q) = \sqrt{(x_2 - x_1)^2 + (y_2 - y_1)^2}$$

In higher dimensions, for two vectors u and v of dimension n, the Euclidean distance is geralized as:

$$d(u,v) = \sqrt{\sum_{i=1}^{n}(u_i - v_i)^2}$$

Where $u_i$ is the $i^{th}$ component of vector u and $v_i$ is the $i^{th}$ component of vector v.

**Manhattan Distance:** Manhattan Distance, also known as the L1 distance, City Block Distance, or Taxicab norm, is a distance metric that calculates the total absolute difference between corresponding coordinates of two points or vectors. It's called the Manhattan Distance because it measures the distance a taxi would drive in a city like New York (where streets form a grid) from one intersection to another, only being able to drive in right-angle paths similar to the city blocks. Mathematically, for two points P(x1,y1) and Q(x2,y2) in a two-dimensional space, the Manhattan distance d between them is given by:

$$d(P,Q) = |x_2 - x_1| + |y_2 - y_1|$$

In higher dimensions, for two vectors u and v of dimension n, the Manhattan distance is generalized as:

$$d(u, v) = \sqrt{\sum_{i=1}^{n} |u_i - v_i|}$$

Where $u_i$ is the $i^{th}$ component of vector u and $v_i$ is the $i^{th}$ component of vector v.

**Jaccard Similarity:** The Jaccard Similarity coefficient, often known simply as the Jaccard Index or Jaccard coefficient, is a measure used to quantify the similarity and diversity of two sample sets. This metric is especially suited for cases where the data is binary in nature, such as presence/absence data. Mathematically, the Jaccard Similarity between two sets A and B is computed as:

$$J(A, B) = |\frac{A \cap B}{A \cup B}|$$

Where $|A \cap B|$ represents the size of the intersection of sets A and B (i.e., the number of elements that both sets have in common). $|A \cup B|$ represents the size of the intersection of sets A and B (i.e., the number of elements that both sets have in common). The Jaccard Similarity returns a value between 0 and 1: A value of 0 indicates that the two sets have no common elements. A value of 1 means the two sets are identical.

**Pearson Correlation Coefficient:** The Pearson Correlation Coefficient, often simply referred to as Pearson's correlation or the correlation coefficient, is a measure of the linear correlation or association between two continuous variables. It's a value between -1 and 1 that represents the strength and direction of the linear relationship. Mathematically, for two variables X and Y, Pearson's correlation coefficient r is given by:

$$r = \frac{\sum_{i=1}^{n}(x_i - \overline{x})(y_i - \overline{y})}{\sqrt{\sum_{i=1}^{n}(x_i - \overline{x})^2}\sqrt{\sum_{i=1}^{n}(y_i - \overline{y})^2}}$$

Where $x_i$ and $y_i$ are individual data points, $\overline{x}, \overline{y}$ are the means of X and Y. The output r can be interpretedas: r = 1: Perfect positive linear correlation. r = −1: Perfect negative linear correlation, r = 0: No linear correlation.

The pre-trained models from tf.keras.applications with pre-trained weights on the 'imagenet' dataset. This means the model is initialized with weights that were learned from training on the ImageNet dataset. This is beneficial as it leverage the knowledge that the VGG19 model has gained from that large-scale dataset, making it useful for various computer vision tasks The include_top = False argument specifies excluding the fully connected layers of the VGG19 model, which are typically used for classification. By setting this to False, pretrained models are removed the classification head, and only keeping the convolutional layers.

The input_shape = (224, 224, 3) argument defines the shape of the input images that will be fed into the VGG19 model. 224 and 224 are the desired width and height in pixel of input images, respectively, and 3 represents the number of color channels (RGB images have 3 channels). Next, base_model.trainable = False freezes the weights of the VGG19 base model, making it non-trainable. This is done to ensure only the layers added after the VGG19 base will be trainable. Freezing the base model's weights can be useful when you have limited data or computational resources and do not want to retrain the entire VGG19 model. Create a feat_extractor2 model using tf.keras.Sequential. It consists of the base_model (VGG19 with frozen weights) followed by a GlobalMaxPooling2D layer. The GlobalMaxPooling2D layer performs global max pooling, reducing the spatial dimensions of the output from the base model to a single vector per channel.

After successfully creating the model, we use this model to extract the feature vector of all the images in the train set and save it as .npy. Finally, when the user imports an image, we use the model to extract the feature vector of that image and compute the cosine between that vector and the train set vectors stored as npy. Get k train images with highest cosine.

## 4.3   Experimental Results

Figures 2, 3, and 4 visualize some sample recommendation results. In these three figures, the left-top image is the input and five other images are the recommendation with similarity values.

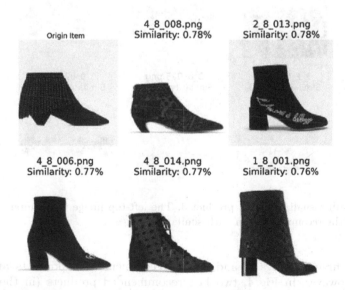

**Fig. 2.** Result visualization for product 1. The left-top image is the input, five other images are the recommendation with similarity values

**Fig. 3.** Result visualization for product 2. The left-top image is the input, five other images are the recommendation with similarity values

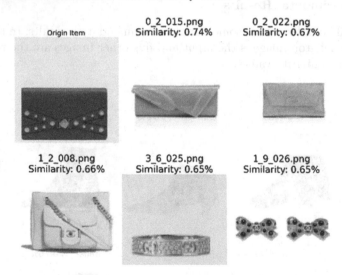

**Fig. 4.** Result visualization for product 3. The left-top image is the input, five other images are the recommendation with similarity values

In the first two Figs. 2 and 3, the recommendation products are almost correct. However, in Fig. 4, two last recommended products (in the bottom-middle and bottom-right images) are not correct. This can be seen clearly by human-eye, since in the original image, there is an X decoration part with some buttons, when using feature extraction, it looks very similar to the X one in the two last images.

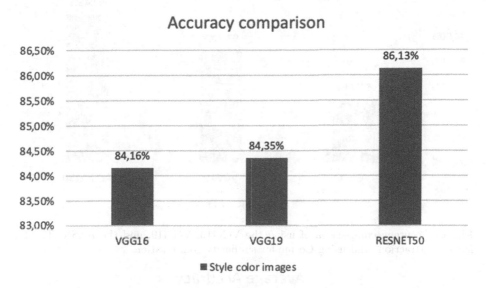

**Fig. 5.** Accuracy comparison of using the VGG16, VGG19, and the RESNET50 for feature extraction, and using Cosine for similarity computation

In the first experiments, we have investigated the similarity between the Cosine and Pearson on the fist data set (the Style Color Images) using the VGG16 model. Results show that the accuracy of using Cosine is 84.16% compared to using Pearson with 83.75%. Thus we have used Cosine similarity for the rest experiments.

Figure 5 and Fig. 6 presents for the accuracy comparison of using the VGG16, VGG19, and the RESNET50 for feature extraction, and using Cosine for similarity computation on data sets of Style color images and Fashion product images. These results show that in the proposed approach, we can use any pre-trained deep learning models since three models can achieve nearly similar results in feature extraction, thus, can get similar accuracy. However, the Style Color Images data set has less images in training/testing than the other one, thus, its accuracy has 84.35% compared to the Fashion Product Images Data with 93.72%, using the VGG19 model.

The second experiments, we have compared the similarity between the Cosine, Pearson, Euclidean, Manhattan and Jaccard on the second data set (the Fashion Product Image) using the VGG16 model. The results show as in Fig. 7.

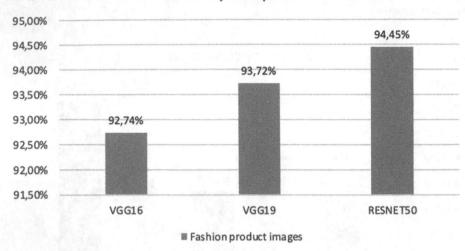

Fig. 6. Accuracy comparison of using the VGG16, VGG19, and the RESNET50 for feature extraction, and using Cosine for similarity computation

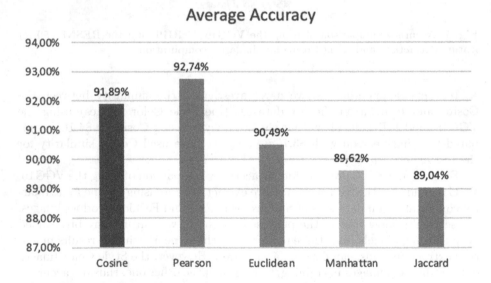

Fig. 7. Accuracy comparison using the VGG16 on different similarities

## 5  Conclusion

This work proposes an approach for recommendation based on product images using pre-trained deep learning models and similarity matching. Specifically, in the proposed model, we have utilized the pre-trained deep learning models (e.g., the VGG16) to extract the image features. Then, based on the image features, we compute similarities between the products (e.g. using Cosine similarity). Experimental results on two public data sets show that the approach can generate good recommendations.

However, this is approach is not personalized for each users. In future work, we try employ personalizing to the proposed approach for better recommendation.

# References

1. Dien, T.T., Thanh-Hai, N., Thai-Nghe, N.: Novel approaches for searching and recommending learning resources. Cybern. Inf. Technol. **23**(2), 151–169 (2023). https://doi.org/10.2478/cait-2023-0019
2. Alamdari, P.M., Navimipour, N.J., Hosseinzadeh, M., Safaei, A.A., Darwesh, A.: Image-based product recommendation method for e-commerce applications using convolutional neural networks. Acta Informatica Pragensia **2022**(1), 15–35 (2022)
3. Ullah, F., Zhang, B., Khan, R.U.: Image-based service recommendation system: a JPEG-coefficient RFs approach. IEEE Access **8**, 3308–3318 (2020). https://doi.org/10.1109/ACCESS.2019.2962315
4. Tuinhof, H., Pirker, C., Haltmeier, M.: Image-based fashion product recommendation with deep learning. In: Nicosia, G., Pardalos, P., Giuffrida, G., Umeton, R., Sciacca, V. (eds.) LOD 2018. LNCS, vol. 11331, pp. 472–481. Springer, Cham (2019). https://doi.org/10.1007/978-3-030-13709-0_40
5. Yin, P., Zhang, L.: Image recommendation algorithm based on deep learning. IEEE Access **8**, 132799–132807 (2020). https://doi.org/10.1109/ACCESS.2020.3007353
6. Sivaramakrishnan, N., Subramaniyaswamy, V., Viloria, A., et al.: A deep learning-based hybrid model for recommendation generation and ranking. Neural Comput. Appl. **33**, 10719–10736 (2021). https://doi.org/10.1007/s00521-020-04844-4
7. Yin, P., Wang, J., Zhao, J., Wang, H., Gan, H.: Deep collaborative filtering: a recommendation method for crowdfunding project based on the integration of deep neural network and collaborative filtering. Math. Probl. Eng. **2022**, 1–15 (2022). https://doi.org/10.1155/2022/4655030
8. Simonyan, K., Zisserman, A.: Very Deep Convolutional Networks for Large-Scale Image Recognition, eprint arXiv:1409.1556 (2015)
9. He, K., Zhang, X., Ren, S., Sun, J.: Deep residual learning for image recognition, pp. 770–778 (2016). https://doi.org/10.1109/CVPR.2016.90
10. Kobyshev, K., Voinov, N., Nikiforov, I.: Hybrid image recommendation algorithm combining content and collaborative filtering approaches. Procedia Comput. Sci. **193**, 200–209 (2021). https://doi.org/10.1016/j.procs.2021.10.020. ISSN 1877-0509
11. Afoudi, Y., Lazaar, M., Al Achhab, M.: Hybrid recommendation system combined content-based filtering and collaborative prediction using artificial neural network. Simul. Model. Pract. Theory **113**, 102375 (2021). https://doi.org/10.1016/j.simpat.2021.102375. ISSN 1569-190X
12. Sulthana, A.R., Gupta, M., Subramanian, S., et al.: Improvising the performance of image-based recommendation system using convolution neural networks and deep learning. Soft Comput. **24**, 14531–14544 (2020). https://doi.org/10.1007/s00500-020-04803-0
13. Liu, K., Xue, F., Guo, D., Sun, P., Qian, S., Hong, R.: Multimodal graph contrastive learning for multimedia-based recommendation. IEEE Trans. Multimedia (2023). https://doi.org/10.1109/TMM.2023.3251108

14. Wu, L., Chen, L., Hong, R., Fu, Y., Xie, X., Wang, M.: A hierarchical attention model for social contextual image recommendation. IEEE Trans. Knowl. Data Eng. **32**(10), 1854–1867 (2020). https://doi.org/10.1109/TKDE.2019.2913394
15. Thai-Nghe, N., Xuyen, N.T.K., Tran, A.C., Dien, T.T.: Dealing with new user problem using content-based deep matrix factorization. In: Fujita, H., Wang, Y., Xiao, Y., Moonis, A. (eds.) IEA/AIE 2023. LNCS, vol. 13926, pp. 177–188. Springer, Cham (2023). https://doi.org/10.1007/978-3-031-36822-6_16

# A Practical Approach to Leverage Knowledge Graphs for Legal Query

Dung V. Dang[1,3], Vuong T. Pham[2,3,4], Thanh Cao[4(✉)], Nhon Do[5],
Hung Q. Ngo[6], and Hien D. Nguyen[1,3]

[1] University of Information Technology, Ho Chi Minh city, Vietnam
{dungdv,hiennd}@uit.edu.vn
[2] University of Science, Ho Chi Minh city, Vietnam
[3] Vietnam National University, Ho Chi Minh city, Vietnam
[4] Sai Gon University, Ho Chi Minh city, Vietnam
{vuong.pham,thanh.cao}@sgu.edu.vn
[5] Hong Bang International University, Ho Chi Minh city, Vietnam
nhondv@hiu.vn
[6] Technological University Dublin, Dublin, Ireland
hung.ngo@tudublin.ie

**Abstract.** The significance of legal information in society stems from
its role in facilitating individuals' access to and comprehension of legal
knowledge. This study introduces a novel methodology for legal querying,
leveraging knowledge graphs as representations of the semantic essence
within legal documents. Empirical investigations validate the efficacy
of knowledge graph-based querying in effectively addressing the diverse
requirements of users on a substantial scale. The success of the practical
legal query application is a testament to the potential of this approach.

**Keywords:** Knowledge graph · Subgraph matching · Legal
document · Knowledge-based system

## 1 Introduction

The process of researching and retrieving legal information plays a crucial role in
understanding and applying the law in everyday life [18,21]. It enables individuals to grasp legal information quickly and accurately, ensuring fairness, effectiveness, and compliance with the law. However, current systems face challenges
in extracting meaning from legal documents and retrieving relevant legal information, requiring precision to meet user needs.

This study focuses on the query processing phase based on the knowledge
representation method presented in a previous study [7]. In this study, the
effectiveness of the knowledge representation method has been identified and

---

Equal contribution by Dung V. Dang and Vuong T. Pham

© The Author(s), under exclusive license to Springer Nature Singapore Pte Ltd. 2024
N. Thai-Nghe et al. (Eds.): ISDS 2023, CCIS 1949, pp. 271–284, 2024.
https://doi.org/10.1007/978-981-99-7649-2_21

demonstrated in [7] Specifically, the study uses a knowledge graph to represent knowledge from legal documents and identify relationships between concepts. It leverages the knowledge representation method and the represented knowledge to optimize the knowledge graph, reduce its complexity and size, and develop an efficient knowledge query method.

This paper proposes an effective query method based on knowledge represented in the form of a knowledge graph, facilitating users to easily and quickly search for legal information. The paper also conducts experiments and evaluates the effectiveness of the new query method on real-world datasets, comparing it with previous query methods.

The next section reviews related work in the field of legal knowledge query. Section 2 presents the process of refining the knowledge graph from legal documents. Next, Sect. 3 describes the development of the query method and the formation of the query module. We describe the process of testing the search capability on the knowledge repository of road traffic laws in Sect. 5. Finally, the paper concludes and provides future research directions in Sect. 6.

## 2   Related Work

In the domain of legal query, significant advancements have been made in the application of artificial intelligence to intelligent query systems [19]. These systems not only facilitate efficient retrieval and comprehension of legal information but also contribute to judicial decision-making. This progress is attributed to the utilization of advanced technologies such as natural language processing (NLP), data mining, and information extraction in the context of query processing and analysis [9,18,21].

The study [6] introduces an approach using BERT for the selection of answers in the field of Vietnamese law. The method is fine-tuned on a dataset of Vietnamese legal questions and answers, achieving an F1 score of 87%. Furthermore, the study conducts pre-training of the original BERT model on a domain-specific Vietnamese legal corpus, resulting in a higher F1-Score of 90.6% on the same task. This approach demonstrates high effectiveness in answer selection. However, it requires a large amount of training data and faces challenges in knowledge updating. Additionally, it encounters difficulties in handling complex relationships.

The study in [20] introduced an unsupervised legal information extraction method using a deep pre-trained language model called LamBERTa to build an information extraction system for Italian legal texts, known as the Unsupervised Law Article Mining System (ULAMS). By utilizing unsupervised learning techniques, this system automatically identifies important topics within the provisions of the Italian Civil Code. The model was evaluated on real-world data from Italian Civil Law and demonstrated high accuracy.

Wang et al. proposed an efficient distributed method to answer subgraph matching queries on large RDF graphs using MapReduce [22]. In this method, the query graph is decomposed into a set of stars, taking advantage of the semantic information embedded in the RDF graph. Additionally, optimization techniques are introduced to enhance the performance of algorithms, such as filtering out invalid input data to reduce intermediate results. In Vietnamese, there is one study on building linked data for legal documents based on RDF graphs [12]. However, this study only builds the dataset as linked entities from unstructured documents without applications exploiting this data source.

Li et al. proposed an approach to approximate subgraph matching in fuzzy RDF graphs [11]. The method utilizes fuzzy graph models to represent RDF data and searches for subgraphs that have a high likelihood of matching the given query graph. By decomposing the query graph into paths, finding candidates for each path, and combining the candidates to construct query answers, the proposed method achieves efficient approximate subgraph matching.

Significant research efforts aim to provide convenient and efficient means for the general public to access and understand legal information [18,21]. However, current methods may not be suitable or require substantial effort to implement effectively. This paper presents a comprehensive solution to designing a robust legal query system harnessing the power of a knowledge base and a knowledge graph. The proposed solution aims to enhance the efficiency and accuracy of legal knowledge retrieval. By leveraging the capabilities of the knowledge base and knowledge graph, users can effectively query and access relevant legal knowledge, reducing the efforts and time required by knowledge system managers.

## 3    Legal Knowledge Base Construction

This section focuses on designing the structure and methods for representing knowledge that are suitable for the legal domain. The aim is to construct a supportive platform for organizing and representing knowledge from legal documents into a knowledge graph automatically, thereby saving time and effort for humans.

### 3.1    Legal Ontology Model and Knowledge Graph

The ontology structure comprises concepts and their relationships, which are utilized to convey the meaning of legal documents. In the context of the Vietnamese legal system, the adoption of the Ontology Rela-model is suitable as it offers a knowledge model of relations [8]. The Ontology Legal-Onto [15,17] is constructed based on the Rela-model and enhanced to meet the requirements of representing legal knowledge. The structure of the Legal-Onto ontology encompasses concepts, relationships, and inference rules, all of which contribute to establishing a comprehensive and precise knowledge model.

The Onto Legal-Onto comprises the following elements:

$$K = (C,\ R,\ Rules) + (Concept,\ Rela) \tag{1}$$

In which, *(C, R, Rules)* represents the structure of the Rela-model used for organizing information. Specifically:

- *C* is a collection of concepts, with each concept in C having an improved internal structure adapted to its organization and information storage.
- *R* is a set of relationships between concepts, key phrases, and a database that stores the content of legal documents.
- *Rules* is a set of inference rules within the knowledge domain.

*(Concept, Rela)* is a knowledge graph that illustrates the relationships between key phrases in legal documents, where:

- *Concept* is a collection of terms used to express concepts present in legal documents.
- *Rela* is a set of arcs, each having a direction and representing a semantic relationship between two concepts.

### 3.2 Building Knowledge Graph and Its Optimization

Building upon the knowledge graph extraction and representation method discussed in study [7], the next crucial phase involves optimizing the knowledge graph for enhanced performance. This optimization process helps reduce the complexity and size of the text by removing meaningless triples from the text containing them. Along with that, we will search for triples with the same meaning to proceed with the reduction.

Knowledge graph optimization involves reducing the complexity and size of the graph by removing meaningless triples specific to the Document Item. Additionally, equivalent triples are identified and combined to reduce redundancy (Fig. 1).

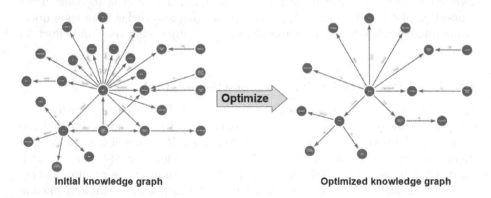

Initial knowledge graph                    Optimized knowledge graph

**Fig. 1.** Illustrate the graph optimization process

**Step 1:** *Identify meaningless triples*
During the triple construction process, especially through automatic extraction, there may be triples with low tf-idf values that appear in multiple Document Items but do not contribute specific content. These tf-idf values serve

as hints for system administrators to decide whether to remove them from the graph representing the passage's meaning. Any triples deemed irrelevant by the administrator are blacklisted and excluded from the graph.

**Step 2:** *Optimize equivalent triples*

Synonymous are key phrases, which have the same or nearly the same meaning.

- Check if two key phrases belong to the same value field in the LDOCE Vietnamese version of the thesaurus. If they do, they are considered synonymous.
- If the key phrases do not belong to any value fields in the thesaurus, Phobert is used to convert them into vectors and calculate the cosine similarity[1]. Two key phrases are considered semantically similar if their similarity score is greater than or equal to a predefined value, denoted as alpha. The value of alpha depends on the specific knowledge domain under investigation. In this study, alpha is set to 0.6, a value determined through the collection of synonymous word clusters within the legal field and verified by domain experts.

Equivalent triples are three-word phrases that have the same or synonymous values for each pair of subject, relation, and object.

To determine equivalence between two triples, all three pairs of subject, relation, and object values are compared for sameness or synonymity. If any component does not match, the comparison stops, and the two triples are not considered equivalent. If there are equivalent triples within the same graph representing the Document Item's meaning, they are omitted to reduce the graph's size and complexity.

## 4  Legal Search Platform

The legal search platform After completing the knowledge base construction for the system, it is necessary to find a solution to accurately answer the questions entered by users within the scope of knowledge represented in the knowledge base. To achieve this, it is essential to design algorithms for accurately querying the knowledge represented in the form of a knowledge graph. In this session, the study will present the steps, algorithms, and techniques used to design the query algorithm for knowledge retrieval from legal documents (Fig.2).

### 4.1  Scenarios for Searching on Legal Documents

To ensure accurate question answering, the study collects actual questions posed by road users. These questions are then subjected to semantic analysis to determine the query requirements for the knowledge base. Semantic analysis can be performed using various methods, such as natural language processing (NLP)

---

[1] https://github.com/VinAIResearch/PhoBERT.

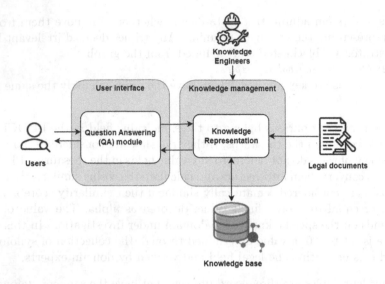

**Fig. 2.** Architecture of Legal Search Platform

tools to extract keywords, relationships, and relevant information from the questions, or employing machine learning techniques to classify and analyze the questions. This analysis helps in accurately understanding the question's characteristics and providing specific solutions to answer user questions.

After organizing and representing knowledge using the Rela-Model model, the system needs to identify the specific problems that need to be solved in order to provide correct answers to user questions. Understanding the user's intent through the query is crucial for providing accurate answers. Therefore, the system needs to classify and define the semantics of the query to comprehend the user's requirements. Once the semantics of the question are determined, the system searches and matches the information stored in the knowledge base to provide the correct answer to the user. Specifically, there are two key problems to be solved:

- Problem 1: Query classification and semantics determination: The system determines the question type and retrieves information about the query's meaning and semantics to understand the user's requirements.
- Problem 2: Finding relevant knowledge: Based on the query's semantics, the system searches and matches the information stored in the knowledge base K to provide the answer to the user.

Solving these two important problems is essential for accurately answering user questions in the knowledge query system, ensuring the system can deliver precise and relevant answers to fulfill user requirements.

## 4.2   Requirements of Platform and Its Architecture

The requirements for a platform used to represent knowledge from legal documents include:

- **Knowledge representation:** The platform needs to have the capability to automatically construct a knowledge graph from input data and efficiently represent information. This knowledge graph should be able to represent concepts and the relationships between them within the data. This function converts the legal documents into a knowledge graph, depicting the semantic relationships within the text. In this step, nodes in the graph are extracted and their corresponding edges are automatically determined. The process of identification and extraction is performed based on the Rela-model and with the support of the VNCoreNLP library.
- **Data storage**: This function takes legal texts in the Vietnamese legal system as input and performs text analysis to identify meaningful sections representing chapters, sections, articles, clauses, and points within legal normative documents. These sections are then stored in a database, ensuring efficient and organized storage of legal information. This facilitates easy access and retrieval of legal information whenever needed, ensuring effective management of legal knowledge.
- **Graph optimization**: This function optimizes the knowledge graph by reducing its complexity and size. This process utilizes the tf-idf method to eliminate redundancy and the LDOCE Vietnamese synonym dictionary combined with PhoBert to remove synonymous triples. By optimizing the graph, the speed and accuracy of legal information retrieval are improved, ensuring optimal performance of the knowledge retrieval system.
- **Scalability:** The platform should have scalability to meet the increasing demands for data and users. It should be able to handle and store new documents or documents from new legal domains, ensuring that the system can expand to accommodate diverse legal knowledge and keep up with the development of the legal field.

With the knowledge organization method and the representation model designed in the previous section. Combined with the requirements of a platform. This study has developed a platform to support the organization of knowledge base from legal texts entered by platform users, following the predefined structure of legal documents (including chapters, sections, articles, clauses, and points).

This platform utilizes user-input legal documents as input and generates a knowledge graph that represents the meaning of those documents. The main functionalities of the platform include:

The architecture in Fig. 3 represents the platform for knowledge representation from legal texts. Upon receiving a legal text, the Data Storage component analyzes its textual components. The Knowledge Representation component is utilized to represent the knowledge embedded within these components. The Graph Optimization component is employed to optimize the knowledge graph and store it in the knowledge base (Fig. 3).

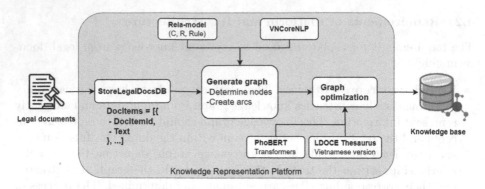

**Fig. 3.** Knowledge Representation Model

With the design and implementation of the platform, the study has created a useful tool for organizing and extracting knowledge from legal documents. The semi-automated process of document analysis, graph construction, and graph optimization helps save time and effort for users. The role of the knowledge manager is to regularly check and update the knowledge graph to ensure accuracy and reliability when storing it in the system's knowledge base.

### 4.3  Algorithms for Searching Legal Knowledge

The legal searching process involves transforming the query into a knowledge graph representation, and then searching knowledge graphs, which match the knowledge graph representation of the input query. There are two types of input queries, including conceptual queries and violation queries.

**Algorithm 4.1**: *Query Classification and Semantic Determination*
**Input:** Query $q$
**Output:**
- Type of question
- Knowledge graph representation (representing the semantics of the question)

**Step 1:** Categorize questions:
- Classify into two types (conceptual/violation queries) by gazette terms:
    ○ Conceptual query: if query $q$ contains question words about definitions, concepts, for example *"what is"*.
    ○ Violation query: Similarly, violations specified in legal documents with corresponding fines if query $q$ has keywords releated to fines.

**Step 2:** Represent the knowledge of the question in a knowledge graph
- Extract subjects, relations, and objects from the query.
- Transform them into triples to represent the knowledge graph.

The knowledge graph that represents the semantics of the question contains multiple triples with different subject types. In general, several questions may not provide enough information to construct a complete triple. Otherwise, if the query does not mention the value of the subject or object, they will be represented with '*' to indicate that those elements are not fully specified in the question.

**Algorithm 4.2**: *Finding Relevant Knowledge*
**Input:** Knowledge graph representation
**Output:** Matched passages (comparing with the knowledge of the question)
**Step 1:** Decompose the question graph
  • The graph is decomposed into star graphs [23].
  • Each star represents triples sharing the same subject.
  • This decomposition helps subdivide the question into subquestions.
**Step 2:** Find knowledge that matches the star graph
  • For each decomposed star graph:
    ○ Search for semantically similar subgraphs in the knowledge base.
    ○ Save the potential answers.
  *Details of the matching and subgraph-finding process:*
    ○ Traverse each triple in the star graph to find semantically equivalent triples.
    ○ Combine semantically equivalent triples to represent the matched star graph.
**Step 3:** Take the intersection set of the answer sets of the star graphs
  • Obtain a list of matching IDs for each star in the question's knowledge representation graph.
  • Calculate the intersection of these sets.
  • The resulting list of IDs represents the answer to the user's input question.

During traversing triples to find semantically equivalent triples, if any component of the triple in the question is missing, the matching process relies on the remaining components. In addition, if the knowledge graph includes components that have an "is-a" relationship with the corresponding elements in the knowledge base, this algorithm also considers those related knowledge parts. For example, "motorbike" may have an "is-a" relationship with "motorcycle", "moped", etc.

### 4.4  Question Answering Module

The question answering module is a crucial component in query systems, and designing its architecture is an important step to ensure the efficiency and accuracy of the system. This module contains 5 components, including question classification, question-to-graph, star decomposition graph, and subgraph matching component (Fig. 4).

To answer user queries based on the represented knowledge base, the study has developed a query module with the following sequential components:

**Question Classification:** To identify the correct user-entered queries, we categorize them into two categories: concept queries and violation queries. To do this, we use the technique of identifying question words in the query. Queries with

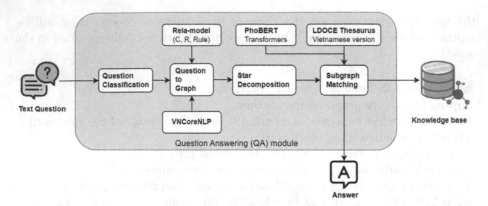

**Fig. 4.** Question Answering module

words like "what is", "what is?" classified as a conceptual question. The remaining cases are considered as queries about violations specified in legal documents, together with corresponding fines.

**Question-to-Graph:** The question is represented in the form of a knowledge graph similar to the way legal documents are represented in the knowledge base. Each triple in the knowledge graph still consists of three components: subject, relation, and object. In some cases, the question does not provide enough information to fully construct a triple. In this case, if the query does not mention the value of the subject or object, an asterisk (*) will be used to represent the elements that are not fully informed in the question.

**Star Decomposition Graph:** Here, if the question entered by the user is a complex question with many different subjects in the question knowledge graph, it proceeds to decompose the knowledge graph of the question into Star Graphs, this is similar to subdividing the question into subquestions with each star being triples that share the same subject.

**Subgraph Matching:** After the question has been decomposed into Star Graphs, for each Star Graph research, search for subgraphs in the knowledge base with semantically similar content and save a list of similar id values response. After collecting a list of ids that match each Star Graph, the research proceeds to combine these values to create a final answer to the user's question. The answers are generated by assigning or combining values from the respective subgraphs.

The query module comprises the following components, which are integrated together into a designed workflow with the following architecture:

This module allows users to query and answer questions based on the represented knowledge. Steps and algorithms are tightly integrated to achieve high performance and accuracy in answering questions in the knowledge base.

# 5 Experiments and Results

## 5.1 User Interface

Users will use the user interface to input their questions into the system. The questions will be recorded by the question and answer module. Subsequently, the questions will be sent to the sentence analysis module for question classification, semantic analysis, and extraction of relevant information. This extracted information will be represented as knowledge graphs.

Next, the system employs the subgraph search component to extract information from the knowledge base that matches the question's knowledge graph. Through matching and searching, the system identifies subgraphs in the knowledge base that correspond to the knowledge graph representation of the question.

Once the search is completed, the system returns the answers to the question and answer module for display on the user interface (Fig. 5). If no answer is found, the system returns no available relevant information.

**Fig. 5.** System query interface

## 5.2 Experiment

The experimental process used a collection of 36 articles, 306 clauses, and 762 points related to traffic violations and their associated penalties in traffic laws [1–4]. From this dataset, the research employed the knowledge representation platform to construct 616 concepts and establish 340 relationships among them within the legal regulatory documents of road traffic. As a result, 1239 triples were generated, forming the foundational structure of the knowledge graph. This knowledge graph effectively represents the content of traffic violations in road traffic laws.

When users input queries into the system, it processes and represents the question's knowledge into a knowledge graph. Once the knowledge graph for the query is constructed, the system performs matching with the knowledge graph of the database to identify relevant points, clauses, and regulations related to the query. The results are presented to the user as a list of the most detailed Points - Clauses - Regulations that answer their query.

This study conducted experiments with two types of common questions regarding road traffic laws in Vietnam. The first type of question involves asking about concepts or definitions stated within the road traffic laws. For example, a question about a concept would be *"What is a dual carriageway?"* The system analyzes the question and provides the following answer:

*"Based on Article 3.11, Point 3 of the National Technical Regulation QCVN 41:2016/BGTVT on Road Traffic Signs and Signals:*

*A dual carriageway refers to roads where the traffic in both directions is separated by a median strip (in the case of separation by markings, it is not considered a dual carriageway)."*

The second type of question involves violations within road traffic laws. For example, a question like *"What is the fine for not wearing a helmet?"* The system analyzes and represents the knowledge of the question, matches it with the knowledge base, and provides the following answer:

*"Based on Clause 6, Article 6 of Government Decree 100/2019/NĐ-CP: Regulations on Administrative Penalties for Road Traffic and Railway Violations:*

*Article 6. Penalties for riders of motorcycles, mopeds (including electric motorcycles), and similar vehicles violating road traffic regulations: A fine ranging from 2,000,000 VND to 3,000,000 VND shall be imposed on riders who commit one of the following violations: b) Riding on expressways, except for vehicles serving the management and maintenance of expressways."*

This system was tested with 180 real-world questions related to road traffic laws, collected from various sources, including district police websites, provincial police websites, legal advice websites, and other relevant data sources. On average, the system took 0.2 s to respond to conceptual questions and 5.6 s to respond to violation questions. The results are presented in the following Table 1:

**Table 1.** Query test results about Road Traffic Law

| Kinds | Quantity | Correct | Accuracy |
|---|---|---|---|
| Concept definition queries | 30 | 24 | 80.0% |
| Violation queries | 150 | 113 | 75.3% |
| **Total** | **180** | **137** | **76.1%** |

## 6   Conclusion and Future Work

This study has developed a platform that extracts knowledge from legal documents and organizes it into a knowledge base using the Legal-Onto ontology and

NLP techniques. The extracted knowledge is represented as a knowledge graph, which is used to search and query legal knowledge based on user-inputted legal documents.

Furthermore, a legal knowledge query model has been developed and tested in the field of traffic law to evaluate the effectiveness of the knowledge representation platform and the accuracy of the algorithm. This query model enables the retrieval of concepts, violations, and corresponding penalties according to the regulations of traffic law.

In the future, as a knowledge-based intelligent system, the project aims to automate ontology construction and ensure its accuracy through validation by knowledge engineers [5,14]. Additionally, advanced NLP techniques will be incorporated to handle complex user queries [12,13]. These advancements will enhance the system's ability to serve various legal domains and meet user requirements more effectively. Moreover, the solution of this study can be research to apply in other legal domains such as land law [15,16], labor law [10].

**Acknowledgments.** This research was supported by The VNUHCM-University of Information Technology's Scientific Research Support Fund.

# References

1. Law on Road Traffic. Law No. 23/2008/QH12. Vietnam National Assembly (2008)
2. Decree on Administrative penalties for road traffic and rail transport offences. No. 100/2019/ND-CP. Vietnam Government (2019)
3. National Technical Regulation on Traffic Signs and Signals. QCVN 41:2019/BGTVT. Vietnam National Assembly (2019)
4. Amending and supplementing some articles of decrees provisions on penalties for administrative violations in the marine region, road, rail traffic, and domestic airline. No. 123/2021/ND-CP. Vietnam National Assembly (2021)
5. Cao, Q., et al.: KSPMI: a knowledge-based system for predictive maintenance in industry 4.0. Rob. Comput. Integr. Manuf. **74**, 102281 (2022)
6. Chau, C.N., Nguyen, T.S., Nguyen, L.M.: VNLawBERT: a Vietnamese legal answer selection approach using BERT language model. In: 2020 7th NAFOSTED Conference on Information and Computer Science (NICS), pp. 298–301 (2020). https://doi.org/10.1109/NICS51282.2020.9335906
7. Dang, D.V., Nguyen, H.D., Ngo, H., Pham, V.T., Nguyen, D.: Information retrieval from legal documents with ontology and graph embeddings approach. In: Proceedings of the 36th International Conference on Industrial, Engineering & Other Applications of Applied Intelligent Systems (IEA/AIE 2023), Shanghai, China, pp. 300–312, July 2023
8. Do, N.V., Nguyen, H.D., Selamat, A.: Knowledge-based model of expert systems using Rela-model. Int. J. Softw. Eng. Knowl. Eng. **28**, 1047–1090 (2018)
9. Governatori, G., Bench-Capon, T., Verheij, B., et al.: Thirty years of artificial intelligence and law: the first decade. Artif. Intell. Law **30**(4), 481–519 (2022)
10. Le, H., Nguyen, T., Ngo, T., et al.: Intelligent retrieval system on legal information. In: Nguyen, N.T., et al. (eds.) Proceedings of 15th Asian conference on Intelligent Information and Database Systems (ACIIDS 2023), Phuket, Thailand, pp. 97–108. Springer, Singapore (2023). https://doi.org/10.1007/978-981-99-5834-4_8

11. Li, G., Yan, l., Ma, Z.: An approach for approximate subgraph matching in fuzzy RDF graph. Fuzzy Sets Syst. **376** (2019). https://doi.org/10.1016/j.fss.2019.02.021

12. Ngo, H.Q., Nguyen, H.D., Le-Khac, N.A.: Building legal knowledge map repository with NLP toolkits. In: Nguyen, N.T., Le-Minh, H., Huynh, C.P., Nguyen, Q.V. (eds.) Conference on Information Technology and its Applications, pp. 25–36. Springer, Cham (2023). https://doi.org/10.1007/978-3-031-36886-8_3

13. Nguyen, H., Huynh, T., Hoang, S.N., et al.: Language-oriented sentiment analysis based on the grammar structure and improved self-attention network. In: Proceedings of 15th International Conference on Evaluation of Novel Approaches to Software Engineering (ENASE 2020), pp. 339–346 (2020)

14. Nguyen, H.D., Do, N.V., Pham, V.T.: A methodology for designing knowledge-based systems and applications. In: Applications of Computational Intelligence in Multi-Disciplinary Research, pp. 159–185. Elsevier (2022)

15. Nguyen, T., Nguyen, H., Pham, V., et al.: Legal-onto: an ontology-based model for representing the knowledge of a legal document. In: Proceedings of the 17th International Conference on Evaluation of Novel Approaches to Software Engineering (ENASE 2022), Online Streaming, pp. 426–434 (2022)

16. Padmanabhan, A., Wadsworth, T.: A common law theory of ownership for AI-created properties. SSRN, p. 4411194 (2023)

17. Pham, V.T., Nguyen, H.D., Le, T., et al.: Ontology-based solution for building an intelligent searching system on traffic law documents. In: Proceedings of the 15th International Conference on Agents and Artificial Intelligence (ICAART 2023), Lisbon, Portugal, pp. 217–224. ScitePress (2023)

18. Sartor, G., Araszkiewicz, M., Atkinson, K., et al.: Thirty years of artificial intelligence and law: the second decade. Artif. Intell. Law **30**, 521–557 (2022)

19. Szostek, D., Zalucki, M.: Legal Tech: Information Technology Tools in the Administration of Justice. Nomos (2021)

20. Tagarelli, A., Simeri, A.: Unsupervised law article mining based on deep pre-trained language representation models with application to the Italian civil code. Artif. Intell. Law **30** (2021). https://doi.org/10.1007/s10506-021-09301-8

21. Villata, S., Araszkiewicz, M., Ashley, K., et al.: Thirty years of artificial intelligence and law: the third decade. Artif. Intell. Law **30**, 561–591 (2022)

22. Wang, X., et al.: Efficient subgraph matching on large RDF graphs using MapReduce. Data Sci. Eng. **4**, 1–20 (2019). https://doi.org/10.1007/s41019-019-0090-z

23. Zhao, Y., Wu, B.: Star decomposition of graphs. Discrete Math. Algorithms Appl. **07**, 1550016 (2015). https://doi.org/10.1142/S1793830915500160

# A Study of the Impact of Attention Mechanisms on Feature Correlation Learning for Building Extraction Models

Nhat-Quang Tau[1,3], Minh-Vu Tran[1,3], Anh-Tu Tran[2,3,4], and Khuong Nguyen-An[1,3(✉)]

[1] Faculty of Computer Science and Engineering, Ho Chi Minh City University of Technology (HCMUT), 268 Ly Thuong Kiet Street, District 10, Ho Chi Minh City, Vietnam
[2] Faculty of Geological and Petroleum Engineering, Ho Chi Minh City University of Technology (HCMUT), 268 Ly Thuong Kiet Street, District 10, Ho Chi Minh City, Vietnam
[3] Vietnam National University Ho Chi Minh City (VNU-HCM), Linh Trung Ward, Thu Duc City, HCMC, Vietnam
nakhuong@hcmut.edu.vn
[4] Ground Data Technology Solution (GDTS), HCMC, Vietnam

**Abstract.** Buildings are the fundamental units of civilization, deeply ingrained into urban elements and synonymous with progress. Therefore, comprehending buildings through extracting their footprint data is vital for assessing and influencing societal advancement. Unlike the direct approach of recent research, we recognize that using identical or related representation is undoubtedly potential through feature correlation learning.

In this study, we examined how plugging the current attention mechanisms in two fundamental building extraction models led to notable enhancements in their performance. Specifically, we achieved a 1.25% increase in the IoU score for CrossGeoNet and a 2.7% improvement for HiSup.

**Keywords:** Building Footprint · Attention Mechanism · Feature Correlation Learning · Building Extraction Models · Remote Sensing Imagery

## 1 Introduction

Several semantic segmentation approaches have been utilized to accurately derive building geometric properties of remote sensing imagery due to its concentration on pixel-level accuracy of distinct objects. Inspired by the performance and proliferation of deep learning, numerous efforts have been made to apply deep

---

N.-Q. Tau and M.-V. Tran—These authors contributed equally to this work.

N. Thai-Nghe et al. (Eds.): ISDS 2023, CCIS 1949, pp. 285–299, 2024.
https://doi.org/10.1007/978-981-99-7649-2_22

neural networks to the semantic segmentation task of remote sensing data. Further post-processing techniques can be used for the segmented building footprints to align the segmentation outlines to the actual building contours described in the intensity image. Besides pixel-wise methods, some polygonization methods are proposed, such as vectorization of grid-like features and direct learning of a vector representation. These concepts can be utilized to generate more regular and cleaner building polygons.

However, most recent research tends to approach pixel-wise or polygonal aspects straightforwardly. Various techniques have been tested without taking advantage of the existing datasets or feature representations to compensate for the performance of the target set and output mutually. With feature correlation learning, these problems can be solved and extended as an open field for further research. Alongside the attention mechanism, feature correlation learning can be improved to achieve higher milestones in building extraction tasks.

In this article, we analyze the effectiveness of existing attention mechanisms [4] on the feature correlation learning modules of building extraction models. We use this formation to determine the best attention mechanisms that should be used in these models.

At a glance, our contributions are the following:

1. We study the learning of correlation between features with identical and related representations. We observe how the model learns the interconnection between the features;
2. We study different channel attention modules and analyze their compatibility with the existing baseline models;
3. We propose the most efficient channel attention module: more precisely, we try replacing the attention block of the baseline with another channel attention module to see if it improves accuracy and verify our improvements through experiments and critical analysis.

The rest of this paper is organized as follows. In Sect. 2, we recall some backgrounds on attention mechanisms and feature correlation learning techniques and conduct a brief literature review on their applications in building extraction tasks. In Sect. 3, we demonstrate our methods for integrating the existing attention mechanisms into feature correlation learning baselines. We then evaluate and discuss our experiments in Sect. 4. Finally, we conclude the paper with Sect. 5

## 2    Preliminaries and Related Works

Feature correlation learning approach developed along with the popularity of deep learning models. These methods use rich contextual feature maps extracted from robust existing backbones to exploit different perspectives of the features. With the presence of attention mechanisms, these methods have been further advanced. However, other feature maps can contradict one another, possibly accumulating irrelevant information for the segmentation task. Therefore, research on these methods has not been generalized, and fusing multiple features effectively is still an open problem to be studied. This section indicates feature correlation learning and how it applies attention mechanisms to building footprint extraction tasks.

## 2.1 Learning Feature Correlation

Intriguingly, the feature maps produced by contemporary convolutional neural networks can be utilized in similar and various representations. To combine the tasks of semantic matching and object co-segmentation, Chen et al. [2] have proposed semantic matching to provide supervision for object co-segmentation by enforcing the predicted object masks to be geometrically consistent across images. The correlation map directs the network based on the feature map pair. Although two feature maps have a similar form, they are retrieved from two input images in a single forward pass. On the other view, Xue et al. [19] have observed that the learned junctions and line segments can be refined to improve the accuracy of line segments. The duality between region representation and boundary contour representation of objects or surfaces is also exploited by Xue et al. [18]. From these observations, although extracted feature maps are in different representations, they are strongly related. Therefore, learning correlation can be applied to two feature representations: identical representation and related representation.

Methods for learning feature correlation of identical representation mainly aim to enhance the correlation between feature maps from different inputs and understand the relationships between other modalities. The feature maps are generated with shared-weight encoders, which contain the most information from the images. This approach is applied to tasks where the underlying correlation between images is essential, like co-segmentation [11] or building change detection from remote sensing imagery [7]. Inspired by the success of these approaches, CrossGeoNet [10] has utilized cross-geolocation attention to learn the correlation of identical feature maps extracted from input images from different datasets for efficient building extraction in scarce label areas.

For related representations, after the feature extraction process from the convolutional backbone, different feature representations are further extracted from the original features to utilize for correlation learning. Motivated to improve accuracy in terms of both semantic mask and boundary, Li et al. [9] utilize an *"attraction field map"* (AFM, introduced in [18]) for reconstructing segmentation masks of buildings in an end-to-end fashion. Recently, Xu et al. [17] applied both AFM and channel attention for solving mask reversibility issues of buildings and set several new state-of-the-art performances.

## 2.2 Feature Fusion with Channel Attention

Humans can easily focus on significant portions of a scene. Inspired by this aspect, attention mechanisms were introduced to attend to essential extracted features. A recent survey [4] has reviewed recent attention mechanisms and demonstrated the dominance of channel and spatial approaches. However, spatial attention can be computationally expensive for high-resolution images and fails to capture global context when the context spans a large area effectively. Meanwhile, channel attention, operating on the channel dimension (feature maps) instead of every spatial location, can capture relationships among channels effectively at a low cost. Therefore, channel attention is generally more compatible with existing CNN architectures, widely used as a mainstream baseline for building extraction tasks.

In deep neural networks, not all channels are essential for scene-parsing tasks. Inspired by this aspect, channel attention adaptively determines which channels will be attended by adjusting their importance. SENet [6] pioneered this concept with a block consisting of a squeeze and excitation module. In the squeeze module, the global average pooling layer (GAP) gathers global spatial information before the excitation module can capture the channel-wise relationships by exploiting the fully connected (FC) layer. However, global average pooling (GAP) cannot capture sophisticated global information, and a fully connected (FC) layer adds extra to the model complexity.

Inspired by the success of SENet [6], several squeeze-and-excitation variants have been proposed. GSoP-Net [3] captures channel correlation through covariance matrix. EncNet [21] uses context encoding to obtain scene context and object categories, therefore enhancing the performance of small object segmentation. SRM [8] reduces computations by replacing the original FC layer with a channel-wise fully connected (CFC) layer. In summary, the Squeeze-Excitation network improvement approach can be squeeze-concentrated, excitation-concentrated, or focusing on both aspects.

## 3 Proposed Method

In this section, different channel attention blocks are introduced to replace the original block in the baseline models to test their effectiveness in terms of accuracy metrics. For the learning of features with identical and related representation, CrossGeoNet co-segmentation and HiSup polygonal mapping framework are selected as baselines, respectively.

### 3.1 Channel Attention Blocks

Given that $x$ is the input feature map and $g$ is a channel attention mechanism. The output feature map $\hat{x}$ can be formatted as follows

$$\hat{x} = g(x) \times x.$$

SENet [6] block enables the network to re-calibrate features, allowing it to learn to emphasize informative features while suppressing less essential ones selectively (Fig. 1b). The formula for the SE attention is

$$g_{SE}(x) = \sigma(W_{FC_2}\delta(W_{FC_1}, GAP(x))), \tag{1}$$

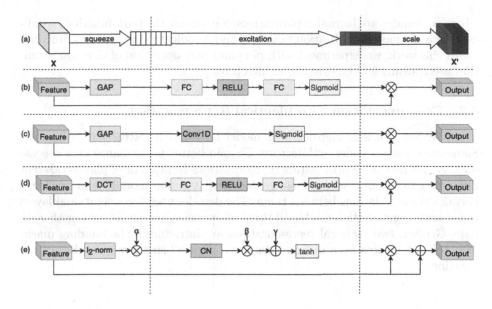

**Fig. 1.** An overview of (a) channel attention mechanisms [4] being experimented in this article, with more detail in (b) SENet [6], (c) ECANet [16], (d) FcaNet [14], and (e) GCT [20]. (Color figure online)

where $GAP$ is the Global Average Pooling layer, $W_{FC_1} \in \mathbb{R}^{(C/r) \times C}$ and $W_{FC_2} \in \mathbb{R}^{C \times (C/r)}$ are the weights of the FC layers, $\delta$ is the ReLU operation, and $\sigma$ is the sigmoid function. The block's final output is generated by rescaling the channel of the input feature.

Despite being widely used, SENet's FC layers require dimensional reduction to control model complexity. To efficiently improve excitation module, ECANet [16] replaced the $FC + ReLU + FC$ layers with a $Conv1D$ layer (Fig. 1c)

$$g_{ECA}(x) = \sigma(W_{C1D}(GAP(x))),$$

where $W_{C1D} \in \mathbb{R}^{C \times C}$ is the weights of the $Conv1D$ layer.

Representing channel information as a scalar can be considered a compression problem, and it is difficult for GAP to collect complicated information for diverse inputs. Therefore, FcaNet [14] replaces the GAP with 2D DCT [1] (Fig. 1d), which is observed as a more general case of GAP

$$g_{FCA}(x) = \sigma(W_{FC_2}\delta(W_{FC_1}, 2DDCT(x))).$$

GCT [20] redesigned architecture for modeling channel relationships with $l^2$-norm in the squeeze module to control channel significance and a $CN$ layer with $tanh$ activation function to create competition and cooperative relationships among channels (Fig. 1e)

$$g_{GCT}(x) = 1 + \theta(\gamma\tau(\alpha\lambda(x)) + \beta),$$

where $\alpha$, $\beta$ and $\gamma$ are learnable parameters, $\theta$ indicates the $tanh$ function, $\tau$ indicates the channel normalization function, and $\lambda$ indicates the $\ell^2$-norm function.

In this work, we experiment with plug-and-play operation of different attention mechanisms for the existing architecture.

## 3.2   Co-segmentation with Identical Representation

CrossGeoNet [10] is a deep learning model for building extraction, utilizing a Siamese encoder-decoder architecture. To gain better performance at a specific dataset, the network learns feature representations from a pair of photos taken at different geolocations, with a cross-geolocation attention module, and explicitly encodes the correlations between them. The decoder uses deconvolutional layers to generate segmentation masks; all three components are optimized together. In CrossGeoNet, two identical representations are introduced, the building masks of two input images from different datasets. Figure 2 shows an overview of the baseline.

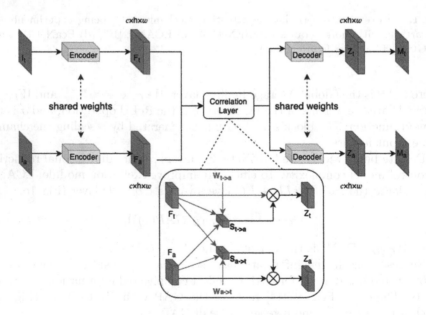

**Fig. 2.** An overview of CrossGeoNet's architecture. This model receives a target image $I_t$, an auxiliary image $I_a$, and outputs two respective masks, $M_t$ and $M_a$. In the original baseline CrossGeoNet [10], the author utilizes a Cross-Geolocation Attention Module for the Correlation Layer (the green box at the center). In this work, we experiment with CAM [5] and its variants for this layer. (Color figure online)

**Revisiting Cross Geolocation Attention.** In CrossGeoNet [10], the Cross-Geolocation Attention Module is utilized for the correlation layer. This module performs multiplication on the flattened feature maps of size $C \times WH$ (originally $C \times W \times H$) and a learnable square matrix. More precisely, given two flattened feature maps $F_t, F_a \in \mathbb{R}^{C \times WH}$, we have

$$S_{t \to a} = F_a^T W_{t \to a} F_t,$$
$$S_{a \to t} = F_t^T W_{a \to t} F_a,$$
$$Z_t = S_{t \to a} F_t^T,$$
$$Z_a = S_{a \to t} F_a^T,$$

where $W_t, W_a \in \mathbb{R}^{WH \times WH}$ are learnable matrices, and $Z_t, Z_a \in \mathbb{R}^{WH \times C}$ are the output feature maps.

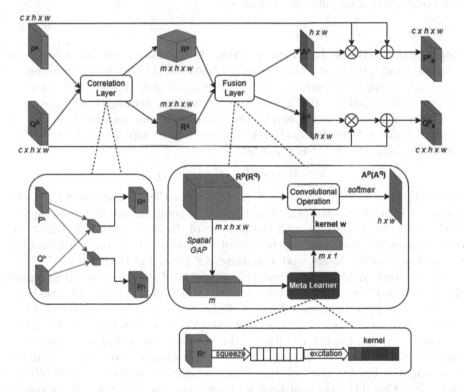

**Fig. 3.** An overview of Cross Attention Module (CAM) [5]. CAM is a few-shot learning module in CAN [5], receiving two input feature maps $P_k$ from the target set and $Q_b$ from the query set. This module consists of two sub-modules: the correlation and fusion layers. As of our observations, SENet [6] inspired the fusion layer (Fig. 1).

**Correlation Layer with Meta Fusion.** It is noticeable that the cross-geolocation attention module in CrossGeoNet [10] is in fact the *correlation layer*

292    N.-Q. Tau et al.

in CAM [5]. However, CrossGeoNet integrates the weight matrix between two feature maps to adjust the correlation matrix directly (Fig. 2). On the other hand, CAM adopts the meta-fusion layer with an adaptive meta-learner to generate attention maps based on the calculated correlation maps without directly injecting the weight matrix (Fig. 3).

The fusion layer receives the correlation map $R^p$ as input. It performs convolutional operation with a $(m \times 1)-$kernel, $w \in \mathbb{R}^{m \times 1}$, to fuse each local correlation vector $\{r_i^p\}$ of $R^p$ into an attention scalar. A softmax function is then applied

$$A_i^p = \frac{\exp((w^T r_i^p)/\tau)}{\sum_{j=1}^{h \times w} \exp((w^T r_j^p)/\tau)},$$

where $\tau$ is the temperature hyperparameter controlling the entropy of distribution concentration on different positions.

**Replacing SE-Block in Meta Fusion.** As mentioned, the *meta-learner* is designed to adaptively generate the kernel based on the correlation between the class and the query features. It is intriguing to observe that meta-learner is a variant of the SE-Block [6] due to its global average pooling (GAP) operation to obtain global information and further enhance feature relationship with the $Conv + ReLU + Conv$ layers. Given the correlation map $R^p$ as the input, the attention scalar $w \in \mathbb{R}^{m \times 1}$ is given by

$$w = W_{C_2}(\delta(W_{C_1}(GAP_s(R^p)))),$$

where $W_{C_1} \in \mathbb{R}^{(m/r) \times m}$ and $W_{C_2} \in \mathbb{R}^{m \times (m/r)}$ are the parameters of the meta-learner, $r$ is the reduction ratio, $\delta$ is the $ReLU$ operation, and $GAP_s$ is a spatial GAP operation. Compared with the formula of the SE attention in SENet (see (1)), this module does not use the sigmoid operation at the end, utilizes Conv layers instead of FC layers, and a spatial-GAP instead of GAP.

In this work, ECANet [16], FcaNet [14], and GCT [20] are adapted to replace the SE-Block [6], which is the meta learner in meta fusion layer. The SE-Block in this architecture leverages $Conv + ReLU + Conv$ layers instead of $FC + ReLU + FC$ in the original work [6], with the $Sigmoid$ layer being replaced by a $Softmax$ to normalize the attention scalar, and without scaling operation being performed. Therefore, we changed to adapt other channel attention mechanisms to this model. For ECANet [6], the $Sigmoid$ layer is replaced with a $Softmax$, and only a $Conv1D$ layer is utilized for excitation. For FCANet [14], we replace $GAP$ with a $2DDCT$ layer for the squeeze sub-module and keep the $Conv + ReLU + Conv$ layers for excitation. For GCT, we replace the $(1+tanh)$ operation with a $Softmax$, while still keeping the learnable $\alpha$, $\beta$ and $\gamma$ parameters. All these modules are adapted without the scaling sub-module.

### 3.3 Polygonal Mapping Using Related Representation

It can be seen that both semantic and geometric features are essential for the representation of a building footprint. While masks can easily handle cases where the object's shape is complex, or the boundaries are not well-defined, polygons can be very accurate if the points are well-placed, capturing the object's shape with high precision. Because both semantic and geometric features are essential, HiSup [17] is selected as a baseline to reduce the gap between the mask predictions and the polygon ones. In HiSup, three related representations, vertices, attraction field maps, and masks, which are highly associated with each other to demonstrate the building footprint, are considered for feature correlation learning. The architecture of this model is shown in Fig. 4.

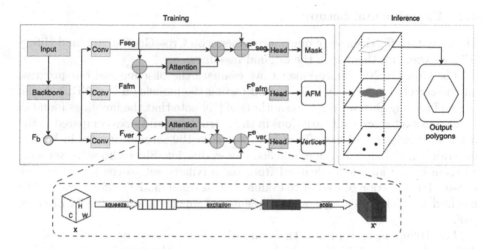

**Fig. 4.** An overview of Hierarchical Supervision Learning. In the original baseline, HiSup [17], the author uses HRNetW48-V2, the second version of HRNet [15] for the backbone, and ECANet [16] as the attention mechanism (the red block in bold). To study the effectiveness of this module, we try utilizing SENet [6], FcaNet [14], and GCT [20].

In the original baseline, HiSup utilizes ECANet [16] for learning Cross-level Interaction (the red block in the figure). ECANet [16] is a channel attention model inspired by the success of SENet [6]. To study the effectiveness of other channel attention modules in this architecture, we experimented with replacing the whole ECA-block [16] with SE-block [6], Fca-block [14], and GCT-block [20]. The result of this experiment is further discussed in Sect. 4.

## 4   Experiments

### 4.1   Dataset

**INRIA Aerial Image Labeling Dataset.** INRIA Aerial Image Labeling dataset [12] comprises images captured by airborne sensors for five cities (Austin, Chicago, Kitsap County, Western Tyrol, and Vienna). This dataset consists of

360 aerial imagery tiles, each with 5000 × 5000 pixels at a spatial resolution of 30 cm/pixel.

**AICrowd Dataset.** AICrowd dataset [13] comprises 300 × 300 pixels images, with 280741 tiles in the training set, 60317 tiles in the validation set, and 60697 tiles in the evaluation set.

In Table 1, we present the statistics of the dataset used for training, evaluating, and testing our two baseline models.

## 4.2   Experimental Setup

To verify the effectiveness of our improvements on CrossGeoNet [10], and HiSup [17], we compare them with the original baselines.

For CrossGeoNet improvements, we evaluate the baseline and our progress on the correlation layer by training and testing the models on the INRIA Aerial Image Labeling Dataset [12]. The authors of [10] note that the buildings found in Vienna exhibit significant variations in their structures and sizes compared to the other four cities mentioned. Therefore, we chose 100 satellite images in Vienna for training, 100 for validation, and 3000 for evaluation. Since CrossGeoNet aims to transfer the knowledge learned from the auxiliary set to the target set, we chose a large portion of INRIA to evaluate the learned model's effectiveness. The method for picking a target and supplemental images is presented in Li et al. [10].

For HiSup improvements, the evaluation is processed on a subset of the AICrowd dataset [13]. The models are trained on the training set and then evaluated on the validation set. We do not use the testing images in this work as they have no annotation yet.

**Table 1.** The statistics of the images used for evaluating the baseline models and their improvements.

| Baseline model | | Number of patches | | |
|---|---|---|---|---|
| | | Train | Validation | Evaluation |
| CrossGeoNet | Target city | 100 | 100 | 3000 |
| | Auxiliary set | 200000 | 10000 | - |
| HiSup | - | 4182 | - | 910 |

The statistics of the data used in this experiment are shown in Table 1. The models are implemented with PyTorch framework and trained on NVIDIA RTX 2080 Super with 8 GB of memory.

## 4.3   Experimental Results

**Co-segmentation on INRIA Dataset.** We train and test the model on the INRIA Aerial Image Labeling dataset [12] to evaluate the performance of the CrossGeoNet [10] model. The model was trained with a batch size of 6 for 50000 iterations. Table 2 shows the result of evaluating the trained model on a subset of INRIA dataset [12], with all 3000 images from Vienna, Austria. Overall accuracy (OA), F1 score, and IoU are the metrics used to compare the models in this evaluation.

**Table 2.** Results of CrossGeoNet on INRIA Aerial Labeling Image Dataset. The best metrics are in bold.

| Methods | Fusion layer | OA | F1-score | IoU |
|---|---|---|---|---|
| CrossGeoNet (*baseline*) | - | 84.61 | 88.43 | 79.25 |
| CrossGeoNet+CAM | SENet | 85.38 | 88.77 | 79.80 |
| CrossGeoNet+CAM | **ECANet** | **85.99** | **89.20** | **80.50** |
| CrossGeoNet+CAM | FcaNet | 85.38 | 88.77 | 79.80 |
| CrossGeoNet+CAM | GCT | 85.22 | 88.78 | 79.83 |

As shown in Table 2, our proposed improvements outperform the baseline. The reason for this is that the cross-geolocation attention has directly eliminated the original information of the correlation map by injecting a weight matrix. On the other hand, CAM [5] adaptive kernel is generated from the original correlation map, resulting in better knowledge transferring between feature maps. Furthermore, replacing SENet [6] in the fusion layer with ECANet [16] shows a slight improvement in all benchmark metrics. However, when this layer is replaced with FcaNet [14] and GCT [20], the model does not improve.

In addition, with the replacement in CAM [5]'s fusion layer, the performance is also significantly affected. The accuracy of FcaNet [14] is not different from that of SENet [6], and that of GCT even decreases. GCT's under-performance can be because the *tanh* function cannot be well replaced with the *sigmoid* function.

Figure 5 compares some qualitative results of CrossGeoNet baseline and our improved models. The improvement in the cross-attention module, with ECANet for the fusion layer, which is expected to perform best in Vienna, shows better accuracy in detecting portions of pictures that are not buildings. Furthermore, the outputs of SENet and FcaNet are similar regarding predictions for each pixel. However, these models fail to differentiate a building-like object from an actual building.

**Polygonal Mapping on AICrowd Dataset.** On a subset of the AICrowd dataset [13], we train the HiSup model [17] with different improvements for 30

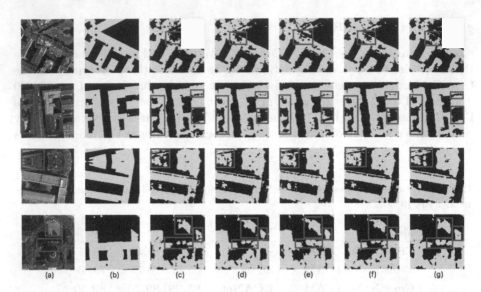

**Fig. 5.** Examples of building extraction results obtained by CrossGeoNet and our improvements on four images from the INRIA dataset in Vienna, Austria. (a) Input image. (b) Ground truth. (c) CrossGeoNet baseline. (d) CrossGeoNet with CAM. (e) CrossGeoNet with CAM and ECANet as the fusion layer. (f) CrossGeoNet with CAM and FcaNet as the fusion layer. (g) CrossGeoNet with CAM and GCT as the fusion layer.

**Table 3.** Results of HiSup on AICrowd dataset. The best metrics are in bold.

| Attention | AP↑ | AP$_{50}$ ↑ | AP$_{75}$ ↑ | AR↑ | AP$^{boundary}$ ↑ | IoU | C-IoU |
|---|---|---|---|---|---|---|---|
| SE | 51.8 | 75.2 | 61.3 | 50.2 | 23.9 | 79.1 | 69.9 |
| ECA (baseline) | 52.0 | 77.0 | 60.8 | 50.4 | 24.9 | 80.3 | 71.8 |
| FCA | 51.2 | 77.0 | 60.1 | 54.5 | 25.7 | 80.7 | 72.0 |
| *GCT* | **53.8** | **78.1** | **63.4** | **57.2** | **27.1** | **81.8** | **72.5** |

epochs. Table 3 demonstrates the quantitative evaluation results on the AICrowd dataset. We compare the original HiSup model with some improvements we made in the model's cross-level interaction module.

As shown in Table 3, the proposed model with the FCA module cannot outperform the baseline on AP↑ and AP$_{75}$ ↑. In addition, the SE replacement even affects the model negatively. In contrast, the proposed model with the GCT channel attention module outperforms the baseline and our other improvements. It can be explained that $l_1$-norm beats GAP in extreme cases (such as networks with Instance Normalization layers) and computation efficiency. However, the baseline HiSup has not yet met the performance shown in its original work [17], and so have our proposed improvements as we did not adapt the whole AICrowd dataset [13] to our model. Instead, we evaluated the model using only a tiny subset of AICrowd, whose amount of images is only 1.5% that of the whole dataset.

Figure 6 illustrates some qualitative results of HiSup and our improved models. The improvement with GCT channel attention gives slightly better results than the other two models in classifying each pixel of an image. However, in some cases, these models fail to detect the boundary between buildings located near each other, and buildings having small areas can be either detected as a part of a large building standing nearby or entirely ignored.

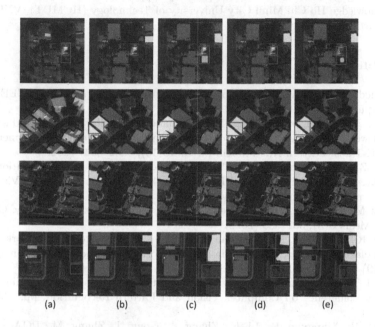

(a)          (b)          (c)          (d)          (e)

**Fig. 6.** Examples of building extraction results obtained by HiSup and its variants on five images chosen from the CrowdAI dataset. (a) Input image. (b) Ground truth. (c) HiSup with SENet. (d) HiSup with ECANet. (e) HiSup with FcaNet. (f) HiSup with GCT

## 5   Conclusion

This paper studies the impact of learning feature correlation on building segmentation models. Several channel attentions have been applied to the feature correlation modules of the networks to improve the proposed baseline model performance. With CrossGeoNet co-segmentation baseline [10], the addition of a fusion layer and replacement of different channel attention in meta-learner of the fusion layer has significantly improved the performance of the model in OA score, F1 score, and IoU.

For the Hierarchical Supervision Learning Baseline [17], try inserting various channel attention mechanisms from ECA [16] to SE [6], FCA [14] and GCT [20] for learning feature correlation has gained positive change in their performance. Especially, GCT [20] outperforms all the other models in all metrics we use for evaluation. In the future, we will conduct further ablation studies to get more critical insights into different existing attention mechanisms and fits HiSup [17] on the complete AICrowd dataset [13].

**Acknowledgements.** This work was partially supported by GDTS, and the authors also acknowledge Ho Chi Minh City University of Technology (HCMUT), VNU-HCM for supporting this study.

# References

1. Ahmed, N., Natarajan, T., Rao, K.R.: Discrete cosine transform. IEEE Trans. Comput. **100**(1), 90–93 (1974)
2. Chen, Y.-C., Lin, Y.-Y., Yang, M.-H., Huang, J.-B.: Show, match and segment: joint weakly supervised learning of semantic matching and object co-segmentation. IEEE Trans. Pattern Anal. Mach. Intell. **43**(10), 3632–3647 (2020)
3. Gao, Z., Xie, J., Wang, Q., Li, P.: Global second-order pooling convolutional networks. In: Proceedings of the IEEE/CVF Conference on Computer Vision and Pattern Recognition, pp. 3024–3033 (2019)
4. Guo, M.-H., et al.: Attention mechanisms in computer vision: a survey. Comput. Vis. Media **8**(3), 331–368 (2022)
5. Hou, R., Chang, H., Ma, B., Shan, S., Chen, X.: Cross attention network for few-shot classification. In: Advances in Neural Information Processing Systems, vol. 32 (2019)
6. Hu, J., Shen, L., Sun, G.: Squeeze-and-excitation networks. In: Proceedings of the IEEE Conference on Computer Vision and Pattern Recognition, pp. 7132–7141 (2018)
7. Jiang, H., Xiangyun, H., Li, K., Zhang, J., Gong, J., Zhang, M.: PGA-SiamNet: pyramid feature-based attention-guided Siamese network for remote sensing orthoimagery building change detection. Remote Sens. **12**(3), 484 (2020)
8. Lee, H., Kim, H.-E., Nam, H.: SRM: a style-based recalibration module for convolutional neural networks. In: Proceedings of the IEEE/CVF International Conference on Computer Vision, pp. 1854–1862 (2019)
9. Li, Q., Mou, L., Hua, Y., Shi, Y., Zhu, X.X.: Building footprint generation through convolutional neural networks with attraction field representation. IEEE Trans. Geosci. Remote Sens. **60**, 1–17 (2021)
10. Li, Q., Mou, L., Hua, Y., Shi, Y., Zhu, X.X.: CrossGeoNet: a framework for building footprint generation of label-scarce geographical regions. Int. J. Appl. Earth Obs. Geoinf. **111**, 102824 (2022)
11. Li, W., Hosseini Jafari, O., Rother, C.: Deep object co-segmentation. In: Jawahar, C.V., Li, H., Mori, G., Schindler, K. (eds.) ACCV 2018. LNCS, vol. 11363, pp. 638–653. Springer, Cham (2019). https://doi.org/10.1007/978-3-030-20893-6_40
12. Maggiori, E., Tarabalka, Y., Charpiat, G., Alliez, P.: Can semantic labeling methods generalize to any city? The Inria aerial image labeling benchmark. In: 2017 IEEE International Geoscience and Remote Sensing Symposium (IGARSS), pp. 3226–3229. IEEE (2017)

13. Mohanty, S.P.: crowdAI mapping challenge 2018 dataset (2019)
14. Singh, V.K.: FCA-Net: adversarial learning for skin lesion segmentation based on multi-scale features and factorized channel attention. IEEE Access **7**, 130552–130565 (2019)
15. Wang, J., et al.: Deep high-resolution representation learning for visual recognition. IEEE Trans. Pattern Anal. Mach. Intell. **43**(10), 3349–3364 (2020)
16. Wang, Q., Wu, B., Zhu, P., Li, P., Zuo, W., Hu, Q.: ECA-Net: efficient channel attention for deep convolutional neural networks. In: Proceedings of the IEEE/CVF Conference on Computer Vision and Pattern Recognition, pp. 11534–11542 (2020)
17. Xu, B., Xu, J., Xue, N., Xia, G.-S.: Accurate polygonal mapping of buildings in satellite imagery. arXiv preprint arXiv:2208.00609 (2022)
18. Xue, N., Bai, S., Wang, F., Xia, G.-S., Wu, T., Zhang, L.: Learning attraction field representation for robust line segment detection. In: Proceedings of the IEEE/CVF Conference on Computer Vision and Pattern Recognition, pp. 1595–1603 (2019)
19. Xue, N.: Holistically-attracted wireframe parsing. In: Proceedings of the IEEE/CVF Conference on Computer Vision and Pattern Recognition, pp. 2788–2797 (2020)
20. Yang, Z., Zhu, L., Wu, Y., Yang, Y.: Gated channel transformation for visual recognition. In: Proceedings of the IEEE/CVF Conference on Computer Vision and Pattern Recognition, pp. 11794–11803 (2020)
21. Zhang, H.: Context encoding for semantic segmentation. In: Proceedings of the IEEE Conference on Computer Vision and Pattern Recognition, pp. 7151–7160 (2018)

# Study and Implementation of AQI Predictive Recommendation System Based on Artificial Intelligence

Nguyen Van Luc[1,2], Le Van Anh Duc[1,2], Nguyen Thi Viet Huong[1,2],
Nguyen Minh Nhut[1,2], and Nguyen Dinh Thuan[1,2(✉)]

[1] University of Information Technology, Ho Chi Minh City, Vietnam
{19521811,19521374,19521374}@gm.uit.edu.vn, thuannd@uit.edu.vn
[2] Vietnam National University, Ho Chi Minh City, Vietnam

**Abstract.** Air quality is increasingly becoming polluted in many major cities worldwide, so updating and keeping track of air quality information is gaining more attention. Based on the Air Quality Index (AQI) of Ho Chi Minh City and the hourly variations in pollution from April to June 2023, using time-series analysis to forecast the AQI using Recurrent Neural Network (RNN), Linear Regression (LR), Long Short-Term Memory (LSTM), and Deep Q-Learning algorithms. From there, we developed a mobile application called SmartAQI as a social platform to suggest locations with good air quality by the hour, display the air quality for the next seven days, and identify the positions of SmartAQI destinations in Ho Chi Minh City. We evaluated the machine learning algorithms using RMSE, MAE, and MAPE to predict the AQI. The experimental results of predicting the AQI demonstrated that the RNN model achieved the highest accuracy in this experiment.

**Keywords:** AQI · Recurrent Neural Network · Linear Regression · Deep Q-Learning · Air pollution · Machine learning

## 1 Introduction

The deteriorating air quality significantly impacts the health of residents in numerous major cities worldwide. In Vietnam, Ho Chi Minh City is considered a city that attracts a large workforce from across the country each year. Health impacts are linked to 81.31% of deaths associated with air pollution due to delicate particulate matter [1]. Currently, the air pollution levels in Ho Chi Minh City exceed the limits set by the World Health Organization (WHO). Consequently, the air quality issue in the city poses a significant concern for human well-being and the local ecosystem [2]. The government considers environmental prevention and control as a top priority, with air quality monitoring being one of the key focuses in Ho Chi Minh City. Therefore, developing an air quality forecasting system is essential to inform residents about the air quality conditions in Ho Chi Minh City.

Currently, various methods are used to predict the Air Quality Index (AQI), including statistical machine-learning techniques such as correlation analysis [3] and regression [4]. Researchers have also employed artificial intelligence algorithms like RNN [5], LR [6], LSTM [7], and ARIMA [8] models. Decision support models like Deep Q-Learning [9] have also been utilized. These studies aim to enable users to analyze environmental information and receive timely and accurate air quality information while providing valuable suggestions and guidance.

In this study, we focus on the following two main issues:

- Firstly, we collected data from the Weather API[1] website from April to June 2023 for Ho Chi Minh City regarding the air quality index (AQI). We used four algorithms: RNN, LR, LSTM, and Deep Q-Learning, to predict the AQI in Ho Chi Minh City. The goal was to find the most accurate prediction algorithm as a prerequisite for implementing the application.
- Secondly, we implemented a SmartAQI App with the following features: viewing weather information and air quality information, predicting the air quality index for the next seven days using machine learning algorithms, posting check-ins at SmartAQI destinations by recognizing the location and name of the place based on the images in the post, and suggesting a list of SmartAQI destinations with favorable conditions (low AQI) that are close to the user's current location. This creates good prerequisites for people to understand the air quality index realistically.

The content of this paper is organized into four sections. Section 2 highlights the essential components of the system, the main processing steps, and the research methodology. The dataset used, and the experimental results will be discussed in Sect. 3. Finally, the achieved results are summarized in the conclusion section.

## 2   Related Works

Climate pollution is one of the topics of great interest today. AQI forecasting is one of the directions to warn and help people control air pollution. The authors [10] used Machine Learning models: support vector regression (SVR) and random forest regression (RFR) to build regression models for predicting the Air Quality Index (AQI) in Beijing and the nitrogen oxides (NOX) concentration in an Italian city. The experimental results showed that the SVR model performed better in predicting AQI (with RMSE = 7.666, $R^2 = 0.9776$, and r = 0.9887), and the RFR model performed better in predicting NOX concentrations (with RMSE = 83.6716, $R^2 = 0.8401$ and r = 0.9180). Authors Kim, Donghyun, et al. [11] studied the Application of Deep Learning Models and Network Methods for Comprehensive Air-Quality Index Prediction. Using LSTM and deep neural network (DNN) applied to six pollutants and predicting a comprehensive air-quality index (CAI) from 2015 to 2020 in Korea. This study shows that combining the DNN model with the network method gives a high CAI prediction ability.

The authors [12] studied Urban Air Quality Prediction Using Regression Analysis. This experiment investigated the effectiveness of several predictive models in predicting

---

[1] https://www.weatherapi.com.

AQI based on the input data of meteorological and pollution information in New Delhi, India. Perform regression analysis on the dataset and obtain results showing which meteorological factors influence AQI more and how helpful the forecasting models are. The authors [13] studied Comparative Analysis of Supervised Machine Learning Techniques for AQI Prediction. This paper focuses on AQI prediction by supervised machine learning technique in Uttarakhand state, India, i.e., Dehradun, based on available pollutants (PM 10, PM 2.5, $SO_2$, $NO_2$). The result shows that the decision tree classifier is more accurate, with an accuracy of 98.63%. In addition, this study also found that the Himalayan drug-ISBT area is in the poor AQI range for the state capital city of Uttarakhand.

## 3 Research Methodology

The process of developing an integrated air quality prediction system with SmartAQI location identification (in Ho Chi Minh City) goes through several main steps, including data collection, data preprocessing, computation of air quality index (AQI) values, construction of machine learning models, and application of the models. The diagram below illustrates the architecture of this research system (Fig. 1).

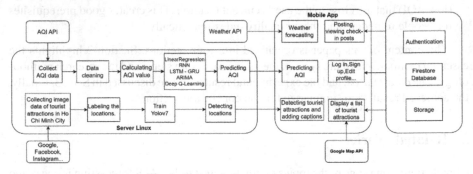

**Fig. 1.** Research System Architecture

### 3.1 Crawl Data

Firstly, regarding the air quality dataset, we have created an automated data collection system using Python to request an API for air quality data once per hour and receive responses containing hourly CO, $SO_2$, $NO_2$, $O_3$, PM 10, and PM 2.5. The website for data collection in this project is Weather API, a reputable website where we can gather the necessary values to compute the AQI from these indices (Fig. 2).

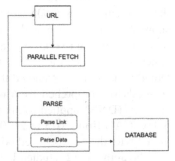

**Fig. 2.** Crawl Data Flow

Regarding the image dataset of SmartAQI locations in Ho Chi Minh City, we have also created a system to collect image data from Facebook, Instagram, and Google Images using the Python language and libraries such as Selenium, Beautifulsoup, and XPath.

## 3.2 Data Pre-processing

After collecting the air quality index data, we checked the dataset for missing or duplicate entries. After this process, we obtained a dataset of 1776 rows collected from April 5, 2023, to June 17, 2023. The dataset contains six attributes: $SO_2$, CO, $NO_2$, $O_3$, PM2.5, and PM10 (Table 1).

**Table 1.** Raw Data Description

|        | CO     | NO2   | O3    | SO2   | PM2.5 | PM10  |
|--------|--------|-------|-------|-------|-------|-------|
| count  | 1776   | 1776  | 1776  | 1776  | 1776  | 177   |
| mean   | 1063.9 | 31.7  | 25.7  | 39.3  | 38.5  | 48.6  |
| std    | 644.8  | 11.8  | 41.5  | 21.4  | 30.7  | 36.4  |
| min    | 363.8  | 8.1   | 0     | 10.1  | 5.6   | 7.2   |
| 25%    | 680.9  | 24    | 0.1   | 26.2  | 19.2  | 25.9  |
| 50%    | 881.2  | 30.3  | 7.9   | 33.9  | 29.1  | 37.9  |
| 75%    | 1215   | 37.4  | 32.2  | 46.7  | 46.8  | 58.9  |
| max    | 5447.4 | 87.7  | 303.3 | 217.4 | 229.2 | 277.7 |

Next, we divide the data into training, testing, and validation sets. After experimentation, we found that splitting the dataset into a ratio of 7:2:1 yielded the best results.

For the image dataset, due to duplicated images from various sources, we removed duplicate images and some images that were mistakenly associated with incorrect locations during the collection process. After preprocessing, we collected 3186 representative images depicting 17 famous SmartAQI locations in Ho Chi Minh City.

Next, we will label each image based on its corresponding location using the LabelImg tool. LabelImg is a software tool used to annotate objects in images. It allows users to draw shapes like rectangles, circles, or polygons around the things of interest in the image. LabelImg is commonly used in machine learning and computer vision projects to create labeled training datasets for object recognition and classification.

The result is a labeled dataset of 3186 images of famous SmartAQI locations in Ho Chi Minh City (Fig. 3).

## 3.3 Calculation and Analysis of Vietnam's Air Quality

According to the "Decision on the issuance of Guidelines for calculating and publishing the Vietnam Air Quality Index (VN_AQI) [15]", this index assesses the current air quality and its impact on human health. After verification, the dataset contains over 1700 valid rows without missing data.

Hourly characteristics of AQI in Ho Chi Minh City from April to June 2023. Based on the data collected from the Weather API website, we have compiled and analyzed

**Fig. 3.** Labeling Result

**Table 2.** AQI Data Description

|       | AQI  |
|-------|------|
| Mean  | 66.7 |
| Std   | 37.2 |
| Min   | 11   |
| Max   | 339  |

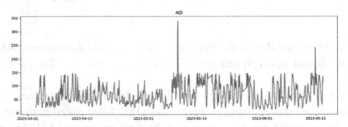

**Fig. 4.** Hourly AQI value visualization chart from April to June

the hourly AQI data illustrated in Fig. 4 below. The average AQI for the three months is 66.7, with the lowest AQI recorded as 11 and the highest as 339 (Table 2).

Related studies have shown that the primary factors influencing AQI are PM2.5, PM10, and $O_3$. During the three-month analysis period, the air quality in Ho Chi Minh City reached its worst level in the early days of May. The main reasons could be the significant increase in ozone concentration during the summer and traffic volume. Air pollution is always a pressing issue in the modern industrial era, so we need preventive measures to reduce air pollution and lower the AQI.

### 3.4  Build and Apply Model to SmartAQI

**Building Model**

The primary purpose of this study is to develop a system for predicting future Air Quality Index (AQI) values based on collected historical AQI values. We used Python and libraries to implement four machine learning algorithms to build a model to achieve this prediction goal:

- Recurrent Neural Network.
- The Long Short-Term Memory.
- The Linear Regression model.
- Deep Q-Learning.

To detect SmartAQI attractions in Ho Chi Minh City and add captions to images based on location and posting date to preserve user memories on social networks. First, we will use YOLOv7 [14] to train the model with labeled datasets.

**Apply Model to SmartAQI**

SmartAQI is an application that uses machine learning algorithms to predict air quality in the next seven days, view the weather of Vietnam's provinces and cities, post check-ins at SmartAQI attractions, and identify locations of SmartAQI attractions through images from the posts and suggest a list of SmartAQI attractions with favorable conditions for SmartAQIs. Here are some pictures of the SmartAQI app (Figs. 5 and 6).

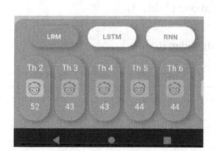

**Fig. 5.** SmartAQI – Predict AQI                **Fig. 6.** SmartAQI – All Posts

In addition, users can visit the air quality page to see the air quality of the province they are in (Fig. 9), and the bottom of the screen will have a prediction of the AQI index and the next seven days of air condition, respectively with each machine learning algorithm.

Posts screen where users can check in to places and post their experiences. Information will be shared with other users; the application will automatically recognize the location and name of the SmartAQI destination the user posts so that others can understand the information and can go there to travel (Fig. 10).

View information on SmartAQI places (Fig. 11) – display information such as weather, air quality index, location image, and location according to the machine learning model results. Users can choose the correct destination for their travel needs depending on each parameter (Fig. 7).

**Fig. 7.** SmartAQI – Recommended Destinations

## 4  Experiments and Result

After completing model training, we use Mean Absolute Error (MAE), Root Mean Square Error (RMSE), and Mean Absolute Percentage Error (MAPE) to evaluate and compare between models.

After using four models: RNN, LSTM, Linear Regression, and Deep Q-Learning, to predict future AQI values, we evaluated and compared the models using three metrics: MAE, RMSE, and MAPE, as mentioned above (Table 3).

**Table 3.** Algorithm Results

| Model | MAE | RMSE | MAPE |
|---|---|---|---|
| RNN | 5.5 | 10.3 | 11.1 |
| LSTM | 6.2 | 10.2 | 12 |
| LinearRegression | 22.2 | 28.8 | 20.5 |
| Deep Q-Learning | 29.9 | 33.5 | 30.2 |

The table above shows that the RNN model gives the best results with a MAPE value of 13.2, the LSTM model with a MAPE value of 18.2, and the linear regression model with a MAPE value of 20.5. The same is the Deep Q-Learning model with a MAPE value of 30.2. After using the labeled dataset to train the YOLOv7 model, we also use the Precision, Recall, and Mean Average Precision indexes to evaluate this model. After training the model, we obtain the following metrics: Precision 0.82, Recall 0.83, and mAP 0.81. From there, the model's accuracy is relatively high, 82%, enough that we can use it to add location detection in the application.

## 5  Conclusions

This study presents and implements algorithms, including machine learning and decision support systems, to predict AQI effectively. These algorithms are also evaluated and compared using MAE, RMSE, and MAPE metrics. Experimental results demonstrate that the RNN model performs better than other algorithms. Therefore, in this study, the RNN model outperforms different algorithms in predicting AQI. Subsequently, it is deployed as the SmartAQI application. This application helps update AQI indices and predict and suggest suitable locations with low AQI. It was built and developed as a social networking platform that supports posting pictures and checking in at various locations, thereby using the images in the posts to identify the location and names of places in Ho Chi Minh City. In the future, we will continue to review and improve the quality of the input dataset by diversifying the data sources. Add more features and experiment with a broader range of algorithms to enhance prediction accuracy.

**Acknowledgment.** This research is funded by Vietnam National University HoChiMinh City (VNU-HCM) under grant number DS2022-26-03.

# References

1. Bui, L.T., Nguyen, P.H.: Evaluation of the annual economic costs associated with pm2. 5-based health damage: a case study in Ho Chi Minh City, Vietnam. Air Qual. Atmos. Health **16**(3), 415–435 (2023)
2. Rakholia, R., Le, Q., Ho, B.Q., Khue, V., Carbajo, R.S.: Multi-output machine learning model for regional air pollution forecasting in Ho Chi Minh City, Vietnam. Environ. Int. **173**, 107848 (2023). https://doi.org/10.1016/j.envint.2023.107848
3. Jiao, Y., Wang, Z., Zhang, Y.: Prediction of air quality index based on LSTM. In: 2019 IEEE 8th Joint International Information Technology and Artificial Intelligence Conference (ITAIC), pp. 17–20. IEEE (2019)
4. Sethi, J.K., Mittal, M.: An efficient correlation based adaptive lasso regression method for air quality index prediction. Earth Sci. Inf. **14**(4), 1777–1786 (2021)
5. Huang, Y., Ying, J.J.-C., Tseng, V.S.: Spatio-attention embedded recurrent neural network for air quality prediction. Knowl. Based Syst. **233**, 107416 (2021). https://doi.org/10.1016/j.knosys.2021.107416
6. Jamal, A., Nodehi, R.N.: Predicting air quality index based on meteorological data: a comparison of regression analysis, artificial neural networks and decision tree. J. Air Pollut. Health **2**(1) (2017)
7. Wang, J., Li, J., Wang, X., Wang, J., Huang, M.: Air quality prediction using CT-LSTM. Neural Comput. Appl. **33**, 4779–4792 (2021)
8. Shishegaran, A., Saeedi, M., Kumar, A., Ghiasinejad, H.: Prediction of air quality in Tehran by developing the nonlinear ensemble model. J. Clean. Prod. **259**, 120825 (2020)
9. Kuan-Heng, Y., et al.: Optimization of thermal comfort, indoor quality, and energy-saving in campus classroom through deep Q learning. Case Stud. Therm. Eng. **24**, 100842 (2021)
10. Liu, H., Li, Q., Dongbing, Y., Yu, G.: Air quality index and air pollutant concentration prediction based on machine learning algorithms. Appl. Sci. **9**(19), 4069 (2019)
11. Kim, D., Han, H., Wang, W., Kang, Y., Lee, H., Kim, H.S.: Application of deep learning models and network method for comprehensive air-quality index prediction. Appl. Sci. **12**(13), 6699 (2022). https://doi.org/10.3390/app12136699
12. Mahanta, S., Ramakrishnudu, T., Jha, R.R., Tailor, N.: Urban air quality prediction using regression analysis. In: TENCON 2019 – 2019 IEEE Region 10 Conference (TENCON), pp. 1118–1123. IEEE (2019)

13. Pant, A., Sharma, S., Bansal, M., Narang, M.: Comparative analysis of supervised machine learning techniques for AQI prediction. In: 2022 International Conference on Advanced Computing Technologies and Applications (ICACTA), pp. 1–4. IEEE (2022)

14. Wang, C.-Y., Bochkovskiy, A., Mark Liao, H.-Y.: Yolov7: trainable bag-of-freebies sets new state-of-the-art for real-time object detectors. In: Proceedings of the IEEE/CVF Conference on Computer Vision and Pattern Recognition, pp. 7464–7475 (2023)

15. Ministry of Natural Resources and Environment. Decision on the Issuance of Technical Guidelines for Calculation and Publication of Vietnam Air Quality Index (VN_AQI) (2019). https://thuvienphapluat.vn/van-ban/Tai-nguyen-Moi-truong/Quyet-dinh-1459-QD-TCMT-2019-ky-thuat-tinh-toan-va-cong-bo-chi-so-chat-luong-khong-khi-Viet-Nam-428215.aspx

# Author Index

N. Thai-Nghe et al. (Eds.): ISDS 2023, CCIS 1949, pp. 309–311, 2024.
https://doi.org/10.1007/978-981-99-7649-2

Printed in the United States
by Baker & Taylor Publisher Services

Printed in the United States
by Baker & Taylor Publisher Services